Plant Protection and Quarantine

Volume II
Selected Pests and Pathogens of Quarantine Significance

Editor

Robert P. Kahn
Senior Staff Officer (Retired)
Plant Protection and Quarantine
Animal and Plant Health Inspection Service
U.S. Department of Agriculture
Hyattsville, Maryland

CRC Press, Inc.
Boca Raton, Florida

Library of Congress Cataloging-in-Publication Data

Plant protection and quarantine / editor, Robert P. Kahn.
p. cm.
Includes bibliographies and index.
Contents: v. 1. Biological concepts — v. 2. Selected pests and pathogens of quarantine significance — v. 3. Special topics.
1. Plant quarantine. 2. Plant quarantine--United States.
I. Kahn, Robert P.
SB980.P58 1989
632'.93--dc19 88-2800
ISBN 0-8493-6650-X (v. 1) CIP
ISBN 0-8493-6651-8 (v. 2)
ISBN 0-8493-6652-6 (v. 3)

Direct all inquiries to CRC Press, Inc., 2000 Corporate Blvd., N.W., Boca Raton, Florida, 33431.

International Standard Book Number 0-8493-6650-X (Volume I)
International Standard Book Number 0-8493-6651-8 (Volume II)
International Standard Book Number 0-8493-6652-6 (Volume III)

Library of Congress Card Number 88-2800
Printed in the United States

PREFACE

Volume II is part of a three-volume series on plant protection and quarantine (PPQ). As such it represents a continuation of the general biological theme of Volume I but along specific lines. In Volume I, biological concepts and the characteristics of 12 groups of pests, pathogens, and weeds are described and related to plant protection and quarantine programs and activities. Most of the organisms mentioned in Volume I as examples or to illustrate points are of quarantine significance to various countries as evidenced by the naming of these organisms in plant quarantine rules and regulations. In addition, Volume I covers the biological aspects and the meaning of "quarantine significance" based on a review of available scientific and regulatory literature.

In Volume II, selected pests and pathogens of quarantine significance are reviewed by contributors in much greater detail than has been possible in the general biological approach of Volume I. The authors in Volume II are either scientists or specialists in the quarantine services of various countries or scientists in nonregulatory agencies that often advise plant protection and quarantine services. Of the 16 contributions, 4 deal with a specific pest or pathogen, 5 with groups of pests or pathogens, 1 with pests and pathogens of one crop, 2 with pathogens of two crops, and 4 with plant quarantine pest and pathogen diagnostic problems.

Many of the authors give the geographical distribution of selected pests and pathogens. However, the information presented can only represent the distribution known to the authors at the time the manuscripts were submitted based on the authors' surveys of information available to them. Consequently, the publication in a book of information about distribution can only represent the past up to a given point in time. Pathogens and pests will continue to move, or be moved, along natural and man-made pathways. A few pests or pathogens may be expected to be eradicated from certain areas or localities where they now occur. Therefore, this volume is not intended to be a source of information for regulatory matters (e.g., export/import activities) that depend on up-to-date knowledge about the world-wide distribution of pests and pathogens.

THE EDITOR

Robert P. Kahn, Ph. D., Consultant, Plant Protection and Quarantine, and Plant Pathology, 14104 Flint Rock Terrace, Rockville, Md., 20853.

Dr. Kahn received his B.A. in botany in 1948 and his Ph.D. in plant pathology in 1951 from the University of Illinois, Urbana, Ill. From 1952 to 1957, he was a Supervisory Plant Pathologist, Crops Division, Ft. Detrick, Frederick, Md. From 1957 to 1970, he was the Plant Virologist and Head of the Plant Quarantine Station, Plant Quarantine Division, Agricultural Research Service, USDA, Glenn Dale, Md. From 1970 to 1972, he was on a 2-year assignment sponsored by U.S. Agency for International Development as the Head of Plant Quarantine, East African Community (Kenya, Tanzania, Uganda) located at the East African Agricultural and Forestry Research Organization, Muguga, Kenya. In 1972, he returned to his former organization which was renamed: Plant Protection and Quarantine (PPQ), Animal and Plant Health Inspection Service (APHIS), USDA. In 1974, he was appointed to the National Program Planning Staff (NPPS), PPQ, APHIS, Hyattsville, Md. In December 1985, he retired as Senior Staff Officer, Biological Assessment Support Staff, NPPS, PPQ, APHIS. Dr. Kahn has worked under short-term plant quarantine assignments in Niger, Nigeria, Zambia, Malaysia, Philippines, India, Columbia, Peru, Mexico, Honduras, Costa Rica, Guatemala, and the People's Republic of China as well as with the Plant Production and Protection Division, Food and Agriculture Organization of the United Nations.

Dr. Kahn is a member of the American Phytopathological Society, British Society for Plant Pathology, Council for Agricultural Research and Technology, American Institute of Biological Science, and the American Association for the Advancement of Science.

Dr. Kahn is the author of 86 papers in the areas of plant virus research, seed pathology, improving the phytosanitary aspects of the international exchange of high risk plant germplasm, postentry quarantine, and exclusion as a plant pest and pathogen control strategy. His current research and regulatory interests are in pest risk analysis and the biological basis of regulatory actions.

CONTRIBUTORS

E. L. Civerolo
Fruit Laboratory
Agricultural Research Service
United State Department of Agriculture
Beltsville, Maryland

Richard A. Frederiksen
Professor
Department of Plant Pathology and Microbiology
Texas A & M University
College Station, Texas

Siew Ming Kang
Head, Plant Quarantine Section
Department of Agriculture
Jalan Gallagher
Kuala Lumpur, Malaysia

Lloyd Knutson
Systematic Entomology Laboratory
Agricultural Research Service
United States Department of Agriculture
Beltsville, Maryland

Jose Carlos Lozano
Cassava Pathologist
Cassava Program
Centro Internacional de Agricultura Tropical
Cali, Colombia

Philip M. D. Martin
Plant Health Directorate
Food Protection and Inspection Branch
Agriculture Canada
Ottawa, Ontario, Canada

Robert R. Martin
Research Scientist
Agriculture Canada Research Station
Vancouver, British Columbia, Canada

A. F. Murant
Department of Virology
Scottish Crop Research Institute
Invergowrie, Dundee, Scotland

Lytton J. Musselman
Professor
Department of Biological Sciences
Old Dominion University
Norfolk, Virginia

Barry L. Nolt
Cassava Virologist
Cassava Program
Centro Internacional de Agricultura Tropical
C.I.A.T.
Cali, Colombia

Luis F. Salazar
Virology Leader
Department of Pathology
International Potato Center
Lima, Peru

Michael J. Shannon
Chief Staff Officer
Animal and Plant Health Inspection Service
United States Department of Agriculture
Hyattsville, Maryland

Kenneth A. Spencer
Department of Biological Sciences
University of Exeter
Exeter, England

Richard Stace-Smith
Research Scientist
Agriculture Canada Research Station
Vancouver, British Columbia, Canada

Tom van der Zwet
Research Plant Pathologist
Host/Parasite Interaction Unit
Appalachian Fruit Research Station
United States Department of Agriculture
Kearneysville, West Virginia

James M. Waller
CAB International Mycological Institute
Kew, Surrey, England

Thomas E. Wallenmaier
Animal and Plant Health Inspection Service
United States Department of Agriculture
Hyattsville, Maryland

PLANT PROTECTION AND QUARANTINE

Volume I

BIOLOGICAL CONCEPTS

Volume II

SELECTED PESTS AND PATHOGENS OF QUARANTINE SIGNIFICANCE

Volume III

SPECIAL TOPICS

TABLE OF CONTENTS

Chapter 1

THE PATHOGENS OF WHEAT

P. M. D. Martin

TABLE OF CONTENTS

I. INTRODUCTION

The determination of the quarantinable organisms of wheat depends directly upon the way in which the pathogen is disseminated. Since wheat is transported by man as seed rather than as grown plants or plantlets, as is the case with some other crops, the seed, soil, and plant refuse in seed lots constitute the most important plant parts by which pathogens may be moved from one place to another. Various related methods comprising passive transport in vegetable material, detritus, soil, and nematode galls that are associated with seeds are significant but can be overcome by proper inspection and use of clean material. Other transmission pathways involving soil in the field and atmospheric currents may eventually distribute pathogens over a wide area but some strategies have been developed to reduce their impact. It is the organisms distributed naturally in live vectors such as insects and mites from one country to another that are difficult to regulate since the latter tend to be widely distributed and their movements cannot be controlled easily. The various means of pathogen dispersal are listed in Tables 1 to 3.

Most of the viruses pathogenic to wheat are carried by insect and mite vectors whereas the fungal and bacterial pathogens are mainly transmitted passively. The majority of pathogens are commonly distributed by one mechanism only and it is less common to find combinations of two or more. It is especially interesting to note that few pathogens are carried both by invertebrate hosts and by seed. The capability of pathogens to disperse naturally by several means enhances their ability to spread and establish themselves. Their potential importance is also increased in some cases by the availability of numerous additional grass and other hosts and by the production of virulent races that compete with the resistant varieties of the host population.

The evolution of pathogenic races is probably best developed and most widespread in the rusts, but several other pathogens such as the bunts and smuts have specialized in this way, and races may be presumed to have developed in many species even though the situation is still imperfectly known. Races often differ in virulence and may not be universally distributed; hence the smallest taxonomic unit rather than the species as a whole is often recognized as the quarantinable organism that needs regulation.

The species that may be regarded as quarantinable represent only a small fraction of the approximately 500 organisms recorded on wheat[73] and predominantly comprise those pathogens which can be transported long distances by air currents and those which are mainly dispersed by seed. The following discussion deals with selected major pathogens, but it should be noted that even apparently minor pathogens could assume a different and unpredictable role if transported into a new area where conditions might encourage their spread and establishment.

II. PRIMARY TRANSPORT OF PATHOGENS

A. Pathogens Primarily Transported by Insect and Mite Vectors

The viral pathogens of cereal crops, including wheat, are usually highly specific with regard to type of invertebrate vector and are not usually transmitted by direct contact (see Table 1). The vectors comprise aphids, leafhoppers (Cicadellidae), planthoppers (Delphacidae), beetles, and eriophyid mites which have been discussed at length elsewhere[71] (see Table 2). Transmission of the pathogen from one generation of wheat to the next can only take place via the invertebrate vector, and it is generally assumed that each virus must survive between crops inside the vectors or in weeds, and perennial or volunteer host plants.[135] In spite of the wide distribution of these pathogens, few devastating epidemics have occurred though considerable losses have been sustained due to wheat streak mosaic virus in Alberta[91] and barley yellow dwarf virus across North America.[23,29,34] Where conditions are favorable

Table 1
METHODS OF TRANSMISSION OF WHEAT PATHOGENS

	Mechanical	Soil	Seed	Insects	Mites	Nematodes	Fungi	Not defined
			Viruses					
Agropyron mosaic virus	+	−	−	−	+	−	−	−
American wheat striate mosaic virus	−	−	−	+	−	−	−	−
Barley mosaic virus[a]	+	−	−[b]	+	−	−	−	−
Barley stripe mosaic virus	+	−	+	−	−	−	−	−
Barley yellow dwarf virus	−	−	−	+	−	−	−	−
Barley yellow striate mosaic virus	−	−	−	+	−	−	−	−
Brome mosaic virus	+	−	+	+[c]	−	+[c]	−	−
Cocksfoot mild mosaic virus	+	−	−	+	−	−	−	−
Cocksfoot mottle virus	+	−	−	+	−	−	−	−
Eastern wheat striate virus	−	−	−	+	−	−	−	−
European wheat striate mosaic virus[a]	−	−	+	+	−	−	−	−
Maize rough dwarf virus	−	−	−	+	−	−	−	−
Maize streak virus	−	−	−	+	−	−	−	−
Panicum mosaic virus	+	−	−	−	−	−	−	−
Tobacco mosaic virus	+	−	−	−	−	−	−	−
Wheat chlorotic streak virus	−	−	−	+	−	−	−	−
Wheat dwarf virus	+	−	−	+	−	−	−	−
Wheat mosaic streak virus	+	−	−	+	−	−	−	−
Wheat rosette stunt virus	−	−	−	+	−	−	−	−
Wheat soilborne mosaic virus	+	+[d]	−	−	−	−	+	−
Wheat soilborne yellow mosaic virus[a]	+	+	−	+	−	−	−	−
Wheat spindle streak mosaic virus (= wheat yellow mosaic virus)	+	+[d]	−	−	−	−	+	−
Wheat spot mosaic agent	−	−	−	−	+	−	−	−
Wheat streak mosaic virus	+	−	−[e]	−	+	−	−	−
Wheat white leaf virus	−	−	−	−	−	−	−	+
Wheat white spike disease agent	−	−	−	−	−	−	−	+
Wheat yellow leaf virus	+	−	−	−	−	−	−	−
Wheat yellow stunt virus	−	−	−	+	−	−	−	−
			Mycoplasma-Like Organisms					
Aster yellows mycoplasma	−	−	−	+	−	−	−	−

[a] Identity as a distinct virus not fully established.
[b] Seed-transmitted at a low rate in *Hordeum vulgare*.
[c] Many of these vectors need confirmation.
[d] Transmitted in soil through the agency of *Polymyxa graminis*.
[e] Seed-transmitted at a low rate in *Zea mays*.

and the crop is susceptible, there may be no effective quarantine measures available to halt the spread of the vector and the disease from one country to another within the same continent since both insects and mites are wind-borne and are carried on vegetative material. However, seed exports are unlikely to be responsible for the dispersal of these pathogens because these particular vectors cannot survive on seed and are likely to desiccate. Thus there is little likelihood of such pathogens as wheat spot mosaic agent[122] borne by *Aceria tulipae*, travelling

Table 2
PATHOGENS TRANSMITTED BY INSECT AND MITE VECTORS

Name Vector	Common name of vector	Ref.
VIRUSES		
Agropyron mosaic virus		
Abacarus hystrix (Nalepa)	Grain rust mite, cereal rust mite	126
American wheat striate mosaic virus		
Amplicephalus (Endria) inimica (Say)	Painted leafhopper	119
Elymana sulphurella Zett (= *Elymana virescens* Fabr.)	Leafhopper	120
Barley mosaic virus		
Rhopalosiphum maidis (Fitch)	Cereal leaf aphid, corn leaf aphid	129
Barley yellow dwarf virus		
Metopolophium dirhodum (Walker)	Rose-grain aphid	45, 128
Myzus persicae (Sulzer)	Green peach aphid	128
Neomyzus circumflexus (Buckton) (= *Myzus circumflexus* (Buckton)	Crescent-marked lily aphid	129
Rhopalosiphum insertum (Walker) (= *Rh. fitchii* (Sanderson)	Apple-grain aphid	128
Rhopalosiphum padi (L.) (= Rh. prunifoliae Fitch)	Bird-cherry aphid	45, 129
Schizaphis (Toxoptera) graminum (Rondani)	Green bug (aphid)	45
Sipha elegans Del Guercio (= *Sipha agropyrella* Hille Ris Lambers)		128
Sitobion avenae Fabricius (= *Macrosiphon avenae* Fabr., *M. granarium* Kirby)	Green aphid	45
Sitobion fragariae Walker	Blackberry-cereal aphid	129
Sitobion miscanthi (Takahashi) (= *Macrosiphum miscanthi* Takahashi)	Grain aphid	129
Barley yellow striate mosaic virus		
Javesella pellucida Fabricius	Leafhopper	25
Laodelphax striatellus (Fallen)	Leafhopper	25
Brome mosaic virus		
Oulema mellanopa (L.)	Cereal leaf beetle	94
Rhopalosiphum padi (L.)[a]	Bird-cherry aphid	141
Xiphinema coxi Tarjan[a]	Dagger nematode	22
Xiphinema diversicaudatum (Micoletzky) Thorne[a]	Dagger nematode	22
Cocksfoot mild mosaic virus		
Myzus persicae (Sulzer)[a]	Green peach aphid Peach potato aphid	118
Oulema cyanella (L.)	Leaf beetle	118
Cocksfoot mottle virus		
Oulema lichenis (Voet.)	Leaf beetle	20
Oulema melanopa (L.)	Cereal leaf beetle Owen green looper	129
Eastern wheat striate virus		
Cicadulina mbila (Naude)	Maize leafhopper	80
European wheat striate mosaic virus		
Javesella (Delphacodes) dubia (Kirschbaum)	Planthopper	14, 111
Javesella obscurella (Boheman)	Planthopper	14
Javesella (Delphacodes) pellucida (Fab.)	Planthopper	14, 111
Psammotettix provincialis (Ribaut)	Leafhopper	4
Maize rough dwarf virus		
Javesella pellucida (Fabricius)	Planthopper	129

Table 2 (continued)
PATHOGENS TRANSMITTED BY INSECT AND MITE VECTORS

Name Vector	Common name of vector	Ref.
Laodelphax striatellus (Fallen) (= *Calligypona marginata* (Fabr.)	Planthopper	26, 43
Maize streak virus		
Cicadulina mbila (Naudé)	Maize leafhopper	27, 35
Wheat chlorotic streak virus		
Laodelphax striatellus (Fallen)	Planthopper	115
Wheat dwarf virus		
Macrosteles laevis Rib.	Leafhopper	89
Psammotettix alienus (Dahlb.)	Leafhopper	3
Psammotettix striatus (L.)	Leafhopper	89
Wheat mosaic virus (Wheat mosaic rosette virus)		
Psammotettix striatus (L.) [= *Deltocephalus striatus* (L.)]	Leafhopper	6, 114
Schizaphis (Toxoptera) graminum (Rondani)	Greenbug (aphid)	8
Wheat mosaic streak virus		
Brachycaudus helichrysi (Kaltenbach)	Leaf-curling plum aphid	10
Macrosiphum avenae (Fabricius)	Grain aphid	10
Rhopalosiphum maidis (Fitch)	Cereal leaf aphid	10
Rhopalosiphum padi (L.)	Bird-cherry aphid	10
Wheat rosette stunt virus		
Laodelphax striatellus (Fallen)	Planthopper	7
Wheat soilborne mosaic virus		
Polymyxa graminis (Ledingham)	Myxomycete fungus	16, 99
Wheat soilborne yellow mosaic virus		
Metopolophium (Acyrthosiphon) dirhodum (Walker)	Rose-grain aphid	137
Rhopalosiphium padi (L.)		137
Schizaphis (Toxoptera) graminum (Rondani)	Greenbug (aphid)	137
Sitobion (Macrosiphum) avenae (Fab.)	Grain aphid	137
Wheat spindle streak mosaic virus (= wheat yellow mosaic virus)		
Polymyxa graminis Ledingham	Myxomycete	9, 31
Wheat spot mosaic agent		
Aceria tulipae (Keifer)	Wheat mite	121
Wheat streak mosaic virus		
Aceria tosichella Keifer	Mite	138
Aceria tulipae Keifer	Wheat mite	88
Wheat yellow leaf virus		
Rhopalosiphum maidis (Fitch)	Cereal leaf aphid, corn leaf aphid	21
Wheat yellow stunt virus		
Psammotettix alienus (Dahlb.)	Leafhopper	97
MYCOPLASMA-LIKE ORGANISMS		
Aster yellows mycoplasma		
Macrosteles quadrilineatus (Forbes) [referred to inaccurately as *Macrosteles fascifrons* (Stahl)]	Aster leafhopper	18

[a] Needs confirmation.

Table 3
PATHOGENS TRANSMITTED BY SEED[103]

Name	Common name	Ref.
VIRUSES		
Barley stripe mosaic virus		30, 77
Brome mosaic virus		22
European wheat striate mosaic virus		1
BACTERIA		
Clavibacter iranicus (Carlson & Vidaver) Davis et al. (=*Corynebacterium iranicum* Scharif)	Yellow slime	
Clavibacter tritici (Carlson & Vidaver) Davis et al.[a] (= Corynebacterium tritici (Hutchinson) Burkholder)	Yellow slime disease, Yellow ear rot, Tundu or Tannau disease in India (in part)	
Erwinia rhapontici (Millard) Burkholder	Pink grain	
Pseudomonas syringae van Hall pv *atrofaciens* (McCulloch) Young et al.	Basal glume rot, Spikelet rot	
Pseudomonas syringae van Hall pv *syringae* van Hall	Leaf necrosis	
Xanthomonas campestris (Pammel) Dowson pv *undulosa* (Smith, Jones & Reddy) Dye[b]	Black chaff	
FUNGI		
Alternaria alternata (Fr.) Keissler	Black point, kernel smudge, leaf spot, *Alternaria* blotch	
Alternaria triticina Prasada & Prabhu		
Aspergillus spp.[c]		
Athelia rolfsii (Curzi) Tu & Kimbrough[d] (= *Sclerotium rolfsii* Sacc.)		
Cladosporium spp.	Black point, black kernel smudge, sooty mold disease	
Claviceps purpurea (Fr. ex Fr.) Tul.	Ergot	
Cochliobolus sativus (Ito & Kuribay) Drechsl. ex. Dastur[e] (= *Helminthosporium sorokinianum* Sacc.) *H. sativum* Pamm. King & Bakke	Spot blotch, seedling blight, common root rot, crown rot, head blight	
Curvularia spp.[f]		
Dilophospora alopecuri (Fr.) Fr.[g]		
Drechslera spp.[h]	Seedling blight, root rot,	
Fusarium spp.[h]	Head blight	
Gaeumannomyces graminis (Sacc.) von Arx & Olivier[i] (= *Ophiobolus graminis* (Sacc.) Sacc.)		
Gibberella zeae (Schw.) Petch[j] (= *Fusarium graminearum* Schwabe)	Scab, head blight	
Hymenula cerealis Ellis & Everhard (= *Cephalosporium gramineum* Nisik. Matsumoto & Yamanti)	Cephalosporium stripe disease	
Leptosphaeria avenaria Weber f.sp. *triticea* Johnson (= *Septoria avenae* f.sp. *triticea* Johnson.)		
Leptosphaeria nodorum Muller (= *Septoria nodorum* Berk.)	Glume blotch, leaf spot	

Table 3 (continued)
PATHOGENS TRANSMITTED BY SEED[103]

Name	Common name	Ref.
Micronectriella nivalis (Schaffn.) Booth (= *Calonectria nivalis* Schaffn., *Griphosphaeria nivalis* (Schaffn.) Muller & Arx, *Fusarium nivale* (Fr.) Ces.)	Snow mold, brown foot rot	
Mycosphaerella graminicola (Fuckel) Schroet. (= *Septoria tritici* Rob. & Desm.)	Speckled leaf blotch	
Nigrospora spp.		
Puccinia graminis Pers. var. *tritici* Erikss. & Henn.[k]	Black stem rust	
Puccinia striiformis Westend[k] (= *Puccinia glumarum* Erikss. & Henn.)	Yellow rust	
Pyrenophora semeniperda (Brittlebank & Adam) Shoemaker (= *Pleosphaeria semeniperda* Brittlebank & Adam, *Drechslera campanulata* (Lev.) Sutton, *D. verticillata* (O'Gara) Shoemaker, *Podosporiella verticillata* O'Gara, *Helminthosporium cyclops* Drechsler)	Sterility disease	
Pyrenophora tritici-repentis (Died.) Drechsl. (= *P. trichostoma* (Fr.) Fuckel, *Helminthosporium tritici-repentis* Died., *Drechslear tritici-repentis* (Died.) Shoemaker)	Yellow spot, leaf blight	
Sclerophthora macrospora (Sacc.) Thirum. (= *Sclerospora macrospora* Sacc.)	Downy mildew	
Selenophoma donacis (Pass.) Sprague & Johnson var. *stomaticola* Bauml. (= *Septoria oxyspora* Penz. & Sacc.)	Halo spot, Selenophoma spot	
Tilletia caries (DC.) Tul.[l] (= *T. tritici* (Bjerk.) Wolff)	Bunt, rough spored bunt (a)	
Tilletia controversa Kuhn[l] (= *T. brevifaciens* G.W. Fisch.)	Dwarf bunt	
Tilletia foetida (Wallr.) Liro[l] (= *T. laevis* Kuhn)	Smooth spored bunt (b)	
Tilletia indica Mitra[l] (= *Neovossia indica* (Mitra) Mundkur)	Karnal bunt	
Urocystis agropyri (Preuss) Schroet. (= *Urocystis tritici* Kornicke)	Flag smut	
Ustilago nuda (Jens.) Rostr. (= *Ustilago tritici* (Pers.) Rostr.)	Loose smut	
NEMATODES		
Anguina tritici (Steinbuch) Chitwood (= *Tylenchus tritici* (Steinbuch) Bastian, *T. scandens* (Schneider) Cobb, *Anguillulina tritici* (Steinbuch) Gervais & V. Beneden)	Ear cockle	

Note: Most of these appear on the quarantine lists of various countries.

[a] Contained in the seed galls of the nematode *Anguina tritici*.

[b] Seed infection partly deep seated but not always transmitted.

Table 3 (continued)
PATHOGENS TRANSMITTED BY SEED[103]

[c] Although these species may produce mycotoxins in stored grain they are too frequent to make quarantine practicable.
[d] Seed-borne mycelium may survive for at least 1 year in the seed; may occur in storage.
[e] Also soil borne.
[f] Field fungi that may occur in storage.
[g] Twist disease especially on *Triticum spelta*
[h] Although these species may produce mycotoxins in stored grain they are too frequent to make quarantine practicable.
[i] Evidence of seed transmission only circumstantial (see Reference 103).
[j] Also soil borne.
[k] Occasionally teleutosori on seed may occur, but pathogen is mostly transported through air currents.
[l] Once teliospores settle, they may be soil borne as well. Reinfection of wheat plants occurs from sporidia produced from germinating teliospores at soil level.

in this way from North America to another continent. Where the vectors are widely dispersed, the best strategy for control is to prevent the build-up of vector populations to develop resistant varieties, together with elimination of volunteer wheat, susceptible grasses and other weeds, and to delay the sowing of winter wheat until after spring wheat has been harvested.[91]

B. Pathogens Primarily Transported in Soil

Important soil-borne pathogens include the nematode *Heterodera avenae* Woll., the fungi *Cochliobolus sativus* (Ito and Kur.) Drechsl and Dastur, *Fusarium* spp., *Gaeumannomyces graminis* von Arx and Olivier, *Pythium* spp. and *Thanatephorus cucumeris* (Frank) Donk (*Corticium solani* [Prill. and Del.] Bourd. and Galz.) *Gaeumannomyces graminis* is one of the most damaging cereal disease agents, particularly in Australia.[28,107] Infective propagules survive in crop residues, principally root fragments and stem bases. Soil-borne wheat mosaic virus remains infective in soil for many years in the absence of susceptible crops or in dry storage in the laboratory although it is unstable in extracted plant juices or in plant tissue at ordinary temperatures.[125]

Movement of contaminated soil should be restricted wherever possible and care should be taken to exclude it from wheat seed.[110,142] Soil clods, plant debris, and inert extraneous matter may contain many nematodes and other pathogens that may be parasitic on other crops apart from wheat.[107] Other strategies include the elimination of grass hosts that serve as natural carriers of these pathogens. Resistance especially to *Gaeumannomyces graminis* has been difficult to develop, chemical control is expensive, and effective control in the future may lie in the use of biological agents.

C. Pathogens Transported Predominantly by Air Currents

The rusts are a recurrent problem to wheat crops, and the damage to them has been well documented.[148] These pathogens owe their success to the production of huge numbers of asexual spores (uredospores) which are light and therefore easily carried. The spores are readily transmitted by air so that these pathogens cannot be controlled merely by restriction of infected plant material. Rusts are not generally regarded as seed borne; and, although spores of two species have been observed on seed and glumes, transmission is doubtful.[103]

The three rust species commonly attacking wheat are *Puccinia graminis* f.sp. *tritici* (black rust or stem rust) *Puccinia recondita* f.sp. *tritici* (brown rust or leaf rust) and *Puccinia striiformis* f.sp. *tritici* (yellow rust or stripe rust). Since these species are universally distributed, the problems of quarantine interest center around the transport of the more virulent races and the prevention of epiphytotics of the known races within a country. The life cycle may be heterocyclic, as in *P. graminis* and *P. recondita* where the asexual phase develops on wheat and an obligatory alternate host is also involved on which genetic recombination can take place. The alternate hosts on which *P. graminis* depends are species of *Berberis* and *Mahonia,*[51] especially *M. magnifolia* and *B. vulgaris* (the common barberry), which are used as ornamentals in Western Europe, and *M. aquifolium* and *M. repens* in North America.[24] The alternate hosts of *Puccinia recondita* comprise several species including *Anemonella thalictroides, Isopyrum fumaroides, Thalictrum speciosissimum,* and various species of *Anchusa* and *Thalictrum.*[108] In *P. striiformis* the life cycle is isocyclic: no alternate host is known and only the asexual phase is apparent. In warm regions, however, the first two species may also subsist without a sexual stage on an alternate host since other grass hosts such as species of *Aegilops, Agropyron, Bromus, Clinelymus, Dactylis, Festuca, Hordeum,* and *Lolium* at a time when wheat is not being cultivated.[17,41,100,121,136,140] If the alternate host has been destroyed or not available as in the Australian mainland[15,69] or in Egypt,[2] it cannot be a factor in rust variation and the main way in which the pathogen may then renew itself is through aerial transport of uredospores of new races from another region, and their establishment on local grass hosts. In cold regions, however, such persistence is impossible and the existence of an alternate host becomes crucial to the continued survival of the pathogen as well as to the production of new races by recombination. Where an alternate host has been allowed to linger, as in Finland[51] and in eastern North America[39,40] new rust races are more easily produced, and a focus is established for the infection of surrounding areas. The potential virulence of these new races made possible by the presence of even a few barberry bushes has been demonstrated.[104,132] Eradication of the barberry has limited the production of widely diverse races.[40]

The question that most concerns the plant health official is the source and nature of the rust spore inoculum. The inoculum has been termed *endemic* in origin when the rust survives both the summer and winter in the area of the epiphytotic and *exodemic* when the inoculum is carried by winds from one area into another.[148] Annual reinfection of the wheat crops from a geographically remote source has been well established, e.g., the dispersal of stem rust from Mexico to Canada via the U.S. each summer,[39,98] from the hills in India to the wheat-growing areas of the plains,[56] the reinfection of Belgian wheat crops from the south,[90] of English wheat crops by spores from continental Europe,[87] and of New Zealand crops from elsewhere in the Pacific area.[76]

Stem rust and leaf rust development in the prairie provinces of Western Canada is initiated by air-borne inoculum from as far as south central Mexico, a distance of about 4000 km. The rust overwinters in Mexico and southern Texas and climatic conditions and the varieties grown in the U.S. influence the kind and amount of the primary inoculum that blows northwards along the "Puccinia Path". This rust population is essentially asexual.[64] Each year the rust population moves northward infecting those cultivars of wheat that it finds susceptible. Recombination on *Mahonia* and *Berberis* (where available) in the U.S. and Canada will produce new races that subsequently develop further on their return southwards.[63] Races found in the annual Canadian rust race survey represent those which have found compatible host varieties and have survived unfavorable climatic conditions during their migration.[98] It is clear that man is able to manage its evolution and allow it to cause only minimum losses most of the time, by manipulating the varieties available to the rust.

Epidemiological zones corresponding to different prevalences of various rust races are likely to occur worldwide. At present the delimitation of these areas is vague but that there are marked differences in race composition between major wheat growing areas is clear from various reports from Europe,[87] North and East Africa,[38,42,72] Asia,[58,59,81,82,103,109] and Australia.[15] Rajaram and Campos[98] have delimited six zones in the Western Hemisphere, three in the North, two in the South and one in Central America from the available data for virulence patterns of the three species of *Puccinia* either singly or in combination. These zones comprise:

1. The Great Plains of North America (itself divided into northern and southern populations[105])
2. The Pacific Northwest
3. The eastern U.S. and Canada
4. Southern Mexico and Guatemala
5. The Andean countries of Colombia, Ecuador, Peru, and Chile
6. Southern Brazil, Argentina, Paraguay, and Uruguay

Further work is needed to distinguish these zones more clearly.

Spores of *Puccinia graminis* and *P. striiformis* may occur on seeds but these pathogens are not generally regarded as seed-borne since transmission does not appear to occur.[103] The main strategies adopted against the spread of *Puccinia* have therefore involved measures other than seed disinfection, as discussed below.

1. Breaking the Life Cycle

The mandatory elimination of the ornamental barberry, *Berberis vulgaris* and other species and hybrids in Canada and the U.S. and elsewhere has taken place since 1916[87,106,133] and has limited the ability of *Puccinia graminis* to produce new genetic combinations. In the absence of this alternate host new races of *P. graminis* can only result by somatic mutation or hybridization in the uredosori on various grasses giving rise to the uredospores.[15,69,144,145] Other campaigns against the barberry, as for example in Bavaria (Federal Republic of Germany) during the 1950s,[46] and in Denmark, Prussia, and France before the end of the 19th century[87] have also had a demonstrable effect. In Denmark for example, the number of epiphytotics has been reduced and the severe epiphytotic of 1951 was clearly of exodemic origin.[148] Where barberry eradication was delayed, as in eastern Canada, *Puccinia graminis* has been correspondingly difficult to control. Eradication of the barberry in Quebec took place as late as 1964 to 1970, but rust incidence was significantly reduced in treated areas.[19] It is also noteworthy that where the barberry was not eliminated as in the Swiss Alpine valleys[148] and in Yugoslavia[131] wheat cultivation has become uneconomic or not feasible.

It follows that importation of susceptible barberry species into areas where it is absent or has been eliminated must be prevented whenever possible. Material from other countries has been intercepted at the Canadian border, often fraudulently mislabeled!

In the case of *Puccinia recondita* however, the alternate hosts comprise a large number of wild herbaceous plants that are harder to eradicate systematically. Although *P. striiformis* is not known to have an alternate host, the list of hosts besides *Triticum* on which the uredial stage develops is so extensive, comprising species of *Agropyron, Bromus, Elymus, Festuca, Hordeum, Lolium,* and *Phalaris,* that eradication of this pathogen is impracticable except near the areas of cultivation.

2. Cultivation Practices

The effect of farm mechanization on rust epidemics is uncertain according to Zadoks.[148] Theoretically, at least, the elimination of volunteer wheat plants and potential grass hosts should diminish the available rust inoculum in areas where the rusts are endemic. The introduction of dwarf wheats has required greatly increased irrigation but this has not been thought to increase diseases in wheat if properly done.[109] Rusts, however, may be dependent on moisture in arid areas such as in Egypt. Fertilizer, on the other hand, favors development of obligate pathogens in general and rusts in particular. Crop rotation, in general, decreases soil-borne disease agents if the alternate crops do not sustain the pathogens.

3. Breeding for Earliness and Escape from Infection

According to Zadoks[148] black rust epidemics in Italy have been halted through breeding new varieties. Breeding for late sowing and early harvesting prolongs the crop-free period and consequently reduces the danger of endemics in the northern countries. Early maturity allows wheat to escape damage by leaf and stem rusts. Formerly, winter-hardy wheat varieties had to be sown early, resulting in the overlapping of growing periods of spring wheat harvested at the end of September with those of winter wheat sown early in September. This situation has changed and it is now common practice in warmer regions such as the Indian subcontinent to harvest two or more crops annually if irrigation is available. Breeding for 'slow rusting' ability which markedly delays the impact of the disease has led to a further decline in disease prevalence and is responsible for the success of such varieties as Thatcher, Warigo, Idared, and Hopps.[79,101,102,143]

4. Breeding for Direct Resistance

Breeding for resistance is probably the most important factor inhibiting the development of rust epiphytotics. In Canada, however, despite constant shifts in the incidence of physiological races only one important stem rust epidemic (1954) has occurred since 1935 when the first varieties with effective resistance were released. Nevertheless, in such countries with an advanced breeding program, the effect of cultivar resistance to races identified preceding the selection period has been sometimes more than counterbalanced by susceptibility appearing after commercialization of these cultivars — as for example the failure of Heines VII to yellow rust in the Netherlands. The greater the success of a resistant variety, the more severe the epidemic when resistance breaks down. Sometimes, also, a variety resistant against one pathogen is highly susceptible to another. When resistance to yellow rust broke down in the variety Kalyansona in India, the intended replacement Arjun also posed a problem because it was highly susceptible to karnal bunt,[55] caused by the smut fungus *Tilletia indica*.

Many varieties of wheat have had a limited usefulness and a short life because they had too narrow a base of resistance. As a result the recent emphasis has been on the production of resistant varieties with a complex resistance genotype. Relatively few varieties are, however, resistant to all known races under all conditions. According to Stakman[133] there are many thousands of rust types paralleled by resistance genes distributed throughout many varieties of *Tricitum*. In addition to resistant varieties, close relatives of *Triticum aestivum* and *T. durum* such as *T. dicoccoides* and *T. spelta* may also be used as a source of resistance.[32,54,116] Using 12 so-called standard differential wheat varieties, 297 races of *P. graminis* var *tritici* were originally identified by Stakman,[134] but the total number is certainly far greater. These races represent groups of biotypes that have some genes for virulence in common and therefore produce similar effects on some varieties of wheat. The component biotypes may differ in respect to some genes so that many races can be subdivided by the use of supplementary differential varieties. New biotypes are continually being produced by mutation and recombination. Thus it is desirable to counteract the effects of changing populations of rust races by changing the components of the varieties accordingly. Varieties should have specific resistances to as many different rust races or rust groups as possible. However composite resistance alone does not guarantee immunity from rust attack since race 15B-1 was able to attack all the resistant varieties in Canada except one during the 1950s.[40]

The success of detection procedures presupposes prior knowledge regarding population trends of races, the early detection of new races, and availability of lines resistant to them. This requires that a detailed survey of races be made in each wheat-growing country during each growing season. In Canada there is an annual survey across the country under the coordination of the Agriculture Canada Federal laboratory on the campus of the University of Manitoba at Winnipeg[37] which aims to determine the prevalence of the main races to detect new and rare races with important combinations of virulent genes.

The detection of rare and new virulent races requires different methods. The detection of a race comprising 1/1000 of the rust population requires several thousand samples and doubling the maximum number of isolates for identification according to Green[39] would not greatly improve the odds. Screening techniques therefore offer the best chances of success. A good method is to plant selective resistant varieties in the field and carefully observe them for infections that only new races could cause. Discovery of a new race does not always mean breakdown of resistance, since many widely virulent races have been found to occur rarely for short periods only. It seems that races 15B and 56 which later became an important part of the rust population in Canada did first appear in readily detectable quantities, whereas the rare races did not increase and cause serious losses though they had the virulence necessary so to do.

Detection of rare virulent races is nevertheless important because the new virulent combinations can be used to produce combinations of resistance genes that will ensure future protection should these virulent combinations become aggressive.

In Canada there are broad similarities in the races found in eastern and western Canada that indicate a common source of inoculum,[64] presumably from the U.S. Races found in British Columbia are distinct from and less widely virulent than those found on the prairies and apparently represent a population that developed independently of the main rust area of North America. Races 15 and 56 have predominated in Canada for the last 50 years. Race 56 has been the most important race of stem rust in central North America, becoming predominant in Canada from 1934 to 1949 despite the widespread use of varieties resistant to it. Race 15B became predominant suddenly in 1950, possibly because of the widespread use of the susceptible variety Austin that was grown in Texas during the previous year[63] and also because of its ability to attack all the commercial varieties of common and durum wheats resistant to race 56 that were grown in North America until 1954, when the resistant variety Selkirk was released. Epiphytotics due to this particular rust race occurred in 1953, 1954, and 1955. Race 15B declined soon after the release of Selkirk and race 56 again became predominant. During the early 1950s, 15B was favored also by prevalent low temperatures, which were unfavorable to race 56, but its distribution diminished in 1956 when the resistant variety became widely grown and temperatures moderated. These temperature preferences were confirmed in the laboratory: race 56 developed best at 25 and 20°C and race 15B-1 at 15°C. Moreover, studies on different densities of infection showed that race 56 had greater competitive ability in light infections but race 15B predominated in heavy infections.[61]

In a review of the physiological races of wheat stem rust in Canada from 1919 to 1969, Green[39] concluded that the main races observed with the exception of race 56 in the 1960s showed increasing ranges of virulence, and that this trend was a cause for concern since it seemed inevitable that resistant varieties would sooner or later become susceptible. A degree of aggressiveness might eventually be reached that could not often be exceeded; however, it also appears that environmental changes also play an important part in changing the prevalence of races so that a fluid situation could still be maintained.

Similar conclusions can be reached regarding the other two species of *Puccinia, P. recondita* and *P. striiformis.* The cultivar Aurora originally showed a high-yield capacity and resistance to all three rust species attacking wheat when it was introduced into Romania[85] but this resistance was broken by a form of race 77 of *P. recondita.* In 1973 outbreaks had extended its range to comprise over 60% of the total leaf rust isolates.

With regard to stripe rust, in spite of its universal distribution, it seems to be a major threat to wheat production mainly in California and the Pacific Northwest. Shaner and Powelson[112] have described three serious outbreaks in Oregon during 1960, 1961, and 1964.

In 1974 one of the most severe epidemics occurred in California[68] when a new virulent race attacked Pitic 62, the major wheat cultivar in the Sacramento Valley. The pathogen appears to have relied on volunteer wheat to build up its inoculum in both cases, although other grass hosts could have played a part also.[113,139]

It is interesting to compare the rusts with another, unrelated group of pathogens that are also effectively disseminated by airborne spores, namely the powdery mildews. *Erysiphe graminis* var. *tritici* is a widely distributed wheat pathogen but it causes less spectacular losses. A number of reasons can be advanced for this: a relatively slow increase during the growing season, infrequent death of the host and availability of better control methods. Other salient epidemiological factors, such as the rapid fall of the conidia, estimated at 1.2 cm/sec., the low relative humidity, and the short time required for germination (within 24 hr at temperatures above freezing), the susceptibility of the fungus to injury due to rain and its preference for low atmospheric humidity for germination and low temperatures (15 to 25°C) for growth and development, are less easy to relate to its importance as a pathogen.[52,86,147] The more fastidious nature of *Erysiphe graminis* may explain why it is less injurious than the rusts, but the most effective strategy against it, nevertheless, involves the monitoring of races on a worldwide basis and the development of varieties resistant against them.[11,146] There is a racial diversity within mildew populations similar to that in the rusts and there is some evidence for the establishment of races in countries far from the source of origin, the best example being that of barley mildew infections in the Faeroes and Denmark due to spores of *Erysiphe graminis* derived from Britain.[44]

The pathogen is also able to withstand periods of adversity, as resistant cleistothecia over hot dry summers and as dormant mycelium in volunteer plants during cold winters. The growth cycle may be broken by removal of volunteer plants and weed hosts during the fallow season.

III. SPREAD OF DISEASES THROUGH SEED

The seed-borne pathogens form the major group of quarantinable organisms and may be spread by regular commercial trade and involuntarily through dissemination of new wheat varieties developed by breeders. During the last 20 years, seed industries in many countries have become more aware of the potential risks of the spread of pathogens through seed transport. Most of the pathogens listed in Table 3 appear in the quarantine regulations of various countries. Small quantities of seed infected superficially with bacteria or fungi can be treated with an appropriate pesticide prior to export; seed infected with viruses or with deep-seated bacterial and fungal pathogens should normally be rejected. The quarantine procedure which is still generally applicable to seed-borne pathogens has been well described by Purdy[95] for *Urocystis agropyri* immediately after flag smut was discovered in the U.S. In 1919 the affected area was quarantined by the state of Illinois, and as the known distribution of the disease increased, the quarantine area was extended accordingly. Movement of wheat and farm machinery from the infested area was prohibited and disinfestation with formalin before harvesting was required. Seed from the infested area had to be treated with formaldehyde to kill all flag smut spores, and stubble had to be burned after harvest. In addition, farmers were required to grow varieties recommended by the Illinois State Department of Agriculture. The U.S. federal quarantine established against *U. agropyri* and *Gaeumannomyces graminis* forbade the importation of cereal seed deemed as unclean from all countries where these pathogens were known to occur, except by special permit which required treatment to destroy all viable flag smut spores. Another quarantine measure regulated the importation of packing material composed of wheat straw, chaff, and hulls which were demonstrated to carry the fungus. Flag smut has since appeared only in the Pacific Northwest, which means that these quarantine measures have been at least partially successful.

In 1945, flag smut appeared in the state of Michoacan in Mexico, never having previously been reported there.[13] The pathogen could have been imported on some commercial wheat from Australia that was grown in the field instead of being milled as intended. An eradication campaign was begun, a domestic quarantine established and mandatory seed treatment required as in the U.S.

Dispersal of seed throughout the world through human agency has had two important effects. The first effect is perhaps not obvious in that the development of new crop varieties potentially alters the host/pathogen equilibrium, making it possible for previously little known local species or their races to flourish if the new varieties are susceptible. A prime example is the development of karnal bunt due to *Tilletia indica* in the Indian subcontinent. In the Punjab the replacement of the varieties Kalyansona and Somalika after 1976 by Arjun, Wa 357, Wa 377, and WL 711 resulted in a greatly increased incidence of Karnal bunt.[60,117] The flowering of new Mexican varieties was uniform and early when compared to other commercial wheat varieties in India and this coupled with high doses of nitrogenous fertilizers was responsible for the sudden success of the pathogen.

The recent green revolution has involved the production of dwarf varieties largely produced by CIMMYT in Mexico and used throughout Asia and Africa. The dwarf wheats were derived from a broad spectrum of parents that had wide adaptability, yield potential, and disease resistance. Concern has been raised about the possible widespread development of diseases especially since the effective resistance of new wheat varieties to rusts was estimated to last only for about 5 years.[12]

In their review of the impact that dwarf wheat varieties have had on the disease situation in Asia and Africa, Saari and Wilcoxson[109] cite evidence that with time new races or biotypes of the three species of wheat rust will develop to attack them, a threat that could be realized particularly in view of the well-known potential of stem rust to reach epidemic proportions in favorable weather. Joshi et al.[57] have indeed reported that introduction of Mexican dwarf varieties into India stimulated the increase of race 12 of brown rust and race 38 of yellow rust. Important cultivars such as Kalyansona, Somalika, and Safed Lerma are susceptible to these races. The introduction of the varieties Soltane, Siété Cerros, and Badri into Tunisia sparked an outbreak of stripe rust in the 1976 to 1977 crop season.[33] The new biotypes of *P. striiformis* had apparently also developed an adaptation to the high spring temperatures which had previously inhibited the rust. Stem rust in East Africa was considered to be especially dangerous[109] because of the occurrence of many different races, rapid shifts in the occurrence of races, presence of inoculum throughout the year, and continuous growing of wheat. Wheat improvement has been hampered by stem rust there and dwarf varieties have failed to successfully establish themselves.

Saari and Wilcoxson[109] further concluded that dwarf wheat varieties were resistant or tolerant to loose smut (*Ustilago tritici*), bunt (*Tilletia* spp. except for *T. indica*), blackpoint (*Alternaria and Helminthosporium*) and bacterial and virus diseases. On the other hand they were susceptible to flag smut (*Urocystis agropyri*), powdery mildew (*Erysiphe graminis*), scab (*Fusarium spp)*, downy mildew (*Sclerophthora macrospora*), and ergot (*Claviceps purpurea*), the last two pathogens being of minor significance. Nematodes (*Anguina tritici* and *Heterodera avenae*) could be a problem but were possible to eliminate by effective management. Progress in obtaining resistance to *Septoria* has been slow: the first dwarf cultivars were susceptible[109] and only later has it been possible to develop resistant varieties.[70]

The second effect of the worldwide movement of germplasm is to facilitate the spread of quarantinable organisms. A good example of this is furnished by the bunts and smuts which form a group of largely universal pathogens allied to the rusts, in which there has also been a fine degree of racial diversification. The smuts comprise three genera, *Tilletia, Urocystis*, and *Ustilago*. The number and distribution of the races of each species have not been as intensively studied as in *Puccinia*; however, the number of races appears to be smaller and more uniformly widespread throughout the world.[47-49,53,62,96] The species that can be effectively quarantined are *Tilletia caries* (DC) Tul., *T. controversa* Kuhn, *T. foetida* (Wr.) Liro, *T. indica* Mitra, *Urocystis agropyri* (Preuss) Schroeter (syn. *U. tritici* Kornicke), and *Ustilago nuda* (Jens.) Rostrup.

Although these species are widely distributed they should still be regarded as quarantinable because of the presence of these races which are potentially dangerous to countries outside their source of origin[84] and because the teliospores may be transported in soil as well as seed. The international spread of *Tilletia indica* on contaminated seed constitutes a spectacular recent case of imperfect quarantine control.[66] Once inside a new area, the teliospores of this fungus, though relatively heavy, may be carried further by air currents over short distances. The pathogen was first discovered in 1930 at Karnal in the Punjab (now Haryana State) in India by Mitra.[78] It does not cause as destructive a disease as other members in the same genus. The prevalence of the disease is usually low although high crop losses have been occasionally reported, and the crop can be spoiled by an offensive fishy odor even with a low incidence of infection.

Sometime before 1970, the disease spread to Pakistan and Afghanistan. From the Indian subcontinent, it appears to have traveled to CIMMYT*, Mexico, inside a seed sample sometime in the late 1960s. The fact that contaminated wheat samples were received back in the National Bureau of Plant Genetic Resources at New Delhi from there, first in 1970 and repeatedly in later years supports the hypothesis that *Tilletia indica* can be moved in wheat germ plasm. Further interceptions by New Delhi were noted in seedlots received from Lebanon (1970), Turkey (1978), Sweden (1979), and also from Syria (1978) where another international cereal station is based, at Aleppo.

Infection of wheat by *Tilletia indica* involves a tortuous route involving both seed and soil.[60] Diseased kernels release teliospores onto the ground where they germinate to produce sporidia, which, together with secondary sporidia, are carried to wheat flowers either by air currents or by splashing water. During anthesis the sporidia germinate and penetrate through the stigma or ovary wall. The pathogen does not invade the host systemically but each kernel is infected separately. Although the pathogen may be distributed through subtropical regions, infection is assisted by cool humid weather. The optimal temperature range for teliospore germination was found to be 15 to 22°C, and teliospores could survive and germinate after long exposure to adverse conditions (4 weeks at 30 or 2°C) once favorable conditions returned.[149]

Despite the rather fastidious nature of the organism which might seem to hinder its dispersal, it has certain features which mitigate against easy control, and these rather than any special adaptation, are responsible for its being strictly considered as a quarantinable organism. The teliospores are extraordinarily long lived (up to 5 years[75]) and customary seed treatments, including pentachloronitrobenzene (PCNB), which are fungistatic rather than fungicidal, lose their activity well before the spores lose their viability. Formalin treatment is the only one reasonably effective for seed in small as well as bulk quantities, since it is a protein denaturant, but it has a serious drawback in that internal infection may not be eradicated without reducing seed viability at the same time. Direct microscopic examination of seed for the presence of teliopspores is a feasible method of detection, providing the quantities are small.

The persistence of the pathogen in adversity[149] suggests furthermore that the organism may survive in most of the winter wheat-growing areas of the world. Third, *T. indica* appears to resist conventional methods of control so that it is hard to eliminate it from areas where it has become firmly established. One of these areas is the southern part of Sonora State near Ciudad Obregon, Mexico, where one of the main breeding stations of CIMMYT is

* Centro Internacional de Mejoramiento de Maiz y Trigo

located. Stringent precautions have been taken in the multiplication of new wheat varieties at CIMMYT, involving treatment of seed with PCNB before planting, addition of PCNB to the irrigation water, and spraying growing wheat plants with mancozeb. Extensive though low level infection (up to 0.04% of the grain) was noted, however, in samples taken from the cultivar multiplication fields at CIMMYT by the writer in 1985 and 1986.

Although the area in Mexico invaded by the organism is still not believed to be extensive, contaminated wheat seed has nevertheless been discovered in railway boxcars at the U.S.-Mexico border. This potentially easy method of entry in conjunction with wind transport of teliospores poses an obvious threat to the wheat growing areas lying to the north.

Of lesser importance perhaps than Karnal bunt is *Alternaria* blight of wheat, principally caused by *Alternaria triticina*. This was first recorded in India in 1924 although the identity of the causal agent was not fully established until 1962.[130] It is now regarded as one of the serious diseases of wheat, particularly of durum varieties.[65,193] Part of this prominence has also been due to the susceptibility of Mexican wheats. Since this pathogen has so far been confined to India it is a potential threat to other wheat-growing areas of the world. Control of primary spread of *Alternaria* blight can be achieved by strict control of seed distribution and by hot water seed treatment at 50 to 54°C for 10 min following presoaking in tap water.[93] The development of resistant varieties shows promise, especially since the dwarf cultivars are generally more resistant than the tall ones. However, breeding programs should include multiple resistance towards *Cochliobolus sativus* and *Puccinia recondita* as well, since there is a synergistic relationship among these three pathogens. Why *Alternaria triticina* has not spread as far as *Tilletia indica* is an interesting question. Although the inoculum required to establish disease for *Tilletia indica* is not known, that for *Alternaria triticina* has been estimated as 40,000 spores per milliliter to produce disease consistently on 5-week-old wheat seedlings[65] after drops were sprayed or placed on the leaf surface. This high number may be the reason why *Alternaria triticina* is still relatively restricted in distribution.

Only 3 of the 29 viruses and mycoplasmata infecting wheat are seed transmitted and most of the others require a live vector. The most important of these is barley stripe mosaic virus whose distribution is directly related to seed transmission, direct contact, and pollen transmission.[125,127] The virus cannot be inactivated by long-term storage by heat treatment, but may be detected by serological methods and seedlings may be grown at warm temperatures (32°C) with adequate lighting to detect mild strains in symptomless plants.[125] Brome mosaic virus is probably not as serious a threat economically as barley stripe mosaic virus but the various insects and nematodes that have been claimed to act as a vector as well as wheat seed make it a potential problem. An apparently seed-borne wheat striate mosaic virus also vectored by *Javesella pellucida* and causing a serious disease of wheat in Egypt[1] would be a threat to other grain-producing countries if imported there. The species of *Javesella* and *Psammotettix* identified as vectors of European wheat striate mosaic virus have not been recorded in North America but other related planthoppers are present as well as *Amplicephalus (Endria) inimica*, the specific vector for American wheat striate mosaic virus.[123,124] Seed-borne viruses thus present an acute problem for seed repositories and gene banks[36] aiming to store seeds over a long period of time for gradual release to research workers and breeders.

Seed-borne bacteria present a similar picture. Although 7 of the 12 bacterial pathogens of wheat are known to be seed transmitted, no definite dispersal mechanism is known for the others, although it can be assumed that vegetative material plays a part. Although bacterial diseases of wheat may be serious, none of them has assumed the importance of those caused by the fungi. Nevertheless, the seed-transmitted bacteria need to be monitored, especially by the countries which do not yet have them.

There is a single seed-borne nematode species, *Anguina tritici*, that causes a serious disease of grain in Europe and North America known as "ear cockle".[67] An interesting association exists between the nematode and *Corynebacterium tritici* which can result in a double infection. The bacterium does not appear to be carried by the nematode itself but persists in seed galls for at least 5 years.[74] However, the problem posed by the wheat gall nematode and its synergist has been greatly reduced by the use of mechanical seed cleaning devices that are able to separate galled from healthy seed, and by the selection of resistant varieties.[50]

IV. STRATEGIES FOR THE CONTROL OF SEED-BORNE PATHOGENS OF WHEAT

The screening of wheat imports should involve a number of different approaches. The first step should be a pest risk analysis to assess the potential impact of foreign pathogens. Wadhi et al.[142] have listed ten pathogens posing a risk to wheat in India, drawn from the records over 3 years of quarantine examinations of a large number of incoming wheat samples. Such lists would obviously differ from one importing country to another. Freedom from serious pathogens such as Karnal bunt may be certified by foreign site field inspections within the exporting countries, which may also have to conform to prescribed growing procedures for the acceptable production of seed destined for breeders and researchers. Detailed inspection of such seed should then be performed at a plant health laboratory within the importing country. As for grapestock (discussed in Chapter 2), it may be necessary also to restrict or prohibit the distribution of seed of wheat cultivars that are susceptible to quarantinable pathogens.

In practice, seed samples should be fairly small. Large commercial shipments risk exclusion because it is difficult for the exporter to guarantee freedom from quarantinable pathogens occurring in the area of production and the problem of handling an adequate number of samples may be too great for the importer.

Inside the plant health laboratory, fumigation with methyl bromide or aluminum phosphide[5,142] has been recommended for elimination of stored grain pests. Elimination of soil clods and plant and other debris will remove nematodes and microbes. Seed washing and culturing the seed microflora on agar media will detect all but the viruses, for which serological methods are available.

The elimination of bacteria and fungi from seed samples may be accomplished normally by soaking them in a dilute formalin solution. However, there is some doubt that internally borne pathogens such as *Tilletia indica* can be eliminated in this way without interfering with the viability of the seed and further research is needed to solve this problem.

In view of the large number of samples that are required by researchers and breeders, this program may present a horrendous challenge to a quarantine facility. Such a program, however, may be the only way to check the spread of the major pathogens as well as those whose behavior is unpredictable.

ACKNOWLEDGMENT

The writer is grateful to Drs. W. P. Campbell, J. W. Martens, and Y. C. Paliwal for assistance and advice.

REFERENCES

1. **Abdel-Hak, T., Ghobrial, E., and Kamel, A. H.,** Studies on striate mosaic of wheat in Egypt, *Agric. Res. Rev.*, 55, 1, 1977.
2. **Abdel-Hak, T. M., Sheheli, A. A., and Nazim, M.,** The source of inoculum of wheat leaf rust in relation to wind directions in Egypt, *J. Phytopathol.*, 6, 17, 1974.

3. **Agarkov, V. O.,** Pale green dwarfing of winter wheat in the Chernigov region, *Visn. Silskogospod.* Nauki, 1965, 53, 1965; *Rev. Appl. Mycol.*, 45(1333), 251, 1966.
4. **Agarkov, V. A.,** Virus diseases of winter wheat at Nemchinovka, Moscow district, *Biol. Nauki,* 10, 126, 1967; *Rev. Appl. Mycol.*, 46(3044), 569, 1967.
5. **Akiyama, H.,** Plant quarantine and fumigation of imported cereals in Japan, *Dev. Agric. Eng.*, 5, 31, 1983.
6. **Alekseev, A. N., Pridantseva, E. A., and Razvyazkina, G. M.,** Individual dose inoculation of *Psammotettix striatus* by feeding with winter wheat mosaic virus in suspension and its transmission to healthy plants, *Biol. Nauki,* 13, 136, 1970; *Rev. Plant Pathol.*, 50(73), 22, 1971.
7. **Anon.,** The pathogen of wheat rosette stunt disease, *Sci. Sin.*, 22, 573, 1979.
8. **Atkinson, R. E.,** Western wheat mosaic in Colorado and its transmission by *Toxoptera gramineum, Phytopathology,* 39, 2, 1949.
9. **Barr, D. J. S. and Slykhuis, J. T.,** Further observations on zoosporic fungi associated with wheat spindle streak mosaic virus, *Can. Plant Dis. Surv.*, 56, 77, 1976.
10. **Basu, A. N. and Raychadhuri, S. P.,** Further studies on transmission of wheat mosaic streak virus, *Indian J. Entomol.*, 34, 115, 1972.
11. **Bennett, F. G. A.,** Resistance to powdery mildew in wheat, *Plant Pathol.*, 33, 279, 1984.
12. **Borlaug, N. E.,** Wheat rust and people, *Phytopathology,* 55, 1088, 1965.
13. **Borlaug, N. E., Harrar, J. G., and Stakman, E. C.,** Flag smut of wheat in Mexico, *Phytopathology,* 36, 479, 1946.
14. **Brcak, J.,** Leafhopper and plant hopper vectors of plant disease agent in Central and Southern Europe, in *Leafhopper Vectors and Plant Disease Agents,* Maramorosch, K. and Harris, K., Eds., Academic Press, 1979, 97.
15. **Brown, A. G. P. and Reeves, J. T.,** Wheat rust epidemics, *J. Agric. West. Aust.*, 16, 72, 1975.
16. **Canova, A.,** Recherche sulle malattie da virus delle Graminacee. III, *Phytopathol. Mediterr.*, 5, 53, 1966.
17. **Chagorova, R. M.,** Some data on the biology of leaf rusts of wheat in the conditions of the Frunze region, in Materials of the First Coordinated Conference of Mycologists of the Republics of Central Asia and Kazakhstan, 155, 1960; *Rev. Appl. Mycol.*, 42, 592, 1963.
18. **Chiykowski, L. N.,** Reaction of some wheat varieties to Aster yellows virus, *Can. J. Plant Sci.*, 47, 149, 1967.
19. **Cloutier, A.,** Éradication de l'épine-vinette au Québec, *Phytoprotection,* 53, 14, 1972.
20. *Commonw. Mycol. Inst. Descr.*, 23, 1970.
21. *Commonw. Mycol. Inst. Descr.*, 157, 1976.
22. *Commonw. Mycol. Inst. Descr.*, 3, 1970; 180, 1977.
23. **Comeau, A. and Dubuc, J. P.,** Observations on the 1976 barley yellow dwarf epidemic in eastern Canada, *Can. Plant Dis. Surv.*, 57, 42, 1977.
24. **Conners, I. L.,** An annotated Index of Plant Diseases in Canada, Canada Department of Agriculture Publ. 1251, 1967, 44.
25. **Conti, M.,** Vector relationships and other characteristics of barley yellow striate mosaic virus (BYSMV), *Ann. Appl. Biol.*, 95, 83, 1980.
26. **Conti, M.,** Epidemiology of maize rough dwarf virus, *Phytoparasitica,* 12, 210, 1984.
27. **Damsteegt, V. D.,** Investigations of the vulnerability of U.S. maize to maize streak virus, *Prot. Ecol.*, 2, 157, 1980.
28. **Davidson, S.,** Countering soilborne cereal diseases, *Rural Res.*, 123, 12, 1984.
29. **Fitzgerald, P. J. and Stoner, W. N.,** Barley yellow dwarf studies on wheat (*Triticum aestivum* L.). I, *Crop. Sci.*, 7, 337, 1967.
30. **Fitzgerald, P. J. and Timian, R. G.,** Effect of barley stripe mosaic on wheat, *Plant Dis. Rep.*, 44, 359, 1960.
31. **Gates, L. F.,** Influence of sowing dates, soil amendments and cultivars on wheat spindle streak mosaic in winter wheat, *Can. J. Plant Sci.*, 55, 891, 1975.
32. **Gerechter-Amitái, Z. K. and Grama, A.,** Inheritance of resistance to stripe rust, *Euphytica,* 23, 387, 1974.
33. **Ghodbane, A.,** Stripe rust on wheat, *FAO Plant Prot. Bull.*, 25, 212, 1977.
34. **Gill, C. C.,** Assessment of losses on spring wheat naturally infected with barley yellow dwarf virus, *Plant Dis.*, 64, 197, 1980.
35. **Graham, C. L.,** Inability of certain vectors in North America to transmit maize streak virus, *Environ. Entomol.*, 8, 228, 1979.
36. **Granhall, I.,** Old and new problems in the field of plant quarantine, *EPPO Bull.*, 11, 139, 1981.
37. **Green, G. J. and Campbell, A. B.,** Wheat cultivars resistant to *Puccinia graminis* in Western Canada, *Can. J. Plant Pathol.*, 1, 3, 1979.
38. **Green, G. J., Martens, J. W., and Ribeiro, O.,** Epidemiology and specialization of wheat and oat stem rust in Kenya in 1968, *Phytopathology,* 60, 309, 1970.

39. **Green, G. J.,** Physiologic races of wheat stem rust in Canada from 1919 to 1969, *Can. J. Bot.*, 49, 1575, 1971.
40. **Green, G. J.,** Virulence changes in *Puccinia graminis* f. sp. *tritici* in Canada, *Can. J. Bot.*, 53, 1377, 1975.
41. **Halisky, P. M., Prato, J. D., Houston, B. R., and Lindt, J. H.,** Stripe rust epidemic in central California, *Calif. Agric.*, 16, 5, 1962; *Rev. Appl. Mycol.*, 41, 449, 1962.
42. **Harder, D. E., Mathenge, G. R., and Mwaura, L. K.,** Physiologic specialization and epidemiology of wheat stem rust in East Africa, *Phytopathology*, 63, 166, 1972.
43. **Harpaz, I.,** *Calligypona marginata* the vector of maize rough dwarf virus, *FAO Plant Prot. Bull.*, 9, 144, 1962.
44. **Hermansen, J. E. and Wiberg, A.,** On the appearance of *Erysiphe graminis* f.sp *hordei* and *Puccinia hordei* in the Faeroes, *Friesia*, 10, 30, 1972.
45. **Herrera, M. G. and Quiroz, E. C.,** Distribución y factores epidemiológicos del virus del enanismo amarillo de la cebada (BYDV) en Chile, *Agric. Tech.*, 43, 121, 1983.
46. **Hinke, F.,** Succcess in black rust control in Bavaria through eradication of Barberry, *Bayer Landwirtsch. Jahrb.*, 41, 221, 1964.
47. **Hoffman, J. A., Kendrick, E. L., and Meiners, J. P.,** Pathogenic races of *Tilletia controversa* in the Pacific Northwest, *Phytopathology*, 57, 279, 1962.
48. **Hoffman, J. A., Kendrick, E. L., and Metzger, R. T.,** A revised classification of pathogenic races of *Tilletia controversa, Phytopathology*, 57, 279, 1967.
49. **Holton, C. S. and Johnson, A. G.,** Physiologic races in *Urocystis tritici, Phytopathology*, 33, 168, 1943.
50. **Ilyas, M. and Hussain, M.,** Screening of genetic stocks of *Triticum aestivum* for resistance to earcockle *(Anguina tritici), J. Agric. Res.*, 21, 23, 1983.
51. **Jamalainen, E. A.,** Black rust and occurrence of Berberidaceae in Finland, *Maataloust Aikak.*, 25, 47, 1953.
52. **Jenkyn, J. F. and Bainbridge, A.,** *Biology and Pathology of Cereal Powdery Mildews*, Academic Press, New York, 288, 1978.
53. **Johnson, A. G.,** Further studies of physiologic races in *Urocystis tritici, Phytopathology*, 49, 299, 1959.
54. **Johnson, R. and Dyck, P. L.,** Resistance to yellow rust in *Triticum spelta* var *album* and bread wheat cultivars Thatcher and Lee, *Les Colloques de l'INRA*, 25, 71, 1984.
55. **Joshi, L. M.,** Recent contributions towards epidemiology of wheat rusts in India, *Indian Phytopathol.*, 29, 1, 1976.
56. **Joshi, L. M.,** Wheat rust management: present knowledge and future prospects, *Kavaka*, 10, 1, 1982.
57. **Joshi, L. M., Goel, L. B., Sharma, R. C., and Mohan, M.,** Prevalence and distribution of physiologic races of wheat rusts on dwarf wheats during the 1967—1968 crop, *Indian Phytopathol.*, 23, 637, 1970.
58. **Joshi, L. M., Nagarajan, S., and Srivastava, K. D.,** Epidemiology of brown and yellow rusts of wheat in north India. I, *Phytopath. Z.*, 90, 116, 1977.
59. **Joshi, L. M. and Palmer, L. T.,** Epidemiology of stem, leaf and stripe rusts of wheat in northern India, *Plant Dis. Rep.*, 57, 8, 1973.
60. **Joshi, L. M., Singh, D. V., Srivastava, K. D., and Wilcoxson, R. D.,** Karnal bunt: a minor disease that is now a threat to wheat, *Bot. Rev.*, 49, 309, 1983.
61. **Katsuya, K. and Green, G. J.,** Reproductive potentials of races 15B and 56 of wheat stem rust, *Can. J. Bot.*, 45, 1077, 1967.
62. **Kendrick, E. L.,** Race groups of *Tilletia caries* and *Tilletia foetida* for varietal resistance testing, *Phytopathology*, 51, 537, 1961.
63. **Knott, D. R.,** Using race-specific resistance to manage the evolution of plant pathogens, *J. Environ. Qual.*, 1, 227, 1972.
64. **Knott, D. R.,** The association and dissociation of genes for virulence in wheat stem rust, *Phytopathology*, 74, 1023, 1984.
65. **Kumar, C. S. K. V. and Rao, A. S.,** Inoculum potential disease development and penetration of host by *Alternaria triticina, Proc. Indian Acad. Sci.*, 88B, 359, 1979.
66. **Lambat, A. K., Nath, R., Mukewar, P. M., Majumdar, A., Rani, I., Kaur, P., Varshney, J. L., Agarwal, P. C., Khatarpal, R. K., and Dev, U.,** International spread of karnal bunt of wheat, *Phytopathol. Mediterr.*, 22, 213, 1983.
67. **Leukel, R. W.,** The nematode disease of wheat and rye, USDA Farmers' Bull. 1607, U.S. Department of Agriculture, Washington, D.C., 1957.
68. **Line, R. F.,** Factors contributing to an epidemic of stripe rust on wheat in the Sacramento Valley of California in 1974, *Plant Dis. Rep.*, 60, 312, 1976.
69. **Luig, N. H. and Watson, I. A.,** The role of wild and cultivated grasses in the hybridization of formae species of *Puccinia graminis, Aust. J. Biol. Sci.*, 25, 335, 1972.

70. **Mann, C. E., Rajaram, S., and Villareal, R. L.,** Progress in breeding for *Septoria tritici* resistance in semidwarf spring wheat at CIMMYT, *Septoria* of Cereals, *Proc. Workshop Montana State University,* Scharen, A. L., Ed., Bozeman, Mont., 1983.
71. **Maramorosch, K. and Harris, K. F., Eds.,** *Leafhopper Vectors and Plant Disease Agents,* Academic Press, New York, 1979.
72. **Martens, J. W.,** Virulence dynamics in the wheat stem rust populations of Kenya, *Plant Dis. Rep.,* 59, 763, 1975.
73. **Martin, P. M. D.,** The Mycoflora of Maize, Sorghum, Wheat and Groundnuts, Res. Bull. No. 1 for 1977, Malkerns Res. Stn., Swaziland, 1977.
74. **Mathur, R. S. and Ahmad, Z. U.,** Longevity of *Corynebacterium tritici* causing tundu disease of wheat, *Proc. Natl. Acad. Sci. India Sect. B,* 34, 335, 1964.
75. **Mathur, S. C. and Ram, S.,** Longevity of chlamydospores of *Neovossia indica* (Mitra) Mundkur, *Sci. Cult.,* 29, 411, 1963.
76. **McEwan, J. M.,** The source of stem rust infecting New Zealand wheat crops, *N. Z. J. Agric. Res.,* 9, 536, 1966.
77. **McNeal, F. H., Berg, M. A., and Carroll, T. W.,** Barley stripe mosaic virus data from six infected spring wheat cultivars, *Plant Dis. Rep.,* 60, 730, 1976.
78. **Mitra, M.,** A new bunt on wheat in India, *Ann. Appl. Biol.,* 18, 178, 1931.
79. **Mortensen, K. and Green, G. T.,** Assessment of receptivity and uredospore production as components of wheat stem rust resistance, *Can. J. Bot.,* 56, 1827, 1978.
80. **Nagaich, B. B. and Sinha, R. C.,** Eastern wheat striate: a new viral disease, *Plant Dis. Rep.,* 58, 968, 1974.
81. **Nagarajan, S. and Joshi, L. M.,** A historical account of wheat rust epidemics in India and their significance, *Cereal Rusts Bull.,* 3, 29, 1975.
82. **Nagarajan, S. and Joshi, L. M.,** Epidemiology of brown and yellow rusts of wheat over north India. II, *Plant Dis. Rep.,* 62, 186, 1978.
83. **Nagarajan, S., Singh, H., and Joshi, L. M.,** Climatic factors in relation to stem rust epidemiology, *Plant Dis. Rep.,* 59, 670, 1975.
84. **Neergaard, P.,** Risks for the EPPO Region from seedborne pathogens, *EPPO Bull.,* 11, 207, 1981.
85. **Negulescu, F. and Ionescu-Cojocaru, M.,** The outbreak of a new form of race 77 of *Puccinia recondita* f.sp. *tritici* on wheat cultivar Aurora in Romania in 1973, *Cereal Rust Bull.,* 2, 19, 1974.
86. **Newton, M. and Cherewick, W. J.,** *Erysiphe graminis* in Canada, *Can. J. Res.,* 25C, 73, 1947.
87. **Ogilvie, L. and Thorpe, G.,** New light on epidemics of black stem rust of wheat, *Sci. Prog.,* 49, 209, 1961.
88. **Oliynik, A. M.,** Transmission of wheat streak mosaic virus by *A. tulipae, Mykrobiol. Zh.,* 29, 338, 1967; *Rev. Appl. Mycol.,* 48(417), 81, 1969.
89. **Onischenko, A. N.,** The effect of different factors on the incubation period of phytopathogenic mycoplasmas in plants and leafhopper vectors, *Mikrobiol. Zh.,* 46, 52, 1984; *Rev. Plant Pathol.,* 64(1535), 154, 1985.
90. **Overlaet, J.,** Opzoekingen in de roestziekten van de Tarwe, *Verh. Acad. Wet. Brussels Class Sci.,* 57, 1, 1958.
91. **Paliwal, Y. C.,** Mite transmitted plant viruses, in *Vistas in Plant Pathology,* Varna, A. and Verma, J. P., Eds., Malhotra, New Delhi, 1986, 567.
92. **Paul, H. L., Huth, W., and Querfurth, G.,** Cocksfoot mild mosaic — Phleum Mottle Virus: a comparison, *Intervirology,* 2, 253, 1973/1974.
93. **Prabhu, A. S. and Prasada, R.,** Evaluation of seed infection caused by *Alternaria triticina* in wheat, *Proc. Int. Seed Test. Assoc.,* 32, 647, 1967.
94. **Proeseler, G.,** Ubertragung des Trespenmosaik Virus durch Getreidehahnchen (Coleoptera Chrysomelidae), *Arch. Phytopathol. Pflanzenschutz,* 14, 267, 1978.
95. **Purdy, L. H.,** Flag smut of wheat, *Bot. Rev.,* 31, 565, 1965.
96. **Rabie, C. J. and Dippenaar, B. J.,** Physiologic races of wheat bunt in South Africa, *S. Afr. J. Agric. Sci.,* 5, 643, 1962.
97. **Radulescu, E. and Munteanu, I.,** Recherches sur la symptomatologie, l'etiologie et la lutte contre le rabougrissement jaune du blé en Roumanie, *Ann. Phytopathol.,* 2, 403, 1970.
98. **Rajaram, S. and Campos, A. V.,** Epidemiology of wheat rusts in the western hemisphere, *CIMMYT Res. Bull.,* 27, 27 pp., 1974.
99. **Rao, A. S. and Brakke, M. K.,** Relation of soilborne wheat mosaic virus and its fungal vector, *Polymyxa graminis, Phytopathology,* 59, 581, 1969.
100. **Rees, R. G.,** *Agropyron scabrum* and its role in the epidemiology of *Puccinia graminis* in North Eastern Australia, *Aust. J. Agric. Res.,* 23, 789, 1972.
101. **Rees, R. G., Thompson, J. P., and Mayer, R. J.,** Slow rusting and tolerance to rusts in wheat. I, *Austr. J. Agric. Res.,* 30, 403, 1979.

102. **Rees, R. G., Thompson, J. P., and Goward, E. A.,** Slow rusting and tolerance to rusts in wheat. II, *Austr. J. Agric. Res.*, 30, 421, 1979.
103. **Richardson, M. T.,** *An Annotated List of Seedborne Diseases,* 3rd ed., Commonwealth Mycological Institute, Kew, 1979, and Supplement 1, Int. Seed Testing Assoc., Zurich, 1981.
104. **Roane, C. W., Stakman, E. C., Loegering, W. Q., Stewart, D. M., and Watson, W. M.,** Survival of physiologic races of *Puccinia graminis* var *tritici* on wheat near barberry bushes, *Phytopathology,* 50, 40, 1960.
105. **Roelfs, A. P.,** Evidence for two populations of wheat stem and leaf rust in the USA, *Plant Dis. Rep.*, 58, 806, 1974.
106. **Roelfs, A. P.,** Effects of barberry eradication on stem rust in the United States, *Plant Dis.*, 66, 177, 1982.
107. **Rovira, A. D. and Ridge, E. H.,** Soil-borne root diseases and wheat, in *Soils: An Australian Viewpoint,* Division of Soils, CSIRO, Melbourne, 1983, 721.
108. **Saari, E. E.,** Susceptibility of North American *Thalictrum* spp. to infection by *Puccinia recondita* f.sp. *tritici, Diss. Abstr.*, 28B, 17, 1967; *Rev. Appl. Mycol.*, 47(2126), 387, 1968.
109. **Saari, E. E. and Wilcoxson, R. D.,** Plant disease situation of high-yielding dwarf wheat in Asia and Africa, *Annu. Rev. Phytopathol.*, 12, 49, 1974.
110. **Sanwal, K. C. and Mathur, V. K.,** Soil as a contaminant of wheat seed, *Indian J. Nematol.*, 5, 118, 1975.
111. **Serjeant, E. P.,** The transmission of European wheat striate mosaic virus by *Javesella pellucida* (Fabr.), Ann. Appl. Biol., 59, 39, 1967.
112. **Shaner, G. and Powelson, R. L.,** Epidemiology of stripe rust of wheat, 1961—1968, Tech. Bull. 117, Agric. Exp. Stn. Oregon State University, Corvallis, 1971.
113. **Shaner, G. and Powelson, R. L.,** The oversummering and dispersal of inoculum of *Puccinia striiformis* in Oregon, *Phytopathology,* 63, 13, 1973.
114. **Shaskol'skaya, N. D.,** Identification of a mosaic disease of millet, *Nauchn. Dokl. Vyssh. Shk. Biol. Nauk:* 1961, 103, 1961; *Rev. Appl. Mycol.*, 40, 531, 1961.
115. **Signoret, P. A., Conti, M., Leclant, F., Alliot, B., and Giannotti, J.,** Données nouvelles sur la maladie des stries chlorotiques du blé, *Ann. Phytopathol.*, 9, 381, 1977.
116. **Silfhout, C. H. and Groenewegen, L. J. M.,** The use of *Triticum dicoccoides* in winter wheat breeding, *Les Colloques de l'INRA,* 25, 129, 1984.
117. **Singh, D. V., Srivastava, K. D., and Joshi, L. M.,** Occurrence and spread of karnal bunt of wheat in India, *Indian Phytopathol.*, 33, 249, 1980.
118. **Sinha, R. C.,** Comparison of the ability of nymph and adult *Delphacodes pellucida* Fabricius to transmit European wheat striate mosaic virus, *Virology,* 10, 344, 1960.
119. **Sinha, R. C.,** Serological detection of wheat striate mosaic virus in extracts of wheat plants and vector leafhoppers, *Phytopathology,* 58, 452, 1968.
120. **Sinha, R. C.,** *Elymana virescens,* a newly described vector of wheat striate mosaic virus, *Can. Plant Dis. Surv.*, 50, 118, 1970.
121. **Skorda, E.,** Studies on the physiologic races of wheat stem rust *(P. graminis tritici)* in Greece during the eight years 1955—1962, *Ann. Inst. Phytopathol. Benaki NS,* 1, 157, 1966; *Rev. Appl. Mycol.*, 46(1531), 291, 1967.
122. **Slykhuis, J. T.,** Wheat spot mosaic caused by a mite-transmitted virus associated with wheat streak mosaic, *Phytopathology,* 46, 682, 1956.
123. **Slykhuis, J. T.,** Wheat striate mosaic in Canada, a disease with international relations, *Proc. Can. Phytopathol. Soc.*, 29, 11, 1962.
124. **Slykhuis, J. T.,** Wheat striate mosaic, a virus disease to watch on the prairies, *Can. Plant Dis. Surv.*, 42, 135, 1962.
125. **Slykhuis, J. T.,** Virus diseases of cereals, *Rev. Appl. Mycol.*, 46, 401, 1967.
126. **Slykhuis, J. T.,** Transmission of *Agropyron* mosaic virus by the Eriophyid mite, *Abacarus hystrix, Phytopathology,* 59, 29, 1969.
127. **Slykhuis, J. T.,** Virus and virus-like diseases of cereal crops, *Annu. Rev. Phytopathol.*, 14, 198, 1976.
128. **Smith, H. C.,** Aphid species in relation to the transmission of barley yellow dwarf virus in Canada, *N.Z. J. Agric. Res.*, 6, 1, 1963.
129. **Smith, K. M.,** *A Textbook of Plant Virus Diseases,* Longman, New York, 1972.
130. **Sokhi, S. S.,** *Alternaria* blight on wheat in India, *Pest Artic. News Summ.*, 20, 55, 1974.
131. **Spehar, V.,** The role of distribution of Barberry in the epidemiology of cereal black rust in Yugoslavia, *Agron. Glasn.*, 12, 357, 1962; *Rev. Appl. Mycol.*, 43(2582), 472, 1964.
132. **Stakman, E. C.,** Relation of Barberry to the origin and persistence of physiologic forms of *Puccinia graminis, J. Agric. Res.*, 48, 953, 1934.
133. **Stakman, E. C.,** Will the fight against wheat rust ever end?, Z. *Pflanzenkr. Pflanzenschutz,* 71, 67, 1964.
134. **Stakman, E. C., Stewart, D. M., and Loegering, W. Q.,** Identification of physiologic races of *Puccinia graminis* var. *tritici,* USDA Bull. E617, U.S. Department of Agriculture, Washington, D.C., 1962, 53 pp.

135. **Sulochana, C. B.,** Leafhopper and plant hopper transmitted viruses of cereal crops, *Proc. Indian Acad. Sci.,* 93, 335, 1984.
136. **Tajimi, A.,** On the host plant of wheat stem rust (*Puccinia graminis* f.sp. tritici), *Bull. Natl. Grassland Res. Inst. (Japan),* 13, 79, 1978; *Rev. Plant Pathol.,* 58(5790), 493, 1979.
137. **Tao, J., Qin, J., Xiao, J., Shen, Y., Zhao, F., Li, T., Xie, Y., He, D., Rao, Y., and Huang, X.,** Studies on the soil borne yellow mosaic virus of wheat in Sichuan, *Acta Phytopathol. Sin.,* 10, 15, 1980; *Rev. Plant Pathol.,* 60(215), 25, 1981.
138. **Tosic, M.,** Prenosenje virusa crticastog mosaika psenice pomocu *Aceria tosichella* na razne domacine, *Zast. Bilja,* 24, 317, 1973; *Rev. Plant Pathol.,* 54(3818), 702, 1975.
139. **Tu, J. C. and Hendrix, J. W.,** The summer biology of *Puccinia striiformis* in southeastern Washington. II, *Plant Dis. Rep.,* 54, 384, 1970.
140. **Turapin, V. P.,** Reservoirs of stem rust of wheat in the northern Kazakhstan Region, *Zashch. Rast.,* 12, 172, 1973; *Rev. Plant Pathol.,* 54(1223), 245, 1975.
141. **Von Wechmar, B., Kauffmann, A., Desmarais, F., and Rybicki, E. P.,** Detection of seed transmitted brome mosaic virus, *Phytopathol. Z.,* 109, 341, 1984.
142. **Wadhi, S. R., Nath, R., and Sanwal, K. C.,** Quarantine problems in the import of wheat seed into India, *Seed Res.,* 6, 85, 1978.
143. **Wahl, I., Wilcoxson, R. D., and Rowell, J. B.,** Slow rusting of wheat with stem rust detected in the glasshouse, *Plant Dis.,* 64, 54, 1980.
144. **Watson, I. A. and Luig, N. H.,** Somatic hybridization in *Puccinia graminis* var *tritici, Proc. Linn. Soc. NSW,* 83, 190, 1958.
145. **Watson, I. A. and Luig, N. H.,** Asexual intercrosses between somatic recombinants of *Puccinia graminis, Proc. Linn. Soc. NSW,* 87, 99, 1962.
146. **Wolfe, M. S. and Schwarzbach, E.,** Patterns of race changes in powdery mildews, *Annu. Rev. Phytopathol.,* 16, 159, 1978.
147. **Yarwood, C. E.,** Powdery mildews, *Bot. Rev.,* 23, 235, 1957.
148. **Zadoks, J. C.,** Epidemiology of wheat rusts in Europe, *FAO Plant Protection Bull.,* 13, 98, 1965.
149. **Zwang, Z., Lange, L., and Mathur, S. B.,** Teliospore survival and plant quarantine significance of *Tilletia indica,* particularly in relation to China, *EPPO Bull.,* 14, 119, 1984.

Chapter 2

THE PATHOGENS OF THE GRAPEVINE

P. M. D. Martin

TABLE OF CONTENTS

I. INTRODUCTION

The quarantine requirements for the importation of grapevine material have become more complex throughout the world as further knowledge of the occurrence and biology of grapevine pathogens has become available. Their control through effective quarantine procedures has progressed a long way since the occurrence of one of the earliest and most famous examples of an epidemic that today would have been controlled by modern expertise. In North America, one of the native vines, *Vitis labrusca*, is immune to *Uncinula necator*, the powdery mildew. This grape species was introduced into Europe in the mid 19th century so that hybridization with *V. vinifera*, the cultivated grape, would increase resistance to the fungus. In this way, grape phylloxera, *Daktulosphaira vitifoliae*, and two fungus diseases, *Plasmopara viticola* (downy mildew) and *Guignardia bidwelli* (black rot) were introduced, causing immense damage and resultant crop losses. The French vineyards were almost entirely destroyed and had to eventually be entirely replanted. Phylloxera alone destroyed one million hectares of grapevines in France between 1860 and 1895[54] and the pest persists in Europe to this day, reducing the output of susceptible cultivars that would otherwise be productive. The threat of phylloxera did, however, stimulate the need for effective plant quarantine services to guard against various noxious organisms from other countries.

Today the number and distribution of grapevine pathogens is gradually becoming better known and a great deal of information on identity, synonymy, host range, and distribution has become available, especially in the last 20 years. Although the increase in knowledge has been helpful to quarantine specialists, the range of quarantine activities has also expanded beyond the mere need to exclude foreign pests and diseases because of their potential threat to productivity.

Each country will have its own quarantine perspectives depending on the perceived threats from outside, and on the size of the country. In general, important pathogens that are absent or not established will be the targets of regulation procedures. In countries of moderate size, such as Australia or South Africa that are wholly or partially isolated geographically from their neighbors, it is relatively easy to control foreign introductions. In large countries like Canada and the U.S. there may be domestic pest risk concerns in addition that determine whether a given region should be the subject of scrutiny. In Canada there are two large grape-growing regions, southern Ontario and the Okanagan valley of British Columbia, separated by several thousand kilometers. The latter is one of the most recently developed in the world, so it is necessary to protect it from pathogens in other parts of the country, whether they actually already invade grapevines or are present in some other host which will act as a reservoir from which some future epiphytotic can be launched. Consequently, there are restrictions on the movement of grapestocks from east to west inside Canada. In the smaller European countries, on the other hand, restrictions of movement of material may be impracticable, so that increased emphasis is placed on the development of resistant varieties.

Another modern quarantine objective is the elimination of pests or diseases inside the same country if their distribution is sporadic or if they have some special biological vulnerability that can be exploited. Even if a pest or pathogen is relatively widespread within a country, it could still be placed on a quarantine list if it is unacceptable to that country's trading partners and hence would imperil the certification of material for export.

The distribution of 42 grapevine pathogens of quarantine importance to Canada is shown in Table 1. It is clear that by far the largest number is still present in Europe and that regions outside Europe may have much to lose by their introduction.

Table 1
DISTRIBUTION OF 42 PATHOGENS OF THE GRAPEVINE

	1	2	3	4	5	6	7	8
Nepoviruses								
Arabis mosaic virus	+	−	−	−	+	−	−	−
Artichoke Italian latent virus	−	(+)	−	−	+	−	−	−
Grapevine Bulgarian latent virus	−	+	−	−	+	−	−	−
Grapevine fanleaf virus and its strains	+	+	+	+	+	+	+	+
Hungarian chrome mosaic virus	−	−	−	−	+	−	−	−
Peach rosette mosaic virus	+	+	−	−	−	−	−	−
Raspberry ringspot virus	(+)	−	−	−	+	−	−	−
Strawberry ringspot virus	−	−	−	−	+	−	−	−
Tobacco ringspot virus	+	+	−	−	(+)	−	−	(+)
Tomato black ring virus	+	+	−	−	+	−	−	−
Tomato ringspot virus	+	+	−	+	+	−	−	−
Other viruses								
Grapevine asteroid mosaic agent	−	+	−	−	+	−	+	−
Grapevine Bratislava mosaic virus	−	−	−	−	+	−	−	−
Grapevine Chasselas latent agent	−	−	−	−	+	−	−	−
Grapevine corky bark "legno riccio" agent	−	+	+	+	+	+	+	+
Grapevine enation agent	−	+	−	+	+	−	+	+
Grapevine leafroll agent	+	+	+	+	+	+	+	+
Grapevine little leaf agent	−	−	−	−	−	−	−	−
Grapevine Rupestris stempitting agent	−	+	+	−	+	−	−	+
Grapevine vein mosaic agent	−	−	−	−	+	−	−	−
Grapevine vein necrosis agent	−	−	−	−	+	−	−	−
Tobacco necrosis virus	(+)	(+)	−	−	+	−	+	−
Tomato bushy stunt virus	(+)	(+)	−	(+)	+	−	−	−
Mycoplasma-like organisms or presumed mycoplasma-like organisms								
[b]Flavescence-dorée agent	−	?+	−	+	+	+	−	−
Black wood agent (bois-noir)	−	−	−	−	+	−	−	−
Bacteria								
Grapevine infectious necrosis bacterium	−	−	−	−	+	−	−	−
Fungi								
Charrinia diplodiella (Speg.) Sacc.	+	+	+	+	+	−	+	+
Elsinoe ampelina (De Bary) Shear	−	+	+	+	+	−	+	+
Eutypa armeniacae Hansford & Carter	+	+	+	−	+	−	(+)	+
Fomes (Phellinus) igniarius L. ex Fr.	(+)	+	+	+	+	−	−	−
Peyronellaea glomerata Ciferri	−	(+)	−	−	+	(+)	−	(+)
Phomopsis viticola Sacc.	+	+	+	−	+	+	+	+
Physopella vitis (Thum.) Arthur	−	+	+	+	−	−	−	−
Phytophthora cinnamomi Rands	−	+	−	+	−	−	+	+
Pseudopeziza tracheiphila Muller Thurgau	−	−	−	−	+	−	−	−
Rhacodiella vitis Shterenberg	−	−	−	−	+	−	−	−
Rosellinia necatrix (Hart.) Berl.	−	(+)	+	+	+	−	+	+
Septoria ampelina Berk. & Curt.	−	+	+	−	+	−	−	−
Stereum hirsutum (Willd.) Pers.	(+)	+	−	+	+	+	+	−
Total incidence	14	27	11	15	37	8	14	12

Note: All 42 pathogens are considered quarantinable by Canada; (− not present; + present; (+) present but not recorded in *Vitis*.) 1 = Canada; 2 = U.S.; 3 = Mexico and Central America; 4 = South America; 5 = Europe and U.S.S.R.; 6 = Israel; 7 = South Africa; 8 = Australia.

[a] To be added to Canada's list of designated pathogens.

[b] See Reference 57. Proof of occurrence in the U.S. is still to be obtained.

II. ANALYSIS OF GRAPEVINE PATHOGENS

Based on their biological attributes, microbial species attacking the grapevine can be arranged in three groups. A list of these is contained in Table 2.

A. Species Dispersed Passively by Mechanical Contact, Propagules, Insects, Nematodes, Weeds, and by Various Natural Media (Air, Soil, Water, etc.) as well as by some Human Agency

1. Species with a Particular Affinity to the Fruit

These organisms comprise various filamentous fungi and yeasts causing 'sour rot' or 'bunch rot' of grapes at or after harvest and others which break down the fruit in storage. They are ubiquitous and are not a major cause of economic loss if picking and storage of grapes is done under controlled conditions. Some of the yeasts participate in the fermentation responsible for wine making.

2. Saprophytes and Pathogens which Mainly Develop on the Roots, Stems, and Leaves

This is the largest group, comprising bacteria, fungi, and a few viruses, with ubiquitous to extremely limited distribution. It is unfortunate that so many of the most serious pathogens, such as *Agrobacterium tumefaciens* (crown gall), *Botrytis cinerea* (grey mold), *Plasmopara viticola* (downy mildew), and *Uncinula necator* (powdery mildew) are now so widely dispersed that in general quarantine measures against them can serve no purpose.

The success of these pathogens is usually due to a complex of factors, and the best that can be hoped for is containment, as far as possible, using hygienic measures in cultivation together with appropriate fungicides.[41] A sound appreciation of the biology of the pathogen may also help in control. Taking *Botrytis cinerea* for example, this pathogen is easily distributed by large numbers of airborne conidia, and infection is aided by the damage caused by pests such as the tortrix moths *Lobesia botrana* and *Eupoecilia ambiguella*.[18] Other pathogens such as *Uncinula necator* may also play a synergistic role. The destructive potential of *Botrytis* is magnified by a humid climate and high summer temperatures, resulting in losses of 13 to 90%.[30,49] A centralized warning system is difficult to set up due to climatic variability and different cultivar susceptibilities, so it is best to treat the fruit three times during growth with an appropriate fungicide.[11] Of the recently developed fungicides available, vinclozolin seems to be preeminent[50,57] and this can be combined with various integrated control measures.[61]

Control of a widespread pathogen can be complicated by the presence of pathotypes with affinities for different hosts, as in the case of the three biovars of *Agrobacterium tumefaciens* that can be differentiated on biochemical grounds.[19] Inoculation with a nonpathogenic strain of *A. tumefaciens*, K84, now marketed commercially, has been shown to inhibit subsequent invasion by most pathogenic strains.[39] Unfortunately, biovar 3, which appears to be the major (though not the only) biovar on grapevines, is resistant to the antibiotic agrocin produced by strain K84, so it cannot be controlled in this way. A good case can be made for the restriction of grape material proved to be infected with biovar 3 until it has been cleaned up by meristem tip culture. There is now considerable interest within the U.S. in incorporating crown gall freedom into grapevine nursery programs. Detection procedures for establishing plants free of the bacterium are being developed using a monoclonal antibody that is apparently specific for biovar-3, and which, therefore, has the potential to be used in a budwood certification program. In this way devastating epidemics of crown gall on grape could be prevented, as well as the selective dissemination of biovar 3 over the other two biovars.[20]

In many other cases where the distribution of a pathogen is more limited, the potential damage caused is often too minor to justify the cost of quarantine measures. It is those

Table 2
MICROFLORA OF *VITIS VINIFERA*

Group 1
SPECIES DISPERSED LARGELY MECHANICALLY BY CONTACT, PROPAGULES, LIVING AGENTS, AND BY NATURAL MEDIA INCLUDING, AIR, SOIL, AND WATER

Species involved in pre- and postharvest diseases of fruit
- Fungi
  - *Aspergillus aculeatus* Iizuka
  - *Aspergillus flavus* Link ex. Fr.
  - *Aspergillus niger* Van Tieghem
  - *Aspergillus ochraceus* Wilhelm
  - *Aspergillus rugulosus* Thom & Raper
  - *Aspergillus terreus* Thom
  - *Aspergillus wentii* Wehm
  - *Candida krusei* Castellani & Berkhout
  - *Candida sorbosa* Hedrick & Burke ex Van Uden & Buckley
  - *Candida steatolytica* Yarrow
  - *Cladosporium cladosporioides* (Fres.) De Vries
  - *Cladosporium oxysporum* (Berk. & Curt.)
  - *Greeneria uvicola* (B. & C.) Punithalingam
  - *Guignardia bidwelli* (Ell.) Viala & Ravaz
  - *Hanseniaspora uvarum* (Niehaus) Shehata et al.
  - *Metschnikowia pulcherrima* Pitt & Miller
  - *Penicillium brevicompactum* Dierckx
  - *Penicillium canescens* Sopp
  - *Penicillium chrysogenum* West
  - *Penicillium frequentens* West
  - *Penicillium stoloniferum* Thom.
  - *Pichia membraneofaciens* Hansen
  - *Rhizopus oryzae* Went & Geerl.
  - *Saccharomycopsis crataegensis* (Kreger-Van Rij) Van Der Walt & Scott
  - *Saccharomycopsis vini* Kreger-Van Rij
  - *Saccharomycopsis ludwigii* Hansen
  - *Sclerotinia fuckeliana* (De Bary) Fuckel
  - *Sclerotinia sclerotiorum* (Lib.) De Bary
  - *Zygophiala jamaicensis* Mason

Species more widely distributed over the growing plant
- Viruses
  - *Tomato bushy stunt virus
- Bacteria
  - *Agrobacterium tumefaciens* (Smith & Townsend) Conn.
  - *Erwinia lathyri* (Manns & Taub.) Magrou (= *E. herbicola [Lohnis] Dye)*
  - *Grapevine yellows bacterium*
  - *Pseudomonas syringae* Van Hall
  - *Pseudomonas viticola* Nayudu
  - *Pseudomonas vitis* (Lev.) Speg.
  - **Xanthomonas ampelina* Panagopoulos
- Fungi
  - *Alternaria alternata (Fr.) Keissler*
  - *Alternaria vitis* Cav.
  - *Armillaria mellea* Vahl ex Fr.
  - *Athelia rolfsii* (Curzi) Tu & Kimbrough
  - *Aureobasidium pullulans* (De Bary) Arn.
  - *Botryodiplodia palmarum* (Cooke) Petrak & Sydow
  - *Botryosphaeria dothidea* (Gross. & Duggar) Arx & Muller
  - *Botryosphaeria obtusa* (Schw.) Shoem.
  - *Botryosphaeria rhodina* Von Arx

Table 2 (continued)
MICROFLORA OF *VITIS VINIFERA*

Group 1
SPECIES DISPERSED LARGELY MECHANICALLY BY CONTACT, PROPAGULES, LIVING AGENTS, AND BY NATURAL MEDIA INCLUDING, AIR, SOIL, AND WATER

Botryosphaeria ribis Grossenbacher & Duggar
Botryosphaeria stevensii Shoemaker
Camarosporium flaccidum Zachos & Tsavella-Klonari
Cercospora riachueli Spegazzini
**Charrinia diplodiella* (Speg.) Sacc.
Cladosporium oxysporum B. & C.
Coryneum microstictum B. & Br.
Cristulariella pyramidalis Waterman & Marshall
Cylindrocarpon obtusisporum Wr.
Cytospora vitis Montagne
Drechslera papendorfii (Van Der Aa) M.B. Ellis
Drechslera rostrata (Drechsl.) Rich. & Fraser
**Elsinoe ampelina* (De Bary) Shear
Erysiphe tuckeri Berk.
**Eutypa armeniacae* Hans. & Carter
Exosporium sultanae Du Plessis
**Fomes igniarius* L. Ex Fr.
Fusarium equiseti (Corda) Saccardo
Fusarium oxysporum Schlecter emend Snyder
Fusarium solani (Mart.) Sacc.
Ganoderma lucidum (W. Curt. ex Fr.) Karst.
Gliocladium roseum (Link) Thom
Glomerella cingulata (Ston.) Spauld. & Schrenk
Greeneria uvicola (B. & C.) Punithalingam
Guignardia bidwellii (Ellis) Viala & Ravaz
Hendersonula toruloidea Nattrass
Laestadia bidwelli (Ellis) Viala & Ravaz
Macrophoma malorum (Berk.) Berl. & Vogl.
Mycosphaerella angulata Jenkins
Mycosphaerella personata Higgins
Mycosphaerella tulasnei (Jancz.) Rathers
Nectria radicicola Gerlach & Nilsson
Oidium tuckeri Berk.
Pestalozzia uvicola Speg.
**Peyronellaea glomerata* (Corda) Goid.
Phialophora parasitica Ajello, Georg, & Wang
Phoma vinifera Cke.
**Phomopsis viticola* Sacc.
Phymatotrichum omnivorum (Shear) Duggar
**Physopella vitis* (Thum.) Arthur
**Phytophthora cinnamomi* Rands
Plasmopara viticola (B. & C.) Berl. & De Toni
Pleospora herbarum (Pers. Ex Fr.) Rabh.
Podosphaera tridactyla (Wallr.) De Bary
Polyporellus rhizophilus (Pat.) Pilat.
Polyporus sterilis Reichert
Psathyrella ampelina Foex & Viala
**Pseudopeziza tracheiphila* Muller Thurgau
Pythium ultimum Trow
Ragnhildiana rosleri (Catt.) Vassil.
**Rhacodiella vitis* Shterenberg
Rhizopus oryzae Went & Geerling

Table 2 (continued)
MICROFLORA OF *VITIS VINIFERA*

Group 1
SPECIES DISPERSED LARGELY MECHANICALLY BY CONTACT, PROPAGULES, LIVING AGENTS, AND BY NATURAL MEDIA INCLUDING, AIR, SOIL, AND WATER

Rhizopus stolonifer (Ehr. Ex Fr.) Lind
Roesleria hypogaea Thum. & Pass.
**Rosellinia necatrix* (Hart.) Berl.
Sclerotinia fructicola (Wint.) Rehm
Sclerotinia fructigena Aderh. & Ruhl
Sclerotinia fuckeliana (De Bary) Fuckel
Sclerotinia laxa Aderh. & Ruhl
Sclerotinia sclerotiorum (Lib.) De Bary
Scolecotrichum vitiphylla (Speschm.) Karakulin & Vassiljevsky
**Septoria ampelina* B. & C.
Septoria buharica Naumov
Septoria melanosa (Viala & Ravaz) Elenk
Sordaria uvicola Viala & Marchais
Sphaeropsis tumefaciens L. Daigle
Stagonospora bulgarica Vanev
**Stereum hirsutum* (Willd.) Pers.
Thanatephorus cucumeris (Frank) Donk
Thielaviopsis basicola (Berk. & Br.) Ferraris
Uncinula necator (Schw.) Burr.
Verticillium albo-atrum Reink. & Berthold
Verticillium dahliae Kleb.

Algae
Cephaleuros parasitica Karst.

GROUP 2
VIRUSES AND MYCOPLASMA-LIKE ORGANISMS WITHOUT A KNOWN VECTOR, TRANSMITTED APPARENTLY BY GRAFTING ONLY (HUMAN AGENCY)

Viruses and viroids
*Grapevine Algerian latent virus
*Grapevine asteroid mosaic agent
*Grapevine Bratislava mosaic agent
*Grapevine Bulgarian latent agent[a]
*Grapevine Chasselas latent agent
*Grapevine corky bark-legno riccio-stempitting complex[b]
*Grapevine enation agent
Grapevine false fanleaf agent
Grapevine infectious chlorosis agent
Grapevine interveinal necrosis agent
Grapevine latent agent
*Grapevine leafroll agent[c]
Grapevine line pattern agent
Grapevine little leaf agent
*Grapevine rupestris stem pitting agent[d]
Grapevine shoot necrosis agent
Grapevine small vein mosaic agent
Graepvine summer mottle agent
*Grapevine vein mosaic agent
*Grapevine vein necrosis agent
Grapevine vein pattern mosaic agent
Grapevine vein yellows and leafroll agent
Grapevine yellow dwarf agent

Table 2 (continued)
MICROFLORA OF *VITIS VINIFERA*

Group 1

SPECIES DISPERSED LARGELY MECHANICALLY BY CONTACT, PROPAGULES, LIVING AGENTS, AND BY NATURAL MEDIA INCLUDING, AIR, SOIL, AND WATER

Grapevine yellow speckle agent
Grapevine yellow vein mosaic agent
Viroid similar to hop stunt viroid[e]
Viroids GV-s and GV-f[f]
Mycoplasma-like organisms (suspected)
Grapevine bois noir agent[g]
Grapevine fleck agent[h]
Australian grapevine yellows agent[g]

GROUP 3

VIRUSES AND MYCOPLASMA-LIKE ORGANISMS TRANSMITTED BY A LIVE VECTOR

Transmitted by nematodes (nepoviruses)
*Arabis mosaic virus
*Artichoke Italian latent virus
*Grapevine chrome mosaic virus
*Grapevine fanleaf virus
*Peach rosette mosaic virus[i]
*Raspberry ringspot virus
*Strawberry latent ringspot virus
*Tobacco ringspot virus
*Tomato black ring virus[j]
*Tomato ringspot virus[k]
Transmitted by leafhoppers, aphids, and other insects
Alfalfa mosaic virus
Broad bean wilt virus
Potato virus X
Potato virus Y
Sowbane mosaic virus
Tobacco mosaic virus
*Flavescence dorée MLO
*Pierce's disease bacterium
Transmitted by fungi
*Tobacco necrosis virus

GROUP 4

NEMATODES

A. Species causing direct damage only
Meloidogyne avenaria (Neal) Chitwood
Meloidogyne hapla Chitwood
Meloidogyne incognita Kofoid & White
Meloidogyne javanica (Treub) Chitwood
Meloidogyne nataliae Golden, Rose & Bird
Paratrichodorus minor (Colbran) Siddiqui
Pratylenchus vulnus Allen & Jensen
B. Species transmitting viruses as well as causing direct damage (usually minor)
Species (viruses transmitted)
**Criconemoides sp.*[l] (Peach rosette mosaic virus)
Longidorus apulus Lamberti & Bleve-Zacheo (Artichoke Italian latent virus)
Longidorus attenuatus Hooper (Artichoke Italian latent virus, tomato black ring virus)
Longidorus caespiticola Hooper (Arabis mosaic virus, raspberry ringspot virus)
Longidorus diadecturus Eveleigh & Allen (?Peach rosette mosaic virus, ?tomato black ring virus)

Table 2 (continued)
MICROFLORA OF *VITIS VINIFERA*

Group 1
SPECIES DISPERSED LARGELY MECHANICALLY BY CONTACT, PROPAGULES, LIVING AGENTS, AND BY NATURAL MEDIA INCLUDING, AIR, SOIL, AND WATER

Longidorus elongata (De Man) Thorne & Swanger (Raspberry ringspot virus, tomato black ring virus)
Longidorus leptocephalus Hooper (Raspberry ringspot virus, tomato black ring virus)
Longidorus macrosoma Hooper (Raspberry ringspot virus, tomato black ring virus)
Siddiqia maxima (Butschli) Khan et al. (Raspberry ringspot virus)
Xiphinema americanum Cobb (Peach rosette mosaic virus, tobacco ringspot virus, tomato ringspot virus)
Xiphinema californicum Lamberti & Bleve-Zacheo (Tomato ringspot virus, Arabis mosaic virus)
Xiphinema coxi Tarjan (Strawberry latent ringspot virus)
Xiphinema diversicaudatum (Micoletzky) Thorne (Arabis mosaic virus, raspberry ringspot virus, strawberry latent ringspot virus)
Xiphinema index Thorne & Allen (Grapevine chrome mosaic virus, grapevine fanleaf virus, ?grapevine yellows bacterium)
Xiphinema italiae Meyl (Grapevine fanleaf virus)
Xiphinema paraelongatum Shuurmans et al. (Arabis mosaic virus)
Xiphinema rivesi Dalmasso (Tomato ringspot virus)
Xiphinema vuittenezi Luc et al. (Grapevine chrome mosaic virus, grapevine fanleaf virus)

Note: Asterisks denote species of quarantine importance.

[a] Seed-borne in grape: considered to be a nepovirus but without proof of nematode transmission.
[b] Isometric virus particles and two closteroviruses: grapevine viruses A and B have been observed in connection with these diseases.
[c] Isometric virus particles (different from those above) and four closteroviruses, including viruses A and B, have been observed in association with the disease.
[d] Latent in *V. vinifera* but is expressed in *V. rupestris*: symptoms are distinct from the corky bark-legno riccio-stempitting complex above and a closterovirus has been observed. Quarantine status still undefined due to recent discovery. This disease has been recently described by Prudencio[58] and Tzeng.[68]
[e] Isolated by Sano et al. (1985) in Japan.[62] Pathology not known.
[f] Isolated by Flores et al. (1986) in California.[29] They may be linked to some grapevine diseases of unknown etiology.
[g] Symptoms are similar to those of Flavescence dorée.
[h] Widely distributed. Although it causes symptoms on vine leaves there is no apparent crop loss and its resistance to heat therapy makes elimination impracticable.
[i] Seed-borne in grape.
[j] Seed-borne in grape.
[k] Seed-borne in grape.
[l] Needs confirmation. See References 40 and 60.

organisms with restricted distribution and a high degree of destructiveness such as the asteroid mosaic virus,[38] which need to be continually monitored although their distinction from other pathogens must always be somewhat arbitrary and must rely on continual evaluation of crop losses where the pathogens are established. Any biological information that allows one to speculate whether any of these pathogens would establish itself in a new area should be examined, including the preferred known climatic range of the pathogen and the agents available to facilitate natural spread.

The organisms in this group are relatively easy to identify because the techniques for isolation and identification are well tried and mostly do not require complex equipment or procedures. In most cases the lesions are obvious so that a qualified person should have no difficulty recognizing, for example, material with galls due to *Agrobacterium tumefaciens*,[3,20,46] excoriosis due to *Peyronellaea glomerata* and *Phomopsis viticola*, cankers due to *Eutypa armeniacae*, or the bronzing and necrosis of leaves due to the *Esca* group of diseases caused by *Fomes igniarius* and *Stereum hirsutum*.[9,10,17,23,45] Appropriate culture techniques should enable confirmation by microscopic examination whether lesions are evident or not and there is an abundance of literature available to facilitate the identification of the pathogens and others discussed in succeeding sections, including the Descriptions of Pathogenic Viruses, Fungi and Bacteria, produced by the Commonwealth Mycological Institute in Kew, England and the recent monographs on grapevine pathogens by Galet,[31] and on virus and virus-like diseases of grapevine by Bovey et al.[15]

B. Viruses and Mycoplasma-Like Organisms Apparently Transmitted by Grafting Only

The pathogens in this group are transmitted by grafting but have no known insect or nematode vector. The natural host range is extremely limited. Consequently they are not known to spread in vineyards* and it follows that man must be chiefly responsible for their widespread distribution, through commercial channels in particular. Adequate indexing and inspection of certified stock before and after transit should theoretically provide total control, but in practice, infected material exported to Canada has frequently escaped detection until indexing was performed after arrival. The amount of time required to test fully incoming material on woody indicators at the Sidney Quarantine Station on Vancouver Island, British Columbia, may be $2^1/_2$ years, due to the long time that grapevine leaf roll may take to express itself. This presents an embarrassing problem for a plant health authority faced with impatience and possibly political pressure from or attempts to smuggle by the grape industry and by individuals.

Grapevine leaf roll, grapevine corky bark, and the recently discovered grapevine rupestris stem pitting diseases are now known to be associated with the presence of one or more closteroviruses[25,28,37,51,52,68] and, in some cases, with isometric particles as well.[8,66] The last disease is so far only known to express itself visibly in *Vitris rupestris* cv St. George though it may be latent in several cultivars of *Vitis vinifera*. The effects of these diseases on those cultivars of *Vitis* that are susceptible are similar in that the growth of the vascular cylinder is disrupted and abnormal growth of tissue takes place that may interfere with water transport. The coloration and early abscission of leaves in plants with leafroll or corky bark have more than a fortuitous resemblance to the natural process of defoliation with the onset of winter.

Leafroll, corky bark, and other diseases in this group appear to be most serious in the warm temperate climates of Italy, Southern France, Spain, California, and South Africa, where they are of significant economic importance.[48] In Canada, however, the distribution of leafroll is limited and, although a systematic survey has not be made, there is no evidence of corky bark in well-established vineyards. Most of the vines in Canada, however, have been derived from imported European stocks before the present quarantine policies came into effect that would have prohibited the entry of diseased material. It is reasonable to suppose that some material has arrived contaminated and that it would either die out or carry the viruses in latent form, and that the cold Canadian climate has had a major influence on symptom expression. As for rupestris stem pitting, its presence in Canada has recently come to light through the routine checking of repository material on the St. George indicator, but little is known about its distribution outside of Western Europe and California.

What should the correct quarantine policy be toward viruses in a latent condition? On one hand, they may not appear to affect production directly but further research is needed to show whether this is true, and whether there could be other more subtle effects on metabolism. Leafroll has already been suspected of delaying maturation of the vines and thus lowering frost resistance later on.[1,55] On the other hand, in the author's judgement, if no ill effects could be demonstrated, vine stocks with these three viruses might be permitted

* Natural spread of grapevine corky bark has been observed on one occasion only, in Mexico, and the agent is still unknown.[67] The grapevine Virus A, one of the closteroviruses implicated in grapevine leafroll, may be spread by species of *Planococcus* (mealy bugs).[48] (See also References 71 and 72.)

to enter importing countries in the future, as long as they were not then used to produce certified material for export.

A particular difficulty about the organisms transmitted only by grafting is the resistance of many of them to heat therapy. The sap-transmissible viruses are normally much more sensitive. In practical terms, the process of sanitizing a particularly desirable grape cultivar may prolong the business of producing virus-free stock considerably after the initial discovery of a pathogen transmissible only by grafting, whereas in the case of sap-transmissible pathogens the time necessary for both activities may be considerably shorter. Prolonged treatment with temperatures of 38°C for 42 days and 35 to 39°C for 28 days has successfully freed mature grapevines from asteroid mosaic and Hungarian chrome mosaic viruses, respectively.[36,64] Longer periods were required for leafroll[35,36] and corky bark[33] and the evidence indicates that although the success rate improved with time, it did not reach 100%. Long periods of exposure (693 days) of shoot tips infected with corky bark to a temperature of 36°C achieved a success rate of 82%.[44] Normally, heat treatment is accompanied by the excision and growth of meristem tips to produce new virus free plants, and extreme care has to be taken to ensure through subsequent indexing that they are indeed healthy. In the case of the agents responsible for grapevine fleck,[33] summer mottle,[5] vein mosaic, vein necrosis,[6] and yellow speckle,[33] the results of thermotherapy have generally been disappointing. The difficulty of elimination plus the fact that the agent causing fleck is not known to cause economic damage supports the hypothesis that this agent should not be subject to quarantine.

C. Viruses and Mycoplasma-Like Organisms Transmitted by a Live Vector

1.Nepovirus Group

These are a well-defined group of viruses distinguished by their polyhedral shape and by their transmission by nematodes, particularly of the genera *Xiphinema* and *Longidorus* — hence the acronym nepo. Actual transport via nematodes between and inside various countries in grape material is fairly remote, since nematodes would normally desiccate on the surface of the bud wood, and finished vines which may be rooted are only rarely sent. Nevertheless, the nepoviruses pose a greater threat than the members of the preceding group because they can become established in a new area in many weed reservoirs through the action of nematode hosts that are already present. Their potential quarantine significance can only be mitigated by the absence of a specific nematode that would act as a vector. The nepoviruses also pose a threat to certification programs developed within various countries responsible for screening a large number of foreign and locally grown cultivars. Nepoviruses usually have a wide host range, including many weeds and widespread herbs. Many nepoviruses are seedborne though not necessarily in grape seed. Recently tomato ringspot virus was demonstrated in the seeds of *Taraxacum officinale*[53] and there seems to be little doubt that this plant could be an important virus reservoir, since it is distributed over most of the northern temperate region.

As Martelli[47] has pointed out, the behavior of grapevine nepoviruses with respect to seed transmission in grape is rather erratic. Grapevine fanleaf virus and the yellow vein strain of tomato ringspot virus are pollen borne but not seed transmitted.[4,12,26] Plants pollinated with GFV-carrying pollen do not become infected. On the other hand, the type strains of tomato ringspot virus, peach rosette mosaic virus, and grapevine Bulgarian latent virus are all seed borne although at a rate not exceeding 10%. Seed transmission in grape is, however, of negligible ecological importance and quarantine significance owing to the low transmission rate and to the fact that grape material is normally transported in the form of woody material. Grape seedlings have no practical use except for breeding work; nevertheless, virus contamination of seeds for the latter purpose is potentitally troublesome.

The grapevine, like the cherry, seems to have a special affinity for nepoviruses since 11 have been recorded on grapevine in contrast to the much lower numbers occurring in pome and stone fruits, small fruits, and various edible berries. Dead or dying vine roots may act as a focus for perpetuation of virus diseases within a vineyard since roots are extraordinarily long lasting and may attract nematodes including viruliferous ones as long as the roots remain alive in the ground.

As with members of the preceding group, the damage that nepoviruses do is often insidious and the systemic diseases that they cause are chronic and longlasting. The harvest is reduced but the vines are not usually killed outright except where cultivation is poor. Thus the viticulturist may be tempted to overlook a mediocre crop corresponding to, for example, a 20% reduction of the normal yield, rather than to fumigate the soil and replant his stock, not being aware that this loss will amount to a full crop over a five year period. Nepoviruses, however, can be diagnosed fairly easily and to a very low level in infected material by serology, particularly by the enzyme linked immunosorbent assay (ELISA) test,[27] so that incoming material can be checked within a matter of days rather than months or years.

Some nepoviruses, particularly grapevine fanleaf (GFV), have been easy to detect serologically during spring and summer but not later as the titer in the sap in the stem and leaves than declines, although rootlets have still been found to be a good source of virus. The most suitable period for detecting GFV in leaf samples from the vineyard was from bud burst to early July in Switzerland[16] and the most suitable leaves were the distal ones. Cuttings taken in winter, stored at cold temperatures, and then forced to sprout in a growth chamber at 18 to 20°C have given reliable results, sometimes even better than those from the field. This is a preferred method of testing vinestock in the quarantine testing station in Sidney, British Columbia.

Thermotherapy for the elimination of nepoviruses is likewise relatively more effective. Successful treatment of shoots together with meristem-tip cultures has been reported for arabis mosaic virus,[64] grapevine fanleaf virus,[32] tomato black ring virus,[64] and tomato ringspot virus.[36] Forcing mother plants at 30°C followed by culturing and development of meristem tips seems to be a satisfactory therapeutic treatment for grapevine fanleaf virus,[65] since ELISA tests performed 7 years after the start of treatment were negative.

2. Mycoplasma-Like Organisms and Bacteria

In contrast to the nepoviruses, transmission of the mycoplasma-like organisms and bacteria rely on aerial vectors such as aphids and leafhoppers rather than nematodes, but their other biological characteristics are similar. The quarantine significance of these pathogens to north temperate countries such as Canada seems to be limited by their apparent sensitivity to cold or warm temperatures and by the availability of their vectors. Exposure to cold temperatures under laboratory conditions has been observed to cause remission or recovery of grapestocks from Pierce's disease.[59] Similar experimental work with flavescence dorée has shown that the mycoplasma is heat sensitive.

There has been no research yet, however, on the behavior of these pathogens in grapestocks maintained under a variety of simulated climates that would enable one to form an opinion on whether they were a threat to grapevines in areas with a cold winter. This is particularly true of Pierce's disease bacterium which occurs in countries in South and Central America, and extend northwards in the U.S. to South Carolina and Tennessee in the east and to mid California, around Chico City, in the west. Theoretically, the disease could extend further north because insect vectors have been found containing the bacterium in New York State, and one of the cyperaceous hosts of the vector and agent, *Eleocharis*, extends into Canada. The bacterium has a natural reservoir in various grasses, some of which are montane, so a cold climate is not necessarily a limiting factor in its distribution. Without further research it is impossible to tell whether the Pierce's disease bacterium could survive and become established if imported on vinestocks into areas further north than its presently known natural habitat. Obviously it has a more complex etiology than most diseases. In addition to the environment, a susceptible host and a vector which together normally determine the range of a pathogen, this disease appears to be dependent on hosts with which it is sufficiently compatible for long-term survival, as well as on certain environmental requirements that are not fully understood. Also, while grapevines may be killed by the disease in the greenhouse, they are difficult to infect, giving the impression that they are not one of the primary hosts. Pierce's disease bacterium is, therefore, like the viruses that can only be mechanically transmitted, perhaps not a major quarantine threat to vines grown in a region with a cold winter.

The same curative methods used for the nepoviruses have been tried for the mycoplasma-like organisms and bacteria. Warm water treatment at 30°C for 3 days resulted in an 83% cure of grapevine shoots infected with flavescence dorée,[21,22] and the treatment could be prolonged with increasing success, but whether the disease agent can be destroyed completely without remission is still questionable. Complete assurance still requires woody plant indexing. Immersion of entire plants in water at 45°C for 3 hr and at higher temperatures for a shorter period of time has successfully eliminated Pierce's disease and likewise dry exposure to temperatures of −8 to −12°C for various periods.[59] Treatment of grape stock with penicillin and oxytetracycline has caused remission of symptoms of grapevine little leaf mycoplasma[63] and grapevine infectious necrosis[69] but permanent cure is also not guaranteed. Consequently thermotherapy and meristem-tip culture still have to be performed to counteract these organisms too.

D. Nematodes

The final group of organisms associated with the grapevine comprises the nematodes. Their main significance is, undoubtedly, as carriers of nepoviruses, but there are other direct pathological effects that may be observed also. Normally nematodes do not survive on dry budwood material in transit from one country to another so that these organisms should not customarily constitute a quarantine problem. The distribution of individual species is, however, of interest in relation to the possible establishment and spread of nepoviruses.

The nematode vectors of grapevine viruses have been extensively reviewed by Martelli,[47] and Lamberti.[42] *Xiphinema index* is by far the most widespread vector throughout Europe, the Mediterranean basin, North and South America, South Africa, and Australia. The origin of this organism is possibly in ancient Persia. *X. americanum* (now considered to be a complex of many species[43]) occurs widely in North America and the Far East, but probably does not occur in Western Europe, its closest record being the Azores Islands. *X. diversicaudatum* and *X. vuittenezi* are frequently found in central Europe in heavy soils with high moisture content and have also been recorded from the U.S. *Xiphinema coxi* has been recorded from North America, Germany, and England.

Conversely, *X. italiae* is restricted almost entirely to the sands and light soils of the Mediterranean coastline of Italy and France. *Longidorus attenuatus* and *X. diversicaudatum* are more northerly species inconsistently associated with the grapevine in nature. *Longidorus macrosoma* and *L. caespiticola* are present in central and southern Europe. No record of these five species exists in vineyards outside Europe.

III. STRATEGIES FOR THE CONTAINMENT OF GRAPEVINE DISEASES

Most major grape-producing countries have developed centers to control and research into the mechanisms governing the spread of grapevine pathogens. Various Canadian and American (federal and state) programs presently in force for the identification and treatment of pathogens on grapevine material imported into and distributed within North America have resulted in extremely rigorous control.[2,34,70]

A. Exclusion

The simplest and most direct way to exclude pathogens is to refuse entry to material which cannot be certified by the country of origin. In the author's judgement it is desirable in practice to try to negotiate an agreement with the government of the exporting country so that only properly certified indexed material is sent. Continual monitoring of budwood for a trial period will establish whether or not the exporting country may be accepted as an approved source and it will be necessary only to audit samples of incoming budwood rather than to examine every piece.

The Federal Quarantine Station, Sidney, British Columbia (assisted by the Vancouver Research Station) has adopted a fairly lengthy procedure for certified and audit material. On arrival each bud stick is rooted directly and buds or scions are budded or grafted onto indicator plants. Both herbaceous and woody indicators are used and the tests proceed simultaneously. The herbaceous indicators are *Cucumis sativus*, (cucumber), *Cucurbita melo* (squash), *Chenopodium quinoa,* and *Gomphrena globosa.* They are inoculated directly with sap from the test material after rubbing the leaves of the indicator plants after dusting with carborundum. Symptoms appear on the leaves usually within 2 weeks. Although the nepoviruses give rise to a characteristic ringspot pattern on leaves, it is only possible to differentiate between them with certainty through serology. The ELISA technique can be used for extensive surveys but not usually for one or a few identifications. When the ELISA and immune electron microscopy (IEM) techniques were compared,[14] both tests detected GFV at higher sap dilutions than mechanical inoculation tests on *Chenopodium quinoa*. IEM has been judged to be more time consuming and unsuitable for routine tests but very useful for research. Indexing on woody indicators is used to confirm the results on herbaceous hosts and to detect additional viruses which are not sap transmissible. Usually the process takes somewhat over 2 years as illustrated in Figure 1.

Imported material that is found to be infected with one or more viruses may either be discarded or heat treated to eliminate the infection. If the importer considers the material to be sufficiently important to his work and if the Plant Quarantine Advisory Committee agrees that it is of potential importance to Canadian agriculture, it is subjected to heat treatment. Young potted vines are exposed to continuous heat at a temperature of about 38°C, in a growth chamber for several weeks. The tips are then removed and wedge-grafted onto virus-free root stocks for meristem culture. Usually only a few grafts grow into virus-free plants which is confirmed by additional testing on woody indicators. Thus, heat therapy is a lengthy and expensive procedure, often requiring 5 to 7 years from the time the plant is tested until virus freedom can be confirmed.

When a cultivar has been declared free of viruses, it is assigned to the Grape Repository. Normally future requests to import this particular cultivar will be refused if material in the Repository is available to the applicant.

The discovery of a latent and possibly serious virus disease in the Repository may present problems if its presence could compromise the production of virus-free certified stock. Such has recently been the case in North America with rupestris stem pitting, a disease latent in *Vitis vinifera* but expressed in *V. rupestris*. This virus has only recently been recognized as distinct from corky bark and its effect on the grapevine are not clear. In the writer's opinion, to deny the use of material in the Repository to customers in Canada could seriously harm the grape industry or generate such antagonism as to prejudice quarantine regulations already in force. On balance, it would seem preferable to continue distributing the best available clones and at the same time embark on a program to clean up existing stocks by therapy and indexing over a period of years. The disease should be added to the quarantine list when healthy domestic material is available to be protected. Healthy stocks fed into expansion and replacement plantings should, over the years, eliminate the disease since it is not known to have a natural vector.

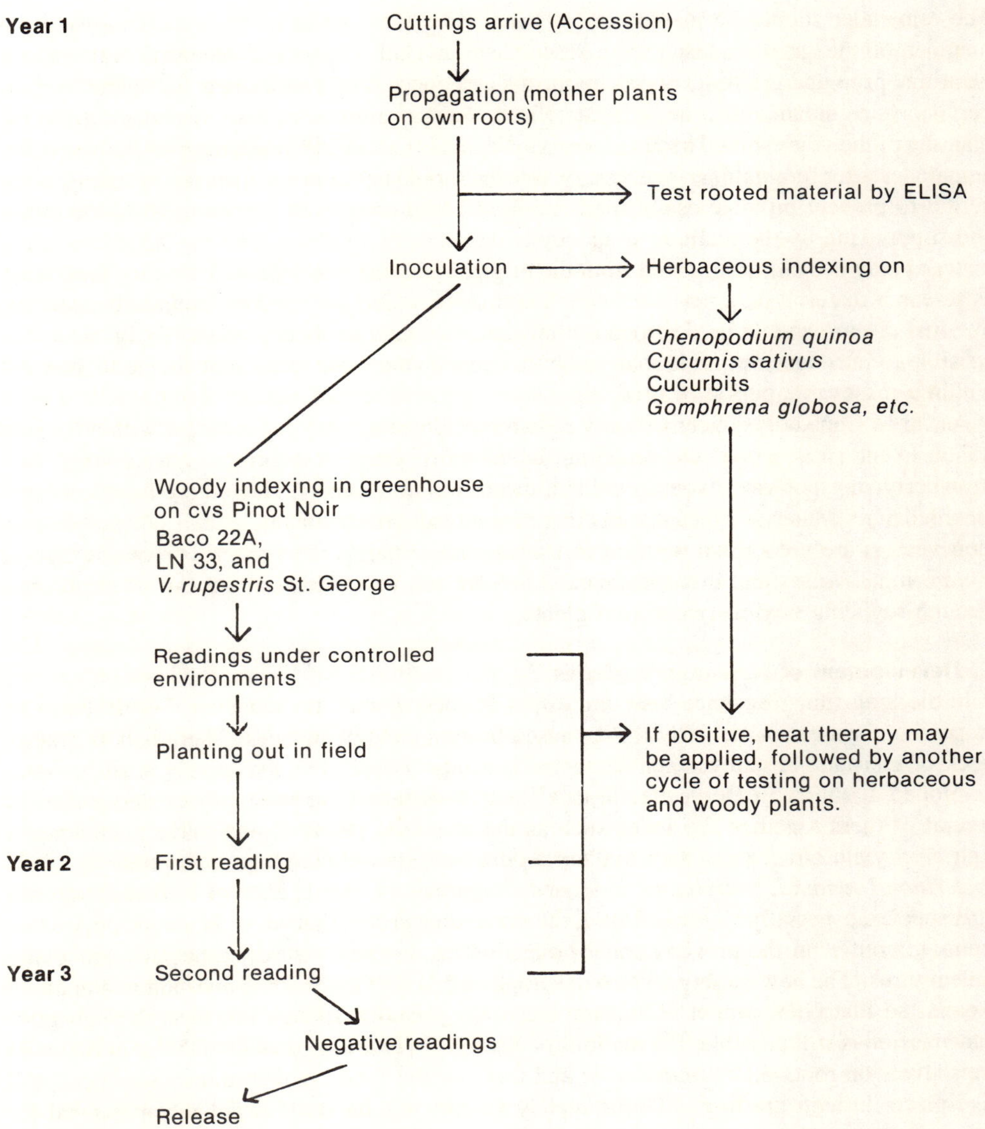

FIGURE 1. Diagram showing the procedures used to determine the health status of imported vegetative propagations of grapevine. Virus indexing tests on woody and herbaceous indicator plants are shown. (Federal Quarantine Station, Agriculture, Canada, Sidney, B. C., in cooperation with the Vancouver Research Station).

B. Importation Agreements

In view of the long time necessary to index grape cuttings, some compromise is often necessary with the industry or with researchers to import uncertified material, provided that supervision can be maintained while samples are being certified at the quarantine station. This involves adherence to a 'Compliance Agreement' agreed upon before the material enters the country, that is mutually acceptable to the importer and to the quarantine agency. Ideally, all material under such an agreement should be screened, as usual, in a recognized official virus testing station in the exporting country. Only material free of nepoviruses would then be eligible for export. On arrival in the host country samples should be checked again for nepoviruses by the designated government authority and submitted to long-term indexing.

The remainder should be planted in isolation under the control of the importer (usually a member of the grape industry or a researcher) and all treatments, including cultivation, planting, pruning, grafting, or selection of budwood, should be carried out under the supervision or guidance of a designated officer of the quarantine agency. Land set aside for planting of the vines should have been surveyed and officially declared free of virus-vectoring nematodes after fumigating, if necessary. Weeds should be suppressed during the compliance period to prevent possible established and seed dissemination of nepoviruses. Cooperation and supervision by the quarantine agency is desirable at all times, and in Canada no grape material may be sold or removed from the original planting site unless cleared for replanting as research material on a new site. Any grape stock visibly affected or found to be infected by virus disease should be destroyed immediately or held for heat therapy. In this way it is possible to introduce desirable material into a new country and make it available to growers within a reasonable period of time.

As far as Canada is concerned, any research or breeding material infected with artichoke Italian latent virus, grapevine Bulgarian latent virus, grapevine chrome mosaic virus, and strawberry ringspot virus would not be allowed into the country, since these have not been recorded here. Material infected with other nepoviruses do not constitute such a threat because their vectors are not known to occur in Canada; nevertheless their progress requires careful monitoring. Viruses and mycoplasmata which are not vector transmitted can be eliminated through rogueing obviously diseased plants.

C. Development of Resistant Varieties

In the long run, resistance breeding offers the best chance for reducing the incidence of grapevine diseases, particularly those caused by fungi and nematodes. There is now a large stock of resistant varieties available against various diseases but a relatively small number developed at INRA in Bordeaux, France[24] and elsewhere have multiple resistance against several of them together. Varieties such as the Russian cultivar Doina[7] have maintained a high crop yield but are resistant to *Plasmopara* and are only slightly susceptible to *Agrobacterium, Uncinula, Botrytis, Elsinoe*, and *Guignardia*. Hybridization of *Vitis vinifera* with *Vitis* species, especially *V. rotundifolia*, the muscadine group, that are resistant to nematodes, seems to confer on the progeny some protection against nepoviruses, especially grapevine fanleaf virus. The new variety 039-16 developed by A. C. Goheen[36a] is resistant to nematode vectors, so that GFV cannot be transmitted in this manner, though infection through graft transmission is still possible. Populations of *Xiphinema index* are unable to thrive or maintain their levels on roots of *V. rotundifolia* and transmit the virus, though transmission can still take place through grafting.[13] Unfortunately *V. rotundifolia* itself is not recommended for use as a rootstock, its resistance to phylloxera and to nematodes being offset *inter alia* by poor rooting capability, lack of graft compatibility with *V. vinifera*, and susceptibility to cold.

The development of resistant grape varieties may have important quarantine implications since, in future, in the author's judgement, it may be desirable to prevent importation of weak or susceptible grape cultivars as well as their pathogens. At present, however, this is not a universally accepted quarantine practice.

ACKNOWLEDGMENTS

The author would like to thank Drs. W. P. Campbell, A. C. Goheen, and R. Stace-Smith for their help in the preparation of this chapter.

REFERENCES

1. **Abracheva, P.,** Grapevine leafroll virus, *Rastit. Zash.*, 25, 32, 1977; *Rev. Plant. Pathol.*, 57(4568), 414, 1978.
2. **Allen, W. R. and von Schagen, J. G.,** The virus-tested grapevine program at Vineland, *Can. Agric.*, 28, 10, 1982.
3. **Amani, B.,** Crown gall of grapevine, *Iran J. Plant Pathol.*, 3, 12, 1966; *Rev. Plant. Pathol.*, 47(1445), 253, 1970.
4. **Balthazard, T.,** Essai de transmission de la dégenerescence infectieuse de la Vigne par le semis, *Prog. Agric. Vitic.*, 83, 84, 1969.
5. **Barlass, M., Skene, K. G. M., Woodham, R. C., and Krake, L. R.,** Regeneration of virus-free grapevines using in vitro apical culture, *Ann. Appl. Biol.*, 101, 291, 1982.
6. **Bass, P. and Legin, R.,** Thermothérapie et multiplication in vitro d'apex de vigne, *C. R. Séances Acad. Agric. Fr.*, 67, 922, 1981.
7. **Bedereu, S. F.,** Investigation of the resistance of some collections and newly selected varieties and forms of grapevine to bacterial canker, *Zash. Rast. Boleznei,* 21, 1978; *Rev. Plant. Pathol.*, 58(5978), 507, 1979.
8. **Belli, G., Faoro, F., Fortusini, A., and Vegetti, G.,** Isometric particles associated with corky bark and legno riccio in grapevine, in *Proc. 7th Meet. ICVG,* McGinnis, A. J., Ed., Niagara Falls, Canada, 77, 1980.
9. **Bellod, M.,** La yesca de la Vid en la región de levante su influencia en la longevidad de las cepas y experiencias sobre su tratamiento, *Bol. Patol. Veg. Entomol. Agric. Madr.*, 15, 223, 1947; *Rev. Appl. Mycol.*, 29, 347, 1950.
10. **Berger, O. and Bouhelier, R.,** Les principales maladies de la Vigne en Chaouia au cours de l'année 1934, *Prog. Agric. Vitic.*, 104, 329, 1935.
11. **Bisiach, M., Zerbetto, F., and Vercesi, A.,** La protection du vignoble contre *Botrytis cinerea* en Italie du Nord et du Centre. Situation actuelle et perspectives d'avenir, *EPPO Bull.*, 12, 87, 1982.
12. **Boubals, D.,** Observations sur la non-transmission par graine de deux viruses chez la Vigne (*Vitis vinifera*), *Ann. Amelior. Plantes,* 27, 213, 1969.
13. **Bouquet, A.,** Resistance to grape fanleaf virus in Muscadine grapes inoculated with *Xiphinema index, Plant Dis.*, 65, 791, 1981.
14. **Bovey, R., Brugger, J. J., and Gugerli, P.,** Detection of fanleaf virus in grapevine tissue extracts by ELISA and IEM, in *Proc. 7th Meet. ICVG,* McGinnis, A. J., Ed., Niagara Falls, Canada, 259, 1980.
15. **Bovey, R., Gartel, W., Hewitt, W. P., Martelli, G. P., and Vuittenez, A.,** *Virus and Virus-Like Diseases of Grapevines,* Payot, Lausanne, Switzerland, 1980.
16. **Bovey, R., Gugerli, P., and Brugger, J. J.,** Some improvements in the detection of grapevine fanleaf virus by enzyme-linked immunosorbent assay (ELISA) and immune electron microscopy, *Phytopathol. Mediterr.*, 24, 208, 1985.
17. **Branas, J.,** Sur l'apoplexie de la Vigne, *Prog. Agric. Vitic.*, 77, 287, 1960.
18. **Bulit, J. and Dubos, B.,** Epidemiologie de la pourriture grise, *EPPO Bull.*, 12, 37, 1982.
19. **Burr, T. K. and Katz, B. H.,** Isolation of *Agrobacterium tumefaciens* biovar 3 from grapevine galls and sap and from vineyard soil, *Phytopathology,* 73, 163, 1970.
20. **Burr, T. K. and Katz, B. H.,** Grapevine cuttings as potential sites of survival and means of dissemination of *Agrobacterium tumefaciens, Plant Dis.*, 68, 976, 1984.
21. **Caudwell, A.,** La biologie de la flavescence dorée et des fondements des mesures preventives, *Bull. Tech. Inf. Ing. Serv. Agric.*, 198, 377, 1965.
22. **Caudwell, A.,** L'inhibition in vivo du virus de la flavescence dorée par la chaleur, *Ann. Epiphyt.*, 17, 61, 1966.,
23. **Chiarappa, L.,** Wood decay of the grapevine and its relationship with black measles disease, *Phytopathology,* 49, 510, 1959.
24. **Clerjeau, M.,** Evolution des principales maladies de la vigne et des méthodes pour les combattre, *Phytoma,* 372, 19, 1985.
25. **Conti, M., Milne, R. G., Luisoni, E., and Boccardo, G.,** A closterovirus from a stem pitting diseased grapevine, *Phytopathology,* 70, 394, 1980.
26. **Doazan, J. P. and Bellegarde, H.,** Essai de transmission par la graine de viroses de la vigne (courtnoué, marbrure), *Ann. Amelior. Plantes,* 27, 245, 1977.
27. **Engelbrecht, D. J.,** Indexing grapevines for grapevine fan leaf virus by enzyme linked immunosorbent assay, in *Proc. 7th Meet. ICVG,* McGinnis, A. J., Ed., Niagara Falls, Canada, 277, 1980.
28. **Faoro, F., Tornaghi, R., Fortusini, A., and Belli, G.,** Association of a possible closterovirus with grapevine leafroll in Northern Italy, *Riv. Patol. Veg.*, 17, 183, 1981.
29. **Flores, R., Duran-Vila, N., Pallas, V., and Semancik, J. S.,** Detection of viroid and viroid-like RNAs from grapevine, *J. Gen. Virol.*, 66, 2095, 1985.

30. **Frazao, A. and Alves, J. A.**, Etude des fongicides utilisés contre la pourriture grise de la vigne au Portugal, *EPPO Bull.*, 12, 117, 1982.
31. **Galet, P.**, Les maladies et les parasites de la Vigne, Tome 1, Paysan du Midi, Montpellier, France, 1977.
32. **Galzy, R.**, Action de la temperature 35 degrees C. sur *Vitis rupestris* atteint de court-noué. *Bull. Soc. Fr. Physiol. Veg.*, 12, 399, 1966; *Rev. Appl. Mycol.*, 48(236), 53, 1969.
33. **Goheen, A. C.**, Virus and virus-like diseases of grapes, *HortScience*, 12, 465, 1977.
34. **Goheen, A. C.**, The California clean grape stock program, *Calif. Agric.*, 34, 15, 1980.
35. **Goheen, A. C. and Luhn, C. F.**, Heat inactivation of viruses in grapevine, *Riv. Patol. Veg.*, 9(Suppl.), 287, 1973.
36. **Goheen, A. C., Luhn, C. F., and Hewitt, W. B.**, Inactivation of grapevine viruses *in vivo, Proc. Int. Conf. Virus Vector Perenn.*, University of California, Davis, 255, 1966.
36a. **Goheen, A. C.**, personal communication.
37. **Gugerli, P., Brugger, J. J., and Bovey, R.**, L'enroulement de la vigne, *Rev. Suisse Vitic. Arboric Hortic.*, 16, 299, 1984.
38. **Hewitt, W. B. and Goheen, A. C.**, Asteroid mosaic of grapevines in California, *Phytopathology*, 49, 541, 1959.
39. **Kerr, A.**, Biological control of crown gall through produciton of Agrocin 84, *Plant Dis.*, 64, 25, 1980.
40. **Klos, E. J., Fronek, F., Knierim, J. A., and Cation, D.**, Peach rosette mosaic transmission and control studies, *Mich. Agric. Exp. Stn. Q. Bull.*, 49, 287, 1967.
41. **Lafon, R.**, Les fongicides viticoles, *Phytoma*, 372, 27, 1985.
42. **Lamberti, F.**, Nematode vectors of grapevine viruses and their spread and control, *Proc. 6th Meet. ICVG*, Madrid-Cordoba, Spain, 239, 1976.
43. **Lamberti, F. and Bleve-Zacheo, T.**, Studies on *Xiphinema americanum sensu lato* with descriptions of fifteen new species *(Nematoda, Longidoridae), Nematol. Mediterr.*, 7, 51, 1979.
44. **Legin, R. and Bass, P.**, Premiers resultats de guérison par thermothérapie et culture in vitro d'une maladie de type cannelure (legno riccio), *Phytopathol. Mediterr.*, 18, 207, 1977.
45. **Lehoczky, J. and Mako, S.**, Premature decay of grapevine caused by *Stereum hirsutum, Kertgazdasag*, 15, 53, 1983; *Rev. Plant. Pathol.*, 63(5061), 483, 1984.
46. **Loubser, J. T.**, Identification of *Agrobacterium tumefaciens* biovar 3 on grapevine in South Africa, *Plant Dis. Rep.*, 62, 703, 1983.
47. **Martelli, G. P.**, Nematode-borne viruses of grapevine: their epidemiology and control, *Nematol. Mediterr.*, 6, 1, 1978.
48. **Martelli, G. P.**, Virus and virus-like diseases of the grapevine in the Mediterranean area, *FAO Plant Prot. Bull.*, 34, 25, 1986.
49. **Mathys, O.**, Importance economique de la pourriture grise de la vigne, *EPPO Bull.*, 12, 3, 1982.
50. **Mihajlovic, D.**, Efficacité de quelques fongicides utilisés dans la lutte contre la pourriture grise des raisins en Macedoine, *EPPO Bull.*, 12, 95, 1982.
51. **Milne, R. G., Conti, M., Lesemann, D. E., Stellmach, G., Tanne, E., and Cohen, J.**, Closterovirus-like particles of two types associated with diseased grapevines, *Phytopathol. Z.*, 110, 360, 1984.
52. **Mossop, D. W., Elliott, D. R., and Richards, K. D.**, Association of closterovirus-like particles and high molecular weight double-stranded RNA with grapevines affected by leafroll disease, *N.Z. J. Agric. Res.*, 28, 419, 1985.
53. **Mountain, W. L., Powell, C. A., Forer, L. B., and Stouffer, R. F.**, Transmission of tomato ringspot from dandelion via seed and dagger nematodes, *Plant Dis.*, 67, 867, 1983.
54. **Nesbitt, W. B.**, Breeding resistant grape rootstocks. *HortScience*, 9, 359, 1974.
55. **Over de Linden, A. J. and Chamberlain, E. E.**, Effect of grapevine leaf roll virus on vine growth and fruit yield and quality, *N.Z. J. Agric. Res.*, 13, 689, 1970.
56. **Pearson, R. C.**, Chemical control of *Botrytis cinerea* on grapes in New York (USA), *EPPO Bull.*, 12, 101, 1982.
57. **Pearson, R. C., Poul, R. M., Gonsalves, D., and Goffinet, M. C.**, Occurrence of *Flavescence dorée*-like symptoms on White Riesling grapevines in New York, USA, *Phytopathol. Mediterr.*, 24, 82, 1985.
58. **Prudencio, S.**, Comparative Effects of Corky Bark and Rupestris Stem Pitting Diseases on Selected Germplasm Lines of Grapes, M.Sc. thesis, University of California, Davis, 1981.
59. **Purcell, A. H.**, Cold therapy of Pierce's disease of grapevines, Plant *Dis. Rep.*, 61, 514, 1977.
60. **Ramsdell, D. C. and Myers, R. L.**, Peach rosette mosaic virus: symptomatology and nematodes associated with grapevine 'degeneration' in Michigan, *Phytopathology*, 64, 1174, 1974.
61. **Safaryan, S. E. and Gambaryan, G. S.**, Grey mould of grapes and systems for its control in Armenia, *EPPO Bull.*, 12, 113, 1982.
62. **Sano, T., Uyeda, F., Shikata, E., Meshi, T., Ohno, T., and Okada, Y.**, A viroid-like RNA isolated from grapevine has high seqeunce homology with hop stunt viroid, *J. Gen. Virol.*, 66, 333, 1985.
63. **Singh, S. J., Sastry, K. S. M., and Sastry, K. S.**, Little leaf disease of grapevine from India, *Curr. Sci.*, 44, 26, 1975.

64. **Stellmach, G.,** Notes on a modification in the technique for inactivating NEPO Viruses in grapes by heat treatment, *Riv. Pathol. Veg.*, 9(Suppl.), 165, 1973.
65. **Stellmach, G.,** ELISA testing of grapevine rootings reared from NEPO virus-infected mother plants forced to rapid growth, *Phytopathol. Mediterr.*, 24, 123, 1985.
66. **Tanne, E. and Givoni, L.,** Serological detection of two different viruses associated with leaf roll diseased grapevines, *Phytopathol. Mediterr.*, 24, 106, 1985.
67. **Teliz, D., Valle, P., Goheen, A. C., and Luevano, S.,** Grape corky bark and stem pitting in Mexico. I, *Proc. 7th Meet. ICVG,* Niagara Falls, Canada, 51, 1980.
68. **Tzeng, H. C.,** Anatomical and Tissue Culture Studies of Corky Bark, Rupestris Stem Pitting and Leafroll-Affected Grapevines, M.Sc. thesis, University of California, Davis, 1984.
69. **Ulrychova, M., Vanek, G., Jokes, M., Klobaska, Z., and Kralik, O.,** Association of rickettsia-like organisms with infectious necrosis of grapevine and remission of symptoms after penicillin treatment, *Phytopathol. Z.*, 82, 254, 1975.
70. **Welsh, M. F. and Hansen, A. J.,** Post Entry Plant Quarantine Station, Sidney, B. C., Facilities and Procedures, Publ. 1661, Agriculture Canada, 1978.
71. **Rosciglione, B. and Gugerli, P.,** Transmission of grapevine leafroll disease and an associated closterovirus to healthy grapevines by the mealybug *Planococcus ficus, Proc. 9th Meet. of the ICVG,* Israel, Sept. 6 to 11, 1987, (Abstr.), Kiryat-Anavim.
72. **Tanne, E., Ben-Dov, Y., and Raccah, R.,** Transmission of closterovirus-like particles by mealybugs in Israel, *Proc. 9th Meet. of the ICVG,* Israel, Sept. 6 to 11, 1987, (Abstr.), Kiryat-Anavim.

Chapter 3

NEPOVIRUSES

A. F. Murant

TABLE OF CONTENTS

I. MEMBERSHIP OF THE NEPOVIRUS GROUP

Most plant viruses can be placed in one or another of 28 taxonomic groups that are presently recognized. The nepovirus group[67] contains viruses that have isometric particles about 28 nm in diameter and bipartite genomes of single-stranded RNA. Members of the group also share other biological and physicochemical properties and some are serologically related to others. Many are transmitted in nature by soil-inhabiting plant parasitic nematodes and in fact the term *nepo*virus was coined[14] to denote viruses that have *ne*matode vectors and *po*lyhedral particles. However, one member, tobacco ringspot virus (TobRV), is reported also to be transmitted by aerial vectors and for a substantial proportion of the 29 viruses now included in the group (Table 1) the vector is unknown. Although undoubtedly nematode vectors will eventually be found for many of them, it remains possible, as will be discussed below, that some of them spread in nature by other means.

II. HOST RANGES

Most nepoviruses infect a very wide range of cultivated and wild plant species. Table 1 lists the most important hosts among cultivated plants and shows that a given nepovirus may infect plants as diverse as monocotyledons and dicotyledons, or annual vegetables and long-lived perennials. Nepoviruses are troublesome especially because of their tendency to infect woody perennial species such as small fruits, grapevines, fruit trees, and forest trees, individual plants of which are valuable and/or expensive to replace.

The experimental host ranges of nepoviruses are also extremely wide. In early studies by Price,[76] TobRV was found to infect 143/178 species in 40/56 families; the known host range of this virus has since been extended considerably by other workers, notably by Hollings,[36] who transmitted it to 58/103 species inoculated in 21/32 families. A series of extensive investigations on the experimental host ranges of nepoviruses was conducted by Schmelzer, who reported that arabis mosaic virus (AMV) infected 93 species in 28 families out of 136 species inoculated,[86] cherry leaf roll virus (CLRV) infected 62 species in 23 families out of 104 species inoculated,[88] strawberry latent ringspot virus (SLRV) infected 126 species in 27 families out of 167 species inoculated[89] and tomato black ring virus (TBRV) infected 76 species in 25 families out of 167 species inoculated.[87] Many other plants susceptible to these viruses have been reported since, and their true host ranges are undoubtedly very wide, as are those of other nepoviruses.

III. SYMPTOMS

The responses of plants to natural infection by nepoviruses are extremely diverse,[67] ranging from symptomless infection, through a variety of different kinds of foliage symptoms, to stunting and death. The nature of the response depends on the virus or virus strain, the host species or cultivar, the age of the plant, the season of the year, and many other factors. Despite this variability, however, some generalizations are possible.

Most wild plants show no symptoms when infected by nepoviruses, an adaptation which suggests that these viruses are long-established pathogens of wild plants. For example, AMV and SLRV occur symptomlessly in wild hedgerow plants,[35,73,102] and artichoke Italian latent virus (AILV), AMV, cherry rasp leaf virus (CRLV), peach rosette mosaic virus (PRMV), raspberry ringspot virus (RRV), SLRV, TobRV, and TBRV occur symptomlessly in many species of arable weed.[24,31,70,74,75,77,78,102,107] Many cultivated plants too are infected symptomlessly but some exhibit severe symptoms. It is when such a sensitive crop species or cultivar is planted and obvious disease results that the infestation of a soil with nepovirus-carrying nematodes is usually first suspected.

Table 1
THE NEPOVIRUSES: CULTIVATED PLANT HOSTS, VECTORS, AND GEOGRAPHICAL DISTRIBUTION

Virus	Cultivated plant hosts[a]	Main natural vector(s)	Geographical distribution	Ref.[b]
Arabis mosaic (AMV)	Blackcurrant, celery, cherry, cucumber, elderberry, forsythia, grapevine **hop,** lettuce, narcissus, **raspberry,** redcurrant, rhubarb, **strawberry,** tamarillo	*Xiphinema diversicaudatum, X. bakeri*	Europe, Japan, New Zealand, U.S.S.R.	Descr. 16
Arracacha A (AVA)	**Arracacha**	Unknown	Peru	Descr. 216
Artichoke Italian latent (AILV)	**Artichoke, chicory,** grapevine, pelargonium	*Longidorus apulus, L. fasciatus*	Bulgaria, Greece, S. Italy	Descr. 176
Artichoke veinbanding (AVBV)	**Artichoke**	Unknown	S. Italy	Descr. 285
Artichoke yellow ringspot (AYRV)	**Artichoke, cardoon,** *Phaseolus* bean, tobacco, *Vicia* bean	Unknown	Greece, Italy	Descr. 271
Blueberry leaf mottle (BLMV)	**Blueberry**, grapevine	Unknown	N. America	Descr. 267
Cherry leaf roll (CLRV)	**Birch**, blackberry, **cherry**, delphinium, dogwood, **elderberry, elm,** lilac, olive, raspberry, rhubarb, **walnut**	Unknown	Europe, N. America, New Zealand, U.S.S.R.	Descr. 80;306
Cherry rasp leaf (CRLV)	**Apple, cherry,** peach, raspberry	*X. americanum sensu lato*	N. America	Descr. 159
Chicory yellow mottle (CYMV)	Celery, **chicory,** lettuce, parsley	Unknown	S. Italy	Descr. 132
Cocoa necrosis (CNV)	**Cocoa**	Unknown	W. Africa	Descr. 178
Crimson clover latent (CCLV)	**Crimson clover**	Unknown	Europe, N. America	49
Grapevine Bulgarian latent (GBLV)	**Grapevine**	Unknown	Bulgaria, Portugal	Descr. 186
Grapevine chrome mosaic (GCMV)	Celery, **grapevine**	Unknown	Hungary, England	Descr. 103
Grapevine fanleaf (GFLV)	**Grapevine**	*X. index, X. italiae*	World-wide	Descr. 28
Hibiscus latent ringspot (HLRV)	**Hibiscus**	Unknown	Nigeria	Descr. 233
Lucerne Australian latent (LALV)	**Lucerne, white clover**	Unknown	Australia, New Zealand	Descr. 225
Mulberry ringspot (MRV)	**Mulberry**	*L. martini*	Japan	Descr. 142
Myrobalan latent ringspot (MLRV)	**Myrobalan plum,** peach, sweet cherry	Unknown	S.W. France	Descr. 160
Olive latent ringspot (OLRV)	**Olive**	Unknown	S. Italy	Descr. 301
Peach rosette mosaic (PRMV)	Blueberry, **grapevine, peach**	*L. diadecturus, S. americanum sensu stricto*	N. America	Descr. 150
Potato black ringspot (PBRV)	**Potato**	Unknown	Peru	Descr. 206
Potato U (PVU)	**Potato**	Unknown	Peru	45

Table 1 (continued)
THE NEPOVIRUSES: CULTIVATED PLANT HOSTS, VECTORS, AND GEOGRAPHICAL DISTRIBUTION

Virus	Cultivated plant hosts[a]	Main natural vector(s)	Geographical distribution	Ref.[b]
Raspberry ringspot (RRV)	Artichoke, blackcurrant, **cherry,** gooseberry, grapevine, narcissus, **raspberry,** redcurrant, **strawberry**	*L. elongatus, L. macrosoma*	Europe, Turkey, U.S.S.R.	Descr. 6; 198
Rubus Chinese seed borne (RCSV)	**Rubus sp.**	Unknown	China	7
Strawberry latent ringspot (SLRV)	Blackcurrant, celery, elderberry, grapevine, narcissus, olive, peach, **raspberry**, redcurrant, rose, **strawberry**	*X. diversicaudatum, X. coxi*	Europe	Descr. 126
Tobacco ringspot (TobRV)	Anemone, ash, apple, blackberry, **blueberry**, cherry, **cucurbits**, dogwood, eggplant, elderberry, gladiolus, grapevine, iris, pelargonium, *Phaseolus* bean, **soybean**, spearmint, **tobacco**, *Vicia* bean	*X. americanum (sensu lato)*	N. America	Descr. 17;309
Tomato black ring (TBRV)	Beet, celery, forsythia, grapevine, leek, lettuce, narcissus, onion, peach, **potato**, privet, **raspberry**, robinia, **strawberry,** swede, turnip	*L. attenuatus, L. elongatus*	Europe	Descr. 38
Tomato ringspot (TomRV)	**Apple**, cymbidium, gladiolus, **grapevine**, hydrangea, narcissus, pelargonium, *Prunus* spp., **Rubus spp.,** *Ribes* spp., strawberry, tobacco	*X. americanum sensu lato* inc., *X. americanum sensu stricto, X. rivesi* and *X. californicum*	N. America, Japan, U.S.S.R.	Descr. 18;290
Tomato top necrosis (TTNV)	Tomato	Unknown	U.S.	6

[a] Hosts most commonly or most seriously affected in bold.

[b] Descr. nos refer to the series CMI/AAB Descriptions of Plant Viruses, (A.F. Murant and B. D. Harrison, Eds.), Commonwealth Mycological Institute, Kew, and Association of Applied Biologists, Wellesbourne.

Where infection of a sensitive crop by a nepovirus is brought about by nematodes, the diseased plants typically are found in patches, which may extend over a few square meters or for a hectare or more. The boundary of the patch delimits the horizontal distribution of the vector in the soil, and when an infected crop is replaced with another that is also sensitive, the pattern of disease in the field is repeated. In contrast, where the disease results from the use of infected planting material, the distribution of affected plants is more or less random.

The types of foliage symptom induced by nepoviruses in sensitive species or cultivars include chlorotic mottling or flecking, mosaic, vein banding, vein yellowing, leaf rolling, and the production of enations.[67] A very characteristic type of symptom is the formation of chlorotic or necrotic ringspots, and this propensity is reflected in the names of several nepoviruses. Plants infected naturally through the roots by nematode feeding may show no obvious reaction in the year of infection, and the foliar symptoms that appear in the following spring may fade in midsummer often to reappear in the autumn. In subsequent years the cycle of symptom expression is repeated, but the plants become progressively debilitated and eventually may die. The seasonal variations in symptom expression may in some instances result from a fluctuation in virus concentration within the plant. Erratic distribution within the plant is also common, especially in woody plants; for example, CLRV is present in only some branches of walnut trees and infection is difficult to detect reliably by graft indexing or enzyme-linked immunosorbent assay (ELISA) unless heroic efforts are made to collect samples from all parts of the tree.[82]

The responses of herbaceous test plants to experimental infection by nepoviruses are as varied as those shown by plants in the field and a more complete account of the types of symptom caused is given elsewhere.[67] Suffice to say here that many species develop chlorotic or necrotic spots or rings in inoculated leaves, and spots, rings, and line patterns or apical necrosis in systemically infected leaves. Many plants, especially species of Solanaceae, show a very characteristic phenomenon known as "recovery": after the acute phase of symptom expression the leaves subsequently produced show few or no symptoms but still contain virus, although in somewhat lower concentration than in symptom-bearing leaves.

IV. ECONOMIC IMPORTANCE

Some nepoviruses are of little or no known economic importance; others are seriously damaging in the areas where they occur. Undoubtedly the most important is grapevine fanleaf virus (GFLV), different strains of which cause "fan leaf", "yellow mosaic" and "vein banding" diseases of grapevine; the virus is found throughout the world wherever the crop is grown and may cause up to 50% loss of yield.[108] In Britain, infections with RRV +/− TBRV, or AMV +/− SLRV, may kill raspberry or strawberry crops or render them worthless.[56,66] In North America, TobRV causes a severe disease ("bud blight") of soybean[93] and is also very damaging in tobacco,[60] blueberry[18,59] and cucurbits.[62,94] Also in North America, fruit trees such as almond, cherry, nectarine, and peach can suffer serious damage from "yellow bud mosaic", "rasp leaf", and "stem pitting" diseases all caused by tomato ringspot virus (TomRV).[85,95] TomRV also causes serious diseases of raspberry[96] and grapevine[22,28] and is thought to be responsible for graft-incompatibility reactions in some rootstock/scion combinations in apple ("union necrosis and decline")[99] and in prune ("brownline").[38] Another graft-incompatibility reaction ("blackline") is caused by CLRV in walnut and is regarded as a serious threat to walnut tree production in some regions of California.[64] Although both TomRV and CLRV have been studied for many years in other crops, their role and importance in causing these conditions in apple, prune, and walnut has been recognized only comparatively recently. Thus, the wide host ranges of the nepoviruses and the great variety of disease syndromes they cause means that their potential for causing serious economic damage in crop species hitherto unsuspected to be hosts, or not previously exposed to infection, should not be underestimated. As a possible example of this, an outbreak of AMV in New Zealand, which was probably a result of an importation, was responsible for disease in tamarillo (*Cyphomandra betacea*), a host that had not been reported previously.[104,105]

V. TRANSMISSION AND ECOLOGY

A. Nematode Vectors

All nematode vectors so far identified are ectoparasitic and belong to the genera *Longidorus* and *Xiphinema* (Table 1). The relation between virus and vector is highly specific: even the major serological variants of each virus usually have different vectors. Thus, the natural vector of the English serotype of RRV, *L. macrosoma*, is a very inefficient vector of the Scottish serotype;[33] the latter is transmitted by *L. elongatus*, a nematode which has never been found associated with the English serotype in the field although it has been shown to transmit it moderately well experimentally.[101] Similarly with TBRV, isolates of the Scottish serotype are transmitted by *L. elongatus* but those of the English serotype by *L. attenuatus*.[33,101] Different biotypes of a nematode also may differ in vector specificity: cultures of *X. diversicaudatum* originating from France, Italy, and Spain were less effective than cultures from Bulgaria, New Zealand, Norway, Switzerland, U.K. (four cultures), and U.S. in transmitting British isolates of AMV and SLRV, but the Italian culture was the only one that transmitted Italian isolates of SLRV.[11-13]

The nematodes in these different *X. diversicaudatum* populations were anatomically similar and are considered all to belong to one species. However, the taxonomy of virus-vector nematodes is still rather fluid. For example, as recently as 1979 *X. americanum* was subdivided into 25 separate species on the basis of morphometric data.[53] The status of most of these individual species as vectors of TomRV and TobRV has yet to be determined, although three of them, *X. americanum sensu stricto, X. rivesi,* and *X. californicum*, have now been shown to transmit TomRV.[9,39,65] Thus, even for those nepoviruses for which nematode vectors have been found, information is often fragmentary or incomplete. Moreover, few attempts have been made to find out whether nepoviruses from one locality or geographical region can be transmitted by "foreign" nematodes of species with which they have had no previous contact, and our ignorance is such that this possibility cannot be excluded.

At least one nepovirus, CLRV, is probably not transmitted by nematodes, despite earlier reports that it is. In very thorough tests,[44] none of ten species of nematode were found able to act as vectors of CLRV and this is in line with the general observation that CLRV does not occur in patches in crops suggestive of infection via the soil. A similar picture is emerging for blueberry leaf mottle virus (BLMV), for which also there is a lack of evidence for spread either by nematodes or through the soil.[16]

B. Aerial Vectors

Transmission by nematodes may therefore not be the only way in which nepoviruses spread. In soybean crops, TobRV is transmitted inefficiently by the nematode *X. americanum sensu lato* and the pattern of spread suggests that an aerial vector may be involved, possibly thrips,[63] although spider mites,[103] grasshoppers,[23] flea beetles[90] and aphids[50] have also been reported to transmit the virus. So far there is no evidence that any other nepovirus is transmitted by such aerial vectors, but the possibility should be borne in mind.

C. Seed and Pollen Transmission

A striking characteristic of nepoviruses is their ability to be transmitted through seed: this is known to occur with at least 19 viruses in the group[69] and the data are so extensive as to suggest that all nepoviruses are probably seed borne in at least some of their hosts. Table 2 gives a selection of examples. In some instances, as with AMV in lettuce, crimson clover latent virus (CCLV) in crimson clover, SLRV in celery and parsnip, and TBRV in *Capsella bursa-pastoris*, the frequency of seed transmission may approach 100%.

Nepoviruses can enter seed through either gamete. CLRV was found to be transmitted to 30.5% of elm seedlings through the pollen and to 75% through the ovule[15] and is also transmitted through both gametes in birch.[20] RRV and TBRV were transmitted to progeny seedlings both through pollen and ovules of raspberry and strawberry;[58] in strawberry, the frequency of transmission of RRV was 5.5% through the pollen, 10.5% through the ovule, and 44% when both parents were infected. In these experiments, no evidence was obtained for spread of the viruses to the plant pollinated and indeed it was shown that pollen from

Table 2
EXAMPLES OF SEED TRANSMISSION OF NEPOVIRUSES

Host species	AMV	CRLV	PRMV	RRV	SLRV	TobRV	TBRV	TomRV
Capsella bursa-pastoris	33[a]	—	—	2—11	4	—	80—100	—
Chenopodium amaranticolor	—	4—25	—	—	—	—	—	57
Chenopodium quinoa	41	—	90	—	63—100	—	78	—
Lactuca sativa	60—100	—	—	—	—	21	3	—
Stellaria media	58	—	—	29	97	—	66	—
Taraxacum officinale	—	30	4	—	—	9—36	—	24

[a] Figures are the percentage of infected seedlings in progeny from a virus-infected plant.

Data from references cited by Murant[69] and from Mountain et al.[65]

raspberry plants infected with AMV, RRV, or TBRV competed poorly with virus-free pollen for fertilization of the ovule, suggesting that, at least in these virus/host combinations, natural transmission through pollen was not important. However, there is strong circumstantial evidence that CLRV spreads to walnut trees pollinated with infected pollen.[64] There is also evidence in birch for infection of trees pollinated with CLRV-infected pollen.[20] Thus, it seems that transmission in seed and pollen may be the main method of spread of CLRV, and this may account for the lack of evidence for transmission of this virus by nematodes. It may also explain the fact that strains of this virus from different host genera are serologically different, this method of spread presumably offering little opportunity for movement of the virus among unrelated species. BLMV is suggested[16] to spread in the same way, and by implication the same may be true for its serological relative, grapevine Bulgarian latent virus (GBLV). A fourth candidate is artichoke yellow ringspot virus (AYRV) which, experimentally, was transmitted by pollination in *Nicotiana clevelandii*.[52] Some of the other viruses listed in Table 1 as "vectors unknown" also might spread in pollen but probably cocoa necrosis virus (CNV) and grapevine chrome mosaic virus (GCMV) (both distantly serologically related to TBRV), potato black ringspot virus (PBRV) (distantly related to TobRV) and rubus Chinese seed-borne virus (RCSV) (distantly related to SLRV) are transmitted by nematodes.

Usually no symptoms are shown by plants infected with nepoviruses via the seed, even in species that do show symptoms when infected by other routes. The presence of virus in progeny seedlings must be detected by mechanical inoculation to indicator plants or by serological or other tests. Seed transmission of nepoviruses is therefore easily overlooked, whether it occurs in arable weeds in field outbreaks, or in commercial seed lots. Symptomlessly infected progeny seedlings in fact behave as though they are in the "recovered" condition. It has been suggested[8] that the phenomenon of recovery might be linked with the invasion of meristematic tissue by the virus and that the ability of nepoviruses to do this might therefore explain both their ready transmissibility through seed and the absence of symptoms in infected progeny seedlings.[58] Experimental evidence for this was obtained[30] by exposing tobacco seedlings that had been infected with TBRV via the seed to partial heat therapy which probably freed their meristems from virus; when they were returned to normal glasshouse conditions they developed symptoms typical of those that appear after manual inoculation of virus-free plants.

D. Dissemination and Survival in Soils

The ecology of the nematode-borne viruses[34] is critically bound up with the biology of their vectors.[100] Virus-vector nematodes migrate relatively slowly through soils; *X. diver-*

sicaudatum was calculated to have moved at only about 30 cm/year over a period of 75 years.[35] They are also easily killed in soil which is allowed to dry and therefore do not readily survive transportation on farm implements, tractor wheels, etc. Thus, although they are effective in transferring nepoviruses from one plant to another, they are inefficient in the other important function performed by aerial vectors, namely the transportation of the virus over a distance. In nature, such spread presumably occurs, not within the nematode but in seed and pollen from infected plants. In commerce, dissemination of the viruses also occurs through the use of infected planting material.

Nepoviruses are transmitted by juvenile as well as by adult nematodes. They do not multiply in their nematode vectors but are carried in some part of the anterior alimentary tract,[100] from where they are released intermittently to infect any healthy plants on which the nematode feeds. Nematode virus-vectors are long lived and have extended life cycles; some nepoviruses are retained by their vectors for a few weeks after removal from the source plant, others for many months. However, in all instances the nematodes cease to transmit virus after molting (in which process the lining of the anterior alimentary tract is shed) and must obtain renewed access to the virus to regain ability to transmit. In arable soils subjected to a period of fallow, the presence of a nepovirus in weed seeds or in other perennating plant parts may therefore be critical for its survival.[70,71] GFLV, which is unusual among nepoviruses in having only one natural host (grapevine), can survive for many years in soils in deeply buried pieces of grapevine root.[81]

More extended discussions of the ecology and epidemiology of nepoviruses and their vectors, including the importance of seed transmission and the role of wild plants, are available elsewhere.[34,68,100]

E. Man as a Vector

The natural means of spread described above enable the nepoviruses to move only over relatively short distances. Their long distance spread, which must occur within infected plant material, has been accomplished very efficiently by man in the pursuit of his agricultural, commercial, and scientific activities. The use of infected planting material (whole rooted plants, cuttings, scion and budwood material, tubers, bulbs, corms, and seeds) is the main way in which nepoviruses have been disseminated, both within the countries where they are endemic and also from one country to another. It has even been suggested that the Romans were responsible for introducing CLRV when they brought walnut planting stock to Britain.[19] Documented examples of man's activity in disseminating nepoviruses around the world are given in Section VII.

If vector nematodes are present in the soil at the new site it is highly probable that the virus will become established there; if they are not, its establishment will depend on whether vector nematodes accompany the importation. Paradoxically, in these times of rapid movement of commodities around the world, the risk of transporting nematode vectors may be less now than formerly because it is easy to keep soil-free plants alive for a brief air journey whereas in the slow days of sail it was presumably necessary to transport plants in tubs or pots of soil — and the vector nematodes came too. Undoubtedly, this must have been the way in which GFLV and its vector, *X. index*, achieved their present worldwide distribution.

Considering the widespread occurrence of nepoviruses in seed of their numerous host plants it is surprising that they have seldom been reported in commercial seed lots. TobRV may occur in soybean seed.[57] SLRV was found in several sources of parsley seed[10,32] and in almost 100% of one stock of parsnip seed,[21] these being the first records of the natural occurrence of the virus in these species. CCLV was found in all of six seed lots of crimson clover and in 97% of all progeny seedlings examined.[49] In fact, CCLV is known at present only from seed and the same is true of RCSV which, however, was found not in commercial seed but in seed of an unknown *Rubus* species imported into Britain from China for scientific purposes.[7] Presumably it was imported as a source of genetic material for breeding and this is a common way in which viruses are introduced into countries. Breeders and indeed scientists of all kinds need to be more aware of the risks of importing pathogens in seed — about 30% of the well-studied viruses are seed transmitted to at least some extent in one or more host species.

In all of the above instances the infected progeny seedlings showed no symptoms and this may explain the lack of other reports of the detection of nepoviruses in commercial seed. Almost certainly, nepoviruses are of much wider occurrence in commercial seed than is presently reported, and this matter deserves investigation.

VI. GEOGRAPHICAL DISTRIBUTION

GFLV is alone among nepoviruses in having worldwide distribution; it is found in nearly all countries where grapevines are grown, together with its main nematode vector, *X. index*, and as already mentioned this is probably because the plants carried by the early colonizers were transported in tubs or pots of soil to keep them alive during prolonged sea voyages. CLRV too is found in many countries; the elderberry and walnut strains (though apparently not other strains, such as those from cherry and elm) occur both in Europe (including U.S.S.R.) and in North America, and the raspberry strain occurs in New Zealand.

However, the majority of the nematode-borne nepoviruses fall into two geographically separated clusters, one from Europe (including U.S.S.R.) and the other from North America. The "European" group comprises AILV, AMV, RRV, SLRV and TBRV. The "North American" group comprises CRLV, PRMV, TobRV and TomRV. This geographical separation of the viruses is probably largely caused by a restricted distribution of their vectors. Although both *L. elongatus* and *X. diversicaudatum*, the major vectors of the European viruses, have been found in North America, and *X. diversicaudatum* is also reported in Australia, New Zealand, and parts of Asia, they are comparatively rare in these places; nevertheless, a population of *X. diversicaudatum* from the U.S. was found able to transmit British isolates of AMV and SLRV experimentally.[11,12] Although the North American viruses are all reported to be transmitted by *X. americanum sensu lato*, the lack of coincidence in the distributions of the viruses within North America suggests that they differ in efficiency of transmission by the individual nematode species within that broad grouping. Thus, TomRV and CRLV occur along the Pacific coastal strip from California to British Columbia where *X. californicum* is prevalent; TomRV (but not CRLV) occurs also around the Great Lakes and in the eastern states of Pennsylvania, Maryland, and New York, where the prevailing vector species are *X. rivesi* and *X. americanum sensu stricto*. In contrast, TobRV is rare along the Pacific coast but has a fairly widespread distribution elsewhere in the U.S. and southern Canada which suggests that *X. californicum* is not a vector of this virus but *X. americanum sensu stricto* and/or *X. rivesi* perhaps are. PRMV is confined to the region around the Great Lakes; however, *L. diadecturus* is considered to be a more important vector of this virus than *X. americanum sensu stricto*, and *X. rivesi* is apparently not a vector.[2,3] *X. americanum sensu stricto* and *X. californicum* are not known to occur outside North America but *X. rivesi* occurs in France and Spain; other species in the broad grouping *X. americanum sensu lato* occur in many countries throughout the world but their ability to act as vectors of the "North American" nepoviruses has not been investigated.

Most of these "North American" and "European" nepoviruses have been reported occasionally from other parts of the world. TomRV has been found in eastern U.S.S.R.,[29] AMV and TomRV in Japan,[40,42] and AMV in New Zealand;[104,105] in all these instances the presence of vector nematodes was reported or may be suspected. In most other instances the "stray" nepoviruses have occurred apparently in imported plant material and in the absence of their vectors. TBRV has been found in narcissus in Japan,[41] in potato tubers imported into Brazil from Poland[51,92] and into Kenya from the German Federal Republic,[47] and in grapevine imported into Ontario, Canada from France.[97] SLRV was found in Ontario, Canada[1] and in New Zealand;[25] in each instance it occurred only in a single sweet cherry tree and was suspected to have resulted from an importation because it was not associated with vector nematode species (*X. diversicaudatum* has been found only in glasshouses in Ontario and is thought not to be able to survive there in the field). In the U.S. SLRV has also been detected in imported parsley seed.[32]

TomRV has been reported in pelargonium in Sweden,[83] U.K.,[37] Netherlands,[61] and Denmark,[72] in *Pentas lanceolata* in Australia,[17] in raspberry in Yugoslavia[46] and Chile,[5] and in redcurrant in New Zealand.[26] Probably all these isolations of TomRV (except possibly that in Chile) were from material imported from North America, as were reports of CRLV from raspberry in U.K.[43] and of TobRV from anemone in U.K.,[36] gladiolus in U.K.,[55] Japan[27] and Australia,[79] iris in the Netherlands,[4] pelargonium in U.K.,[98] and grapevine in New Zealand.[48] Among numerous other reported isolations of TobRV, those that have been identified serologically include isolates from eggplant[84] and petunia[80] in India, from papaya[54] in Nigeria, and, inexplicably, from plants of the weed *Senecio vulgaris* growing on plots of the National Vegetable Research Station in U.K.[106]

VII. QUARANTINE SIGNIFICANCE

From the foregoing account of the properties of nepoviruses it is obvious that they present a considerable potential quarantine problem. To summarize:

1. They infect a wide range of cultivated and wild plant species and cause serious disease in many cultivated ones, including valuable perennial species such as small fruits, grapevines, and fruit and forest trees.
2. They may infect a species not previously known to be a host. This makes it difficult to exclude a virus from a country by banning importation of other than certified virus-free plants of named host species. It also allows the possibility that an imported virus could prove troublesome in a cultivated species not previously exposed to it.
3. They may occur in plants without inducing symptoms, which makes it impossible to detect infected material reliably by inspection.
4. They may be absent from some parts of an infected plant; this makes it difficult to detect them with certainty even by the most sensitive diagnostic methods.
5. They may be present in seed (often in a high proportion) of many or most host species, and the progeny seedlings are commonly infected symptomlessly.
6. Some vector nematode species may occur in countries where the nepovirus they can transmit is not endemic.
7. The possibility cannot be excluded that an exotic nepovirus may be transmissible by indigenous nematodes belonging to a different species from that which transmits it in its country of origin.
8. One nematode-borne nepovirus (TobRV) is reported also to be transmitted by aerial vectors; this could perhaps happen also with other members of the group.
9. Some nepoviruses may be spread not by nematodes but in pollen, with the infection not only of the progeny seedlings but also of the plant pollinated. There would seem to be little hope of preventing the escape of such a virus once infected material has been planted in the field and allowed to flower.

A group of viruses with such properties might seem to present a daunting problem for those engaged in plant quarantine. However, the extent to which nepoviruses pose an actual rather than a potential quarantine problem is not clear. GFLV is the only nepovirus so far to have achieved world-wide distribution, and this occurred before the era of plant quarantine. Other nepoviruses, such as AMV, SLRV, TBRV, TomRV, and TobRV have been detected as importations in various countries but no disease outbreaks seem to have become established as a result, with the possible exception of the AMV outbreak reported in New Zealand.[104,105] Presumably this is because vector nematodes have usually not been imported along with the virus and no indigenous nematodes able to transmit it were present in the soil.

The need for constant vigilance is, however, illustrated by the results of a survey of pelargonium nurseries in England and Wales conducted in 1979 to 1980 by the Plant Pathology Laboratory of the Ministry of Agriculture, Fisheries and Food.[91] This survey was undertaken to make sure that British exporting nurseries would be able to comply with the Import and Export (Plant Health) (Great Britain) Order, 1980. This order prohibits the importation of TomRV and also "other harmful viruses and virus-like pathogens" of the genera *Cydonia, Fragaria, Malus, Prunus, Pyrus, Ribes*, and *Rubus* that are not known to occur in the countries of the European Economic Community (EEC). The order also states that plants of the genus *Pelargonium* (a known host of TomRV) may be imported only if accompanied by an official statement on the Phytosanitary Certificate that the plants were produced at a place free from TomRV and were derived within a specified period from mother plants that had been tested by appropriate virological methods and found to be free from TomRV. These regulations are in accordance with the provisions of the EEC Plant Health Frontier Directive (77/93/EEC) and are intended to protect the European fruit industries. During the pelargonium survey, in which the plants were tested by inoculation of sap to herbaceous test plants and any resulting isolates identified serologically, TomRV was detected in 8/21 nurseries and 19/589 clones examined. Moreover, TobRV (which is not specifically covered by the pelargonium regulations but is able to infect *Malus* and *Prunus* species) was found in 4/21 nurseries and 5/589 clones. Both viruses were presumably imported originally in pelargonium material from North America. It is evident that had nematodes with vector ability been present in the soils the viruses might well have become established in Britain. Fortunately, no evidence of this has been obtained. Following this survey, infected pelargonium stocks were removed from propagation and stock mother plants on all exporting pelargonium nurseries continue to be indexed regularly to ensure that they remain free from TomRV and TobRV.

VIII. SAFEGUARDS

Obviously it is impossible to close all the possible avenues by which nepoviruses may obtain entry into a country. However, much may be achieved by applying the following quarantine regulations.

1. Banning totally the importation of soil
2. Banning the importation of rooted plants of all species with soil attached
3. Identifying the species in which individual foreign nepoviruses are most likely to be imported, whether in propagating material or in seed and prohibiting the importation of such material unless it has been specifically tested for, and is accompanied by a certificate of freedom from infection by, named nepoviruses
4. Monitoring such certified material after importation to confirm that it is indeed free from infection

A battery of regulations such as these would seem practical to enforce and should go a long way towards preventing the movement and establishment of nepoviruses in new areas.

REFERENCES

1. **Allen, W. R., Davidson, T. R., and Briscoe, M. R.,** Properties of a strain of strawberry latent ringspot virus isolated from sweet cherry growing in Ontario, *Phytopathology,* 60, 1262, 1970.
2. **Allen, W. R., Van Schagen, J. G., and Ebsary, B. A.,** Comparative transmission of the peach rosette mosaic virus by Ontario populations of *Longidorus diadecturus* and *Xiphinema americanum* (Nematoda: Longidoridae), *Can. J. Plant. Pathol.,* 6, 29, 1984.
3. **Allen, W. R., Van Schagen, J. G., and Eveleigh, E. S.,** Transmission of peach rosette mosaic virus to peach, grape, and cucumber by *Longidorus diadecturus* obtained from diseased orchards in Ontario, *Can. J. Plant Pathol.,* 4, 16, 1982.
4. **Asjes, C. J.,** Viruses and virus diseases in Dutch bulbous irises (*Iris hollandica*) in the Netherlands, *Neth. J. Plant Pathol.,* 85, 269, 1979.
5. **Auger, J. and Converse, R. H.,** Raspberry bushy dwarf and tomato ringspot viruses in chilean red raspberries, *Acta Hortic.,* 129, 9, 1982.
6. **Bancroft, J. B.,** Tomato top necrosis virus, *Phytopathology,* 58, 1360, 1968.
7. **Barbara, D. J., Ashby, S. C., and McNamara, D. G.,** Host range, purification and some properties of rubus Chinese seed-borne virus, *Ann. Appl. Biol.,* 107, 45, 1985.
8. **Benda, G. T. A. and Naylor, A. W.,** On the tobacco ringspot disease. III. Heat and recovery, *Am. J. Bot.,* 45, 33, 1958.
9. **Bonsi, C., Stouffer, R., and Mountain, W.,** Efficiency of transmission of tomato ringspot virus by *Xiphinema americanum* and *Xiphinema rivesi, Phytopathology,* 74, 626, 1984.
10. **Bos, L., Huttinga, H., and Maat, D. Z.,** Parsley latent virus, a new and prevalent seed-transmitted, but possibly harmless virus of *Petroselinum crispum, Neth. J. Plant Pathol.,* 85, 125, 1979.
11. **Brown, D. J. F.,** The transmission of two strains of arabis mosaic virus from England by populations of *Xiphinema diversicaudatum* (Nematoda: Dorylaimoidea) from ten countries, *Rev. Nématol.,* 9, 83, 1986.
12. **Brown, D. J. F.,** The transmission of two strains of strawberry latent ringspot virus by populations of *Xiphinema diversicaudatum* (Nematoda: Dorylaimoidea), *Nematol. Mediterr.,* 13, 217, 1985.
13. **Brown, D. J. F. and Trudgill, D. L.,** Differential transmissibility of arabis mosaic and strains of strawberry latent ringspot viruses by three populations of *Xiphinema diversicaudatum* (Nematoda: Dorylaimida) from Scotland, Italy and France, *Rev. Nematol.,* 6, 229, 1983.
14. **Cadman, C. H.,** Biology of soil-borne viruses, *Annu. Rev. Phytopathol.,* 1, 143, 1963.
15. **Callahan, K. L.,** Pollen transmission of elm mosaic virus, *Phytopathology,* 47, 5, 1957.
16. **Childress, A. M. and Ramsdell, D. C.,** Lack of evidence for a nematode vector of blueberry leaf mottle virus (BBLMV), *Acta Hortic.,* 186, 87, 1986.
17. **Chu, P. W. G., Francki, R. I. B., and Hatta, T.,** Some properties of tomato ringspot virus isolated from *Pentas lanceolata* in South Australia, *Plant Pathol.,* 32, 353, 1983.
18. **Converse, R. H. and Ramsdell, D. C.,** Occurrence of tomato and tobacco ringspot viruses and of dagger and other nematodes associated with cultivated highbush blueberries in Oregon, *Plant Dis.,* 66, 710, 1982.
19. **Cooper, J. I. and Edwards, M.-L.,** Cherry leaf roll virus in *Juglans regia* in the United Kingdom, *Forestry,* 53, 41, 1980.
20. **Cooper, J. I., Massalski, P. R., and Edwards, M.-L.,** Cherry leaf roll virus in the female gametophyte and seed of birch and its relevance to vertical virus transmission, *Ann. Appl. Biol.,* 105, 55, 1984.
21. **Cooper, P.,** Strawberry latent ringspot virus in parsnips, *Plant Pathol.,* 30, 59, 1981.
22. **Dias, H. F.,** Incidence and geographic distribution of tomato ringspot virus in DeChaunac vineyards in the Niagara peninsula, *Plant Dis. Rep.,* 61, 24, 1977.
23. **Dunleavy, J. M.,** The grasshopper as a vector of tobacco ringspot virus in soybean, *Phytopathology,* 47, 681, 1957.
24. **Frazier, N. W., Yarwood, C. E., and Gold, A. H.,** Yellow bud virus endemic along California coast, *Plant Dis. Rep.,* 45, 649, 1961.
25. **Fry, P. R. and Wood, G. A.,** Further viruses of *Prunus* in New Zealand, *N.Z. J. Agric. Res.,* 16, 131, 1973.
26. **Fry, P. R. and Wood, G. A.,** Two berry fruit virus diseases newly recorded in New Zealand, *N.Z. J. Agric. Res.,* 21, 543, 1978.
27. **Fukumoto, F., Ito, Y., and Tochihara, H.,** Viruses isolated from gladiolus in Japan, *Ann. Phytopathol. Soc. Jpn.,* 48, 68, 1982.
28. **Gilmer, R. M. and Uyemoto, J. K.,** Tomato ringspot virus in "Baco Noir" grapevines in New York, *Plant Dis. Rep.,* 56, 133, 1972.
29. **Gordejchuk, O. G., Krylov, A. V., Krylova, N. V., and Samonina, I. N.,** Virus diseases of berry crops in the Soviet Far East. I. Identification of some mechanically transmitted viruses, detected in Primorye Territory, *Zentralbl. Bakteriol. Parasitenkd. Infektionskr. Hyg. Abt.* 2, 132, 686, 1977.
30. **Hanada, K. and Harrison, B. D.,** Effects of virus genotype and temperature on seed transmission of nepoviruses, *Ann. Appl. Biol.,* 85, 79, 1977.

31. **Hansen, A. J., Nyland, G., McElroy, F. D., and Stace-Smith, R.,** Origin, cause, host range and spread of cherry rasp leaf disease in North America, *Phytopathology,* 64, 721, 1974.
32. **Hanson, C. M. and Campbell, R. N.,** Strawberry latent ringspot virus from "Plain" parsley in California, *Plant Dis. Rep.,* 63, 142, 1979.
33. **Harrison, B. D.,** Specific nematode vectors for serologically distinctive forms of raspberry ringspot and tomato black ring viruses, *Virology,* 22, 544, 1964.
34. **Harrison, B. D.,** Ecology and control of viruses with soil-inhabiting vectors, *Annu. Rev. Phytopathol.,* 15, 331, 1977.
35. **Harrison, B. D. and Winslow, R. D.,** Laboratory and field studies on the relation of arabis mosaic virus to its nematode vector *Xiphinema diversicaudatum* (Micoletzky), *Ann. Appl. Biol.,* 49, 621, 1961.
36. **Hollings, M.,** Anemone necrosis, a disease caused by a strain of tobacco ringspot virus, *Ann. Appl. Biol.,* 55, 447, 1965.
37. **Hollings, M., Stone, O. M., and Dale, W. T.,** Tomato ringspot virus in pelargonium in England, *Plant Pathol.,* 21, 46, 1972.
38. **Hoy, J. W. and Mircetich, S. M.,** Prune brownline disease: susceptibility of prune rootstocks and tomato ringspot virus detection, *Phytopathology,* 74, 272, 1984.
39. **Hoy, J. W., Mircetich, S. M., and Lownsbery, B. F.,** Differential transmission of Prunus tomato ringspot virus strains by *Xiphinema californicum, Phytopathology,* 74, 332, 1984.
40. **Iwaki, M. and Komuro, Y.,** Viruses isolated from narcissus (*Narcissus* spp.) in Japan. II. Tomato ringspot virus and its transmission by *Xiphinema americanum, Ann. Phytopathol. Soc. Jpn.,* 37, 108, 1971.
41. **Iwaki, M. and Komuro, Y.,** Viruses isolated from narcissus (*Narcissus* spp.) in Japan. 4. Tomato black ring virus, *Ann. Phytopathol. Soc. Jpn.,* 39, 279, 1973.
42. **Iwaki, M. and Komuro, Y.,** Viruses isolated from narcissus (*Narcissus* spp.) in Japan. V. Arabis mosaic virus, *Ann. Phytopathol. Soc. Jpn.,* 40, 344, 1974.
43. **Jones, A. T. and Badenoch, S. J.,** Recent studies on virus and virus-like diseases of *Rubus* in Scotland, *Acta Hortic,* 129, 49, 1982.
44. **Jones, A. T., McElroy, F. D., and Brown, D. J. F.,** Tests for transmission of cherry leaf roll virus using *Longidorus, Paralongidorus* and *Xiphinema* nematodes, *Ann. Appl. Biol.,* 99, 143, 1981.
45. **Jones, R. A. C., Fribourg, C. E., and Koenig, R.,** A previously undescribed nepovirus isolated from potato in Peru, *Phytopathology,* 73, 195, 1983.
46. **Jordović, M., Dimitrijević, B., and Ranković, M.,** Identifikacija virusa prstenaste pegavosti paradajza (tomato ringspot virus) na malini u Jugoslaviji, *Zašt. Bilja,* 23, 147, 1972.
47. **Kaiser, W. J., Bock, K. R., Guthrie, E. J., and Meredith, G.,** Occurrence of tomato black ring virus in potato cultivar Anett in Kenya, *Plant Dis. Rep.,* 62, 1082, 1978.
48. **Kearns, C. G. and Mossop, D. W.,** Detection of nepoviruses of *Vitis vinifera* in New Zealand using enzyme-linked immunosorbent assay (ELISA), *N.Z. J. Agric. Res.,* 27, 431, 1984.
49. **Kenten, R. H., Cockbain, A. J., and Woods, R. D.,** Crimson clover latent virus — a newly recognised seed-borne virus infecting crimson clover *(Trifolium incarnatum), Ann. Appl. Biol.,* 96, 79, 1980.
50. **Komuro, Y. and Iwaki, M.,** Bean yellow mosaic virus and tobacco ringspot virus isolated from crotalaria *(Crotalaria spectabilis), Ann. Phytopathol. Soc. Jpn.,* 34, 7, 1968.
51. **Kudamatsu, M., Barradas, M. M., and Alba, A. P. C.,** Characterization of the "bouquet" strain of tomato black ring virus from imported seed potato in Brazil, *Turrialba* 31, 195, 1981.
52. **Kyriakopoulou, P. E., Rana, G. L., and Roca, F.,** Geographic distribution, natural host range, pollen and seed transmissibility of artichoke yellow ringspot virus, *Ann. Inst. Phytopathol. Benaki (Athens),* 14, 139, 1985.
53. **Lamberti, F. and Bleve-Zacheo, T.,** Studies on *Xiphinema americanum sensu lato* with descriptions of fifteen new species (Nematoda, Longidoridae), *Nematol. Mediterr.,* 7, 51, 1979.
54. **Lana, A. F.,** Transmission and properties of viruses isolated from *Carica papaya* in Nigeria, *J. Hortic. Sci.,* 55, 191, 1980.
55. **Lister, R. M.,** Nematode-borne viruses, *Rep. Scott. Hortic. Res. Inst., 1961—1962,* 66, 1962.
56. **Lister, R. M.,** Nematode-borne viruses as pathogens in strawberry, in *Virus Diseases of Small Fruits and Grapevines,* Frazier, N. W., Ed., University of California Press, Berkeley, 1970, 32.
57. **Lister, R. M.,** Application of the enzyme-linked immunosorbent assay for detecting viruses in soybean seed and plants, *Phytopathology,* 68, 1393, 1978.
58. **Lister, R. M. and Murant, A. F.,** Seed-transmission of nematode-borne viruses, *Ann. Appl. Biol.,* 59, 49, 1967.
59. **Lister, R. M., Raniere, L. C., and Varney, E. H.,** Relationships of viruses associated with ringspot diseases of blueberry, *Phytopathology,* 53, 1031, 1963.
60. **Lucas, G. B.,** Ringspot, in *Diseases of Tobacco,* Lucas, G. B., Ed., Biological Consulting Associates, Raleigh, N.C., 1975, 523.
61. **Maat, D. Z. and Van Hoof,** Virus diseases of pelargonium, *Rep. Inst. Phytopathol. Res. Wageningen 1973,* 31, 36, 1974.

62. **McLean, D. M. and Meyer, H. M.,** A survey of cucurbit viruses in the Lower Rio Grande Valley of Texas: preliminary report, *Plant Dis. Rep.*, 45, 137, 1961.
63. **Messieha, M.,** Transmission of tobacco ringspot virus by thrips, *Phytopathology*, 59, 943, 1969.
64. **Mircetich, S. M., Sanborn, R. R., and Ramos, D. E.,** Natural spread, graft-transmission and possible etiology of walnut blackline disease, *Phytopathology*, 70, 962, 1980.
65. **Mountain, W. L., Powell, C. A., Forer, L. B., and Stouffer, R. F.,** Transmission of tomato ringspot virus from dandelion via seed and dagger nematodes, *Plant Dis.*, 67, 867, 1983.
66. **Murant, A. F.,** Soil-borne viruses and diseases in *Rubus*, in *Virus Diseases of Small Fruits and Grapevines*, Frazier, N. W., Ed., University of California Press, Berkeley, 1970, 132.
67. **Murant, A. F.,** Nepoviruses, in *Handbook of Plant Virus Infections and Comparative Diagnosis*, Kurstak, E., Ed., Elsevier/North Holland, Amsterdam, 1981, 197.
68. **Murant, A. F.,** The role of wild plants in the ecology of nematode-borne viruses, in *Pests, Pathogens and Vegetation*, Thresh, J. M., Ed., Pitman, London, 1981, 237.
69. **Murant, A. F.,** Seed and pollen transmission of nematode-borne viruses, *Seed Sci. Technol.*, 11, 973, 1983.
70. **Murant, A. F. and Lister, R. M.,** Seed-transmission in the ecology of nematode-borne viruses, *Ann. Appl. Biol.*, 59, 63, 1967.
71. **Murant, A. F. and Taylor, C. E.,** Treatment of soils with chemicals to prevent transmission of tomato blackring and raspberry ringspot viruses by *Longidorus elongatus* (de Man), *Ann. Appl. Biol.*, 55, 227, 1965.
72. **Paludan, N.,** Virus diseases in *Pelargonium hortorum* specially concerning tomato ringspot virus, *Acta Hortic.*, 59, 119, 1976.
73. **Pitcher, R. S. and Jha, A.,** On the distribution and infectivity with arabis mosaic virus of a dagger nematode, *Plant Pathol.*, 10, 67, 1961.
74. **Powell, C. A., Forer, L. B., and Stouffer, R. F.,** Reservoirs of tomato ringspot virus in fruit orchards, *Plant Dis.*, 66, 583, 1982.
75. **Powell, C. A., Forer, L. B., Stouffer, R. F., Cummins, J. N., Gonsalves, D., Rosenberger, D. A., Hoffman, J., and Lister, R. M.,** Orchard weeds as hosts of tomato ringspot and tobacco ringspot viruses, *Plant Dis.*, 68, 242, 1984.
76. **Price, W. C.,** Comparative host ranges of six plant viruses, *Am. J. Bot.*, 27, 530, 1940.
77. **Quacquarelli, A., Rana, G. L., and Martelli, G. P.,** Weed hosts of plant pathogenic viruses in Apulia, *Poljopr. Znan. Smotra*, 39, 561, 1976.
78. **Ramsdell, D. C. and Myers, R. L.,** Epidemiology of peach rosette mosaic virus in a Concord grape vineyard, *Phytopathology*, 68, 447, 1978.
79. **Randles, J. W. and Francki, R. I. B.,** Some properties of a tobacco ringspot virus isolate from South Australia, *Aust. J. Biol. Sci.*, 18, 979, 1965.
80. **Rani, S., Verma, H. N., and Verma, G. S.,** A virus disease of *Petunia hybrida*, *Plant Dis. Rep.*, 53, 903, 1969.
81. **Raski, D. J., Hewitt, W. B., Goheen, A. C., Taylor, C. E., and Taylor, R. H.,** Survival of *Xiphinema index* and reservoirs of fanleaf virus in fallowed vineyard soil, *Nematologica*, 11, 349, 1965.
82. **Rowhani, A., Mircetich, S. M., Shepherd, R. J., and Cucuzza, J. P.,** Serological detection of cherry leafroll virus in English walnut trees, *Phytopathology*, 75, 48, 1985.
83. **Rydén, K.,** *Pelargonium* ringspot — a virus disease caused by tomato ringspot virus in Sweden, *Phytopathol. Z.*, 73, 178, 1972.
84. **Sastry, K. S. and Nayudu, M. V.,** Ringspot symptoms of eggplant, incited by tobacco ring spot virus, *Phytopathol. Mediterr.*, 15, 60, 1976.
85. **Schlocker, A. and Traylor, J. A.,** Yellow bud mosaic, in Virus Diseases and Noninfectious Disorders of Stone Fruits in North America, Agric. Handbook 437, U.S. Department of Agriculture, Washington, D.C., 1976, 156.
86. **Schmelzer, K.,** Untersuchungen an Viren der Zier- und Wildgehölze. 2. Mitteilung. Virosen an *Forsythia, Lonicera, Ligustrum* und *Laburnum*, *Phytopathol. Z.*, 46, 105, 1963.
87. **Schmelzer, K.,** Untersuchungen an Viren der Zier- und Wildgehölze. 3. Mitteilung. Virosen an *Robinia, Caryopteris, Ptelea* und anderen Gattungen, *Phytopathol. Z.*, 46, 235, 1963.
88. **Schmelzer, K.,** Untersuchungen an Viren der Zier- und Wildgehölze. 5. Mitteilung. Virosen an *Populus* und *Sambucus*, *Phytopathol. Z.*, 55, 317, 1966.
89. **Schmelzer, K.,** Das latente Erdbeerringflecken-Virus aus *Evonymus, Robinia* und *Aesculus*, *Phytopathol. Z.*, 66, 1, 1969.
90. **Schuster, M. F.,** Flea beetle transmission of tobacco ringspot virus in the Lower Rio Grande Valley, *Plant Dis. Rep.*, 47, 510, 1963.
91. **Sharp, C. M.,** personal communication.
92. **Silberschmidt, K. M.,** Sobre uma doença de vírus da batatinha veiculada em tubérculos importados, *Biológico*, 31, 176, 1965.

93. **Sinclair, J. B. and Shurtleff, M. C.,** Bud blight, in *Compendium of Soybean Diseases,* Sinclair, J. B. and Shurtleff, M. C., Ed., American Phytopathological Society, St. Paul, Minn., 1975, 44.
94. **Sinclair, J. B. and Walker, J. C.,** A survey of ring spot on cucumber in Wisconsin, *Plant Dis. Rep.,* 40, 19, 1956.
95. **Smith, S. H., Stouffer, R. F., and Soulen, D. M.,** Induction of stem pitting in peaches by mechanical inoculation with tomato ringspot virus, *Phytopathology,* 63, 1404, 1973.
96. **Stace-Smith, R.,** Red raspberry virus diseases in North America, *Plant Dis.,* 68, 274, 1984.
97. **Stobbs, L. W. and Van Schagen, J. G.,** Occurrence of tomato black ring virus on grapevine in southern Ontario, *Can. Plant Dis. Surv.,* 64, 3, 1984.
98. **Stone, O. M.,** Nine viruses isolated from pelargonium in the United Kingdom, *Acta Hortic.,* 110, 177, 1980.
99. **Stouffer, R. F., Hickey, K. D., and Welsh, M. F.,** Apple union necrosis and decline, *Plant Dis. Rep.,* 61, 20, 1977.
100. **Taylor, C. E.,** Nematodes, in *Vectors of Plant Pathogens,* Harris, K. F. and Maramorosch, K., Eds. Academic Press, New York, 1980, 375.
101. **Taylor, C. E. and Murant, A. F.,** Transmission of strains of raspberry ringspot and tomato black ring viruses by *Longidorus elongatus* (de Man), *Ann. Appl. Biol.,* 64, 43, 1969.
102. **Taylor, C. E. and Thomas, P. R.,** The association of *Xiphinema diversicaudatum* (Micoletzky) with strawberry latent ringspot and arabis mosaic viruses in a raspberry plantation, *Ann. Appl. Biol.,* 62, 147, 1968.
103. **Thomas, C. E.,** Transmission of tobacco ringspot virus by *Tetranychus* sp., *Phytopathology,* 59, 633, 1969.
104. **Thomas, W. and Procter, C. H.,** Arabis mosaic virus in *Cyphomandra betaceae* Sendt., *N.Z. J. Agric. Res.,* 15, 395, 1972.
105. **Thomas, W. and Procter, C. H.,** Further studies on arabis mosaic virus in *Cyphomandra betacea* Sendt., *N.Z. J. Agric. Res.,* 20, 543, 1977.
106. **Tomlinson, J. A. and Carter, A. L.,** Occurrence and seed transmission of tobacco ringspot virus in *Senecio vulgaris, Rep. Nat. Vegetable Res. Stn., 1970,* 109, 1971.
107. **Tuite, J.,** The natural occurrence of tobacco ringspot virus, *Phytopathology,* 50, 296, 1960.
108. **Vuittenez, A.,** Fanleaf of grapevine, in *Virus Diseases of Small Fruits and Grapevines,* Frazier, N. W., Ed., University of California Press, Berkeley, 1970, 217.
109. **Xu, Z., Polston, J. E., and Goodman, R. M.,** Identification of soybean mosaic, southern bean mosaic and tobacco ringspot viruses from soybean in the People's Republic of China, *Ann. Appl. Biol.,* 108, 51, 1986.

Chapter 4

DOWNY MILDEWS

R. A. Frederiksen

TABLE OF CONTENTS

I. INTRODUCTION

The downy mildews are important plant diseases. These fungi are members of the Peronosporales, they are biotropic, obligate parasites of many of the more significant agricultural and horticultural plant species.

Eighteen specific downy mildew pathogens of over 40 major crop plants are reviewed. This large array of pathogens represents an awesome body of information to be summarized in one chapter, consequently, only a cursory review of each species is presented.

A. Symptoms

The downy mildew fungi cause two type of symptoms; systemic and local lesion, for most of the affected hosts. While there are considerable variations in symptoms, patterns of disease caused by these fungi are quite similar.

For most annual plant species, oospores are the primary inoculum. These spores may be soil borne, seed borne, or associated with infected host tissue. Generally, seedlings infected by oospores develop systemic infection. These plants are dwarfed or stunted, and they produce abundant secondary inoculum in the form of conidia or sporangia. These asexual spore forms represent the source of inoculum for the local lesion phase of these downy mildews. For perennial species, oospores are not necessary for initial inoculum because remaining systemically infected plants provide a source of asexual inoculum. For some downy mildews such as *Peronospora destructor* on onion, the systemic phase begins from infected bulbs.

Most species with systemic infection follow a colonization pattern described by Williams[44] for *Sclerospora graminicola*. Partially differentiated tissues become progressively more diseased with each succeeding leaf. Areas of leaf tissue colonized by the downy mildew pathogen are markedly more chlorotic than healthy tissue. Asexual sporulation is associated with the initial systemic infection, followed in succeeding tissue by the formation of the sexual spores. Systemically stricken plants are usually barren and dwarfed.

B. Taxonomic Concerns

The identification of pathogens causing downy mildew has rarely been easy. The descriptions of pathogens vary in adequacy and biological accuracy. The extent of natural variation within a species is rarely addressed yet may constitute a major consideration in the description of numerous species. Even the number of samples from a collection for comparison may be limited. Comparisons of host ranges of the downy mildews species is limited even under the most ideal conditions because, in general, collections of living specimens are frequently prohibited by regulation and where permitted the facilities are limited in space or environment to adequately understand the biology of the pathogen. The reader is referred to Williams[44] for a detailed review of the taxonomic problems in the downy mildews of the Graminae.

C. The Pathogens

Lists of downy mildew fungi, common names, and hosts are summarized in Tables 1 and

Table 1
THE DOWNY MILDEWS OF THE GRAMINEAE, THEIR HOSTS, AND COMMON NAMES

Pathogen	Crops hosts reported seriously diseased	Common disease name	Other reported hosts or host genera[a]
Sclerospora graminicola (Sacc.) Schroet.	Pearl millet, foxtail millet	Green-ear Graminicola DM	Several *Setaria* spp., *Echinochloa, Zea mays, (Saccharum, Sorghum), Z. mexicana*
Peronosclerospora			
P. heteropogoni Siradhana, Dange, Rathore, and Singh	Maize	—	*Heteropogon contortus, Z. mexicana*
P. maydis (Racib.) C. G. Shaw	Maize	Java DM	*(Z. mexicana, Tripsacum)*
P. philippinensis (Weston) C. G. Shaw	Maize	Philippine DM	*Saccharum, Sorghum, Tripsacum, Z. mexicana (Miscanthus, Avena)*
P. sacchari (T. Miyake) Shirai and K. Hara	Maize, sugarcane	Sugarcane DM	*Miscanthus, Sorghum, Tripsacum, Z. mexicana, (Andropogon, Botriochloa, Schizachyrium)*
P. sorghi (Weston and Uppal) C. G. Shaw	Sorghum, maize	Sorghum DM	*Panicum (Andropogon) Z. mexicana*
P. spontanea (Weston) C. G. Shaw	Maize	Spontaneum DM	*Saccharum, Z. mexicana, Miscanthus, Sorghum*
P. noblei (Weston) C. G. Shaw	—	—	*Sorghum plumosum, Sorghum leiocladum*
P. miscanthi T. Miyake and Sacc. C. G. Shaw	—	Leaf-splitting DM	*Miscanthus japonicus, Saccharum, Sorghum, (Zea mays)*
P. westonii Srin., Naras., and Thirum. C. G. Shaw	—	—	*Iseilema laxum*
P. globosa Kenneth and Kubicek	—	—	*Eriochloa contracta* Hitchcock
Sclerophthora			
S. macrospora (Sacc.) Thirum, Shaw and Naras.	Maize, rice, wheat	Crazy top, yellow wilt	140 species in many genera
S. rayssiae var. *rayssiae* Kenneth, Koltin, and Wohl	Barley	Blotch DM	—
S. rayssiae var. *zeae* Payak and Renfro	Maize	Brown stripe DM	*Digitaria sanuinalis*

[a] Hosts in parentheses indicate infection by inoculation.[44]

Adapted from Kenneth, R. G., *The Downy Mildews*, Spencer, D. M., Ed., Academic Press, London, 1981, 367.

2. The downy mildews of the Graminae are separated in part because of the extraordinary confusion regarding the identification of these species and the fact that maize is a host for most of the more commonly observed species. The review by Kenneth[20] and a recent report by Kenneth and Kubicek[21] discuss in detail species relations among these pathogens. Listed in Table 1 are 21 species, however, only the seven most important downy mildew species attacking the major agricultural crops are reviewed in the disease section. Host ranges of 12 other downy mildew pathogens attacking many of the worlds important agricultural and horticultural crops are summarized in Table 2. The biology of each of these species is summarized in the section on downy mildew diseases.

Table 2
DOWNY MILDEW PATHOGENS OF ECONOMICALLY IMPORTANT PLANT SPECIES OTHER THAN THE GRAMINEAE

Pathogen	Common host
Bremia lactucae Regel	
Lactuca sativa L.	Lettuce
Agoseris Rafin.	Mountain dandelion
Andryala L.	
Arctium L.	Burdock
Carduus L.	Plumeless thistle
Carlina L.	Herb
Carthamus L.	Herb
Centaurea L.	Knapwood
Cicerbita Wallr.	
Cichorium L.	Quinine
Cirsium Mill.	Thistle
Crepis L.	
Cynara L.	Herb
Dendroseris D. Don.	
Dimorphotheca Moench.	Cape marigold
Erechtites Rafin.	
Gaillardia Fougeroux	Blanket flower
Helichrysum Mill.	Everlasting
Hemistepta Bunge	
Hieraceum L.	Hawkweed
Inula L.	Herb
Krigia Schreb.	Dwarf dandelion
Lapsana L.	
Launaea Cass.	
Leontodon L.	
Mycelis Cass.	
Onopordum L.	Herb
Parthenium L.	Shrub herb
Prenamthes L. (syn. *Nabalus* Cass.)	Rattlesnake root
Picris L. (syn. *Helminthia* Juss.)	Bitterweed
Senecio L. (syn. *Jacobaea* Mill and *Cineraria* L.)	Groundsel
Solidago L.	Goldenrod
Sonchus L.	
Taraxacum Weber	Dandelion
Tragopogon L.	Goatsbeard
Venidium Less.	Namaqualand daisy
Peronospora destructor (Berk.) Casp.	
Allium cepa	Onion
A. altaicum Poll.	
A. fistulosum L.	Welsh onion
A. flavum L.	
A. galanthum Kar. and Kir.	
A. obliquum L.	
A. ochroleucum L.	
A. schoenoprasum L.	Chive
A. tuberosum Rottl. ex. K. Sprong	Chinese chive
Peronospora farinosa (Fr.) Fr.	
Spinacia oleracea L.	Spinach
Beta vulgaris L.	Beet
Peronospora manshurica (Naum.) Syd, ex. Gaum.	
Glycine max (L.) Merrill	Soybean

Table 2 (continued)
DOWNY MILDEW PATHOGENS OF ECONOMICALLY IMPORTANT PLANT SPECIES OTHER THAN THE GRAMINEAE

Pathogen	Common host
Peronospora parasitica (pers. ex. Fr.) Fr.	
Armoracia rusticana P. Gaertin., B. Mey. and Scherb.	Horseradish
Brassica alboglabra L. H. Bayley	Chinese kale
Brassica campestris L.	Chinese cabbage Turnips
Brassica napus L. var. *napobrassica*	Swedes
Brassica napus L. var. *napus* f. *annua*	Rape
Brassica natus L. var. *napus* f. *biennis*	Colza
Brassica oleracea var. *acephala* D.C. sub var. *medullosa*	Marrow-stem kale
Brassica oleracea var. L. *Botrytis* sub var. *cauliflora* (Gars) DC.	Cauliflower
Brassica oleracea L. var. *botrytris* sub var. *cymosa* Duchesne	Broccoli
Brassica oleracea var. L. *capitata*	Cabbage
Brassica oleracea var. L. *gemmifera*	Brussel sprout
Brassica oleracea L. var. *gongyloides*	Kohlrabi
Cheiranthus cheiri L.	Wallflower
Crambe maritima L. gongyloides	Seakale
Matthiola incana (L.) R. Br.	Stock
Nasturtium officinale R. Br.	Watercress
Raphanus sativus L.	Radish
Peronospora tabacina Adam.	
Nicotiana tabacum L.	Tobacco
Peronospora trifoliorum de Bary (*forage legumes*)	
Coronilla varia L.	Crown vetch
Lotus corniculatus L.	Birds-foot trefoil
L. uliginosus Schk.	
Medicago sativa L.	Alfalfa
M. lupulina L.	Black medic
Melilotus alba Desr.	Bukhara clover
M. officinalis (L.) Lam.	Yellow sweet clover
Trifolium agrarium L.	Hop clover
T. incarntum L.	Crimson clover
T. pratense L.	Red clover
Peronospora viciae (Berk.) de Bary	
Lathyrus odoratus L.	Sweet pea
L. sativus	
Pisum arvense L.	Field pea
P. sativum L.	English pea
Vicia faba L.	Broad bean
V. sativa	Spring vetch
Plasmopara halstedii (Farl.) Berk. and de Toni	
Helianthus annus L. and numerous species in 30 genera of compositae	

II. DOWNY MILDEW OF LETTUCE

Causal agent — *Bremia lactucae* Regel (*Peronospora ganglioniformis* Dby., *Botrytis ganglioniformis* Berk.).

Host range — *B. lactucae* has an extremely large host range, 36 genera of Compositae. However, the pathogen population is composed of several *formae speciales*; each having a relatively narrow host range confined to one host species or at most to one genus. Some biotypes of *B. lactucae* which attack *Lactuca sativa* attack other species of *Lactuca*.[10,26]

Geographic range — *Bremia lactucae* is found in most countries of the world. It is present throughout Europe and North America, the Mediterranean Basin, Asia Minor, and the Middle East. Reports from other regions include the countries of Afghanistan, Argentina, Australia, Boliva, Brazil, Chile, Colombia, Cuba, Ecuador, Ethiopia, India, Japan, Kenya, New Zealand, Pakistan, Paraguay, Peru, South Africa, Tanzania, Tunisia, Turkey, Uruguay, Venezuela, Zaire, and Zambia.[10]

Symptoms — Pale yellow lesions of variable size and shape appear on upper surfaces of the leaves. Downy masses of spores and sporophores of the pathogen are formed on the lower surfaces of infected leaves. The lesions may enlarge and coalesce. The lesions become brown as they age.[10]

Disease cycle — *Bremia lactucae* survives between lettuce crops in infected seed and as oospores in soil and plant debris. Lettuce seedlings infected by overwintering inoculum produce sporangia which are disseminated by wind and splashing rain to initiate the secondary cycle of disease.[10]

Disease spread — The disease is spread over short distance by sporangia during periods of cool weather and high humidity. Long-distance spread results from distribution of infected seeds, transport of oospores in infested soil, and the transport of diseased lettuce through commercial produce channels.[10]

III. ONION DOWNY MILDEW

Causal agent — *Peronospora destructor* (Berk.) Casp. (*Peronspora schleideniana* W. G. Sm.).

Host range — Only *Allium* sp. are reported as hosts to *P. destructor*. Both cultivated and wild hosts are attacked,[43] Table 2.

Geographic range — Onion downy mildew is, for all practical purposes, worldwide in distribution.

Symptoms — Pale green, oval to cylindrical lesions found on leaves and stems. Under humid conditions, lesions will be covered by violet, downy fructification of the pathogen. Systemic infection induces dwarfing and distorted growth. Leaves of systemically infected plants are covered with the violet down of pathogen sporulation. Growth and quality of bulbs are reduced.[43,47]

Disease cycle — The pathogen over-winters as mycelium in bulbs ore perennial onion plants and as oospores in the soil. Secondary infection caused by conidia produced on diseased plants and disseminated to healthy plants by air currents under humid conditions.[43,47]

Disease spread — Short range spread is by airborne conidia. The conidia may retain viability for a few days after maturation if relative humidity and temperature are favorable. Long-range spread is caused by transport of oospore-infested soil, infected bulbs, and oospore-infested seed. Seed transmission of the disease has not been demonstrated.[29,43,47]

IV. DOWNY MILDEW OF BEET AND SPINACH

Causal agent — *Peronospora farinosa* (Fr.) Fr.; *P. effusa* (Grev.) Rabh. (= spinacea Laub.; *P. schachtii* Fuckel). Downy mildew of beet, and downy mildew of spinach are caused by two *formae specialis* of *P. farinosa*, *P. farinosa betae*, and *P. farinosa spinaceae*.[5]

Host range — *Beta vulgaris* and *Spinacia oleracea*.[5]

Geographic range — Downy mildew of beet is widespread in temperature regions of the Northern Hemisphere. The disease has been reported in countries throughout Western Europe as well as Argentina, Egypt, U.S., and U.S.S.R. Downy mildew of spinach occurs in most countries where the crop is grown.[24]

Symptoms — In systemically infected beets the young leaves are yellowish green, stunted, curled, and thickened. Under humid conditions, pathogen sporulation covers leaves and stems with a purplish-gray down. The flower shoots are dwarfed and distorted. Excessive leaf development and compacted appearance of the inflorescence often gives a witches-broom effect. Localized infection of leaves produces light green spots up to 4 cm in diameter under humid conditions, spores are produced on the lower surface of the leaf spot. Spinach downy mildew is characterized by pale yellow spots on leaves of infected plants. The leaves are curled and distorted, and under humid conditions, spores are produced on leaf surfaces and sometimes the petiole also. Infected leaves die.[24]

Disease cycle — In beet downy mildew, the pathogen survives between cropping seasons as mycelium in seeds, roots, crowns, or living plants and as oospores in the soil, and in the seed. Secondary infection is induced by conidia produced on diseased plants. In spinach downy mildew, the pathogen overwinters as mycelium in the seed coat, in perennial plants, as oospores in the soil or on the seed. Secondary inoculum is made up of wind-borne conidia.[24]

Disease spread — Conidia disseminated by wind under humid conditions are the agents of short-range spread in downy mildew of beet and spinach. Long-range spread of beet downy mildew is by means of oospore-infested soil, oospore-infested seed, mycelium in roots, and mycelium and oospores in the seed. Long-range spread of spinach downy mildew may occur through transport of oospore-infested soil, diseased plant material, and oospore-infested seed. Mycelium of the pathogen has been found in the seed coat but seed transmission of the disease has not been demonstrated.[25,30]

V. DOWNY MILDEW OF SOYBEAN

Causal agent — *Peronospora manshurica* (Naum.) Syd. ex Guam. (*P. sojae* Lehman and Wolf; *P. trifoliorum* var. *manshurica* Naum.).

Host range — *Glycine max*.

Geographic range — Bermuda, China, Czechoslovakia, Denmark, England, Hungary, India, Latvia, Rumania, Ryukyus, South Africa, Sweden, Taiwan, U.S., U.S.S.R., and Yugoslavia.[12]

Symptoms — Early symptoms consist of large or small chlorotic spots on the upper surface of infected leaves. A violet to gray down composed of spores and sporophores of the pathogen develops on the lower surface of the lesion. Lesions spread rapidly and the leaves turn more yellow than brown. Oospores of the pathogen form in the brown leaf tissue. Pod infection causes no visible exterior symptoms but the interior of the pod and the seed coat are covered with gray mycelium and oospores.[12]

Disease cycle — The pathogen survives between crops as oospores in plant debris or on the seeds. Soybean seedlings are infected by oospores. The diseased seedlings produce conidia which are disseminated by wind, invade healthy leaves, and start the secondary cycle of the disease.[12,19]

Disease spread — The disease is spread for short distances by airborne conidia. Long-range dissemination of the pathogen occurs with the transport of oospore-infested soil or oospore-infested seed.[12,19,30,45]

VI. DOWNY MILDEW OF BRASSICA

Causal agent — *Peronospora parasitica* (Pers. ex Fr.) Fr. (*Botrytis parasitica* Pers.; *P. brassicae* Gaum.).

Host range — *Peronospora parasitica* attacks a wide range of crucifers and possesses great variability for virulence, at both the interspecific and intraspecific level. Major economic hosts are cabbage, brussel sprouts, and cauliflower (see Table 2).

Geographic range — Angola, Argentina, Australia, Austria, Bermuda, Brazil, Brunei, Canada, Chile, China, Colombia, Costa Rica, Cuba, Cyprus, Czechoslovakia, Denmark, Dominica, Egypt, Ethiopia, Fiji, Finland, France, Germany, Greece, Guatemala, Haiti, Hong Kong, India, Iran, Iraq, Israel, Italy, Jamaica, Japan, Kampuchea, Kenya, Korea, Libya, Malawi, Malaysia, Malta, Mauritius, Mexico, Morocco, Mozambique, Nepal, Netherlands, New Zealand, Norway, Panama, Pakistan, New Guinea, Phillipines, Poland, Portugal, Puerto Rico, Samoa, South Africa, Spain, Sri Lanka, Sweden, Switzerland, Taiwan, Tanzania, Thailand, Trinidad, Turkey, Uganda, U.S.S.R., Venezuela, Vietnam, and Yugoslavia.[6]

Symptoms — The initial symptoms on brassica seedlings are discolored spots on cotyledons. Infected cotyledons frequently turn yellow and die; and seedling death may follow. Sporulation of the pathogen produces a gray down on the hypocotyl and lower surface of the cotyledons. Later infection of developing plants produces small, discolored spots on the leaves. Pathogen sporulation produces a gray down on the under surface of the leaves. Under humid conditions, the infection spreads rapidly and the leaves die. In addition to the leaves, the fungus may colonize the curds of cauliflower and broccoli, cabbage heads, and turnip roots. The affected tissues are frequently rotted by the secondary invasion of other fungi or bacteria.[6]

Disease cycle — The pathogen survives between host crops as oospores infesting seed, soil, or plant debris. Host seedlings are invaded by oospores and systemic infection of the seedling occurs. Under cool moist conditions, the pathogen produces conidia on the diseased seedling. These spores are spread to healthy leaves by air currents and induce the secondary disease cycle.[6,30]

Disease spread — Short-range spread during the growing season is accomplished by conidia. Conidia of *P. parasitica* are comparatively durable, surviving for 110 days in dry soil under winter conditions.[22] Long-range spread results from the transport of oospore-infested soil, oospore-infested seed, diseased seedlings, and fresh vegetable market produce.[6,30]

VII. DOWNY MILDEW OF TOBACCO (BLUE MOLD OF TOBACCO)

Causal agent — *Peronospora tabacina* Adam (*Peronospora hyoscyami* deBary; *Peronospora nicotiana* Speg.).

Host range — *Nicotiana tabacum, Nicotiana* sp., Other species reported susceptible to artificial inoculation with *P. tabacina* are *Capsicum annum, Hyoscyamus muticus, H. niger, Petunia hybrida, Physalis alkekangi, P. peruviana, Solanum lycopersicum esculentum*, and *S. melongena*.[2,36]

Geographic range — Argentina, Australia, Brazil, Canada, Cambodia, Chile, Cuba, U.S., Europe, North Africa, and the near East.[36]

Symptoms — Pale green or yellow lesions appear on upper surface of the leaves. A grayish brown downy mass composed of conidia and conidiophores of the pathogen develop on the undersurface of the leaves. As the disease progresses, the infected leaves turn brown.[36]

Disease cycle — The pathogen overwinters in disease, second growth tobacco plants, in perennial feral species of tobacco, and as conidia and oospores in plant debris and soil. The new tobacco seedlings are infected by oospores in the soil and by conidia from surviving diseased plants of the previous season. The secondary cycle of disease is initiated by the dissemination of conidia produced on the diseased tobacco seedlings.[35,45]

Disease spread — Conidia of *P. tabaci* play a major role in the spread of this disease both short- and long distance. The conidia are resistant to drying and retain viability for 4 to 5 months in dry soil. This durability permits the spores to survive travel distances of up to 1600 km. Other means of long-distance spread are transport of diseased seedlings, oospore-infested soil, and plant debris.[30,36,46]

VIII. DOWNY MILDEW OF ALFALFA AND CLOVER

Causal agent — *Peronospora trifoliorum* de Bary (*Peronospora savulescu* Rayss; *Peronospora romanica* Sav. and Rayss; *Peronospora aestivalis* f. *medicaginis-falcata* (Therem.) Sav. and Rayss; *Peronospora trifolii alpestris* Gaum.; *Peronospora aestivalis* Syd.). An extensive list of earlier synonyms is provided by Chilton.[7]

Host range — *Peronospora trifoliorum* attacks the major forage legume genera. The most economically important host species is *Medicago sativa* L. (Alfalfa). Other genera reported as host are *Coronilla, Lotus, Melilotus*, and *Trifolium*.

Geographic range — The pathogen is essentially global in distribution, corresponding with the distribution of the host.[1]

Symptoms — Leaves are chlorotic and frequently malformed. Stems are smaller than normal, swollen, and have shorter intermodes. A grayish white mycelial mat develops on the surface of leaves and stems. A downy mass of conidophores and violet conidia is produced on the undersurface of the leaves under humid conditions. Oospores are formed in leaves and in the crown of diseased plants.[40]

Disease cycle — The pathogen overwinters as mycelium and as oospores in the crowns of infected alfalfa plants and as oospores in plant debris. Conidia are produced on newly infected plants, and on systemically infected shoots from diseased crowns. The conidia are disseminated to the leaves of healthy plants by air currents and initiate the secondary cycle of infection.[40]

Disease spread — Short-range spread is by means of conidia. Oospores in hay and oospore-infested seed are the most probable means of long distance spread.[40]

IX. DOWNY MILDEW OF PEAS AND BEANS

Causal agent — *Peronospora viciae* (Berk.) de Bary (*Botrytis viciae* Berk.; *P. faba; P. pisi* Syd. emend. Campbell). *Peronospora viciae* is the causal agent of downy mildew of peas and beans. It is a severe disease in areas where cool, humid conditions prevail.

Host range — *Pisum arvense* L., *P. sativum* L., *Lathyrus odoratus* L., *L. sativus* L., *Vicia faba* L. and *V. sativa* L.[11]

Geographic range — Argentina, Australia, Canada, Chile, Denmark, France, Germany, India, Italy, Kenya, Netherlands, New Zealand, Rumania, Spain, Tazmania, Tunisia, U.K., U.S., U.S.S.R., and Zimbabwe.[11]

Symptoms — Yellowish-green to brown spots appear on the upper surface of leaves. A white to bluish-white downy mass of spores and sporophores of the pathogen are produced on the lower surface of the leaves. Systemically infected seedlings are stunted with swollen stems. Under humid conditions, all parts of seedling are covered with downy fructification of the pathogen. Plant tissues are filled with oospores. Systemically infected seedlings frequently die before maturity. Infection of pods induces yellow spot and blotches on the pods. These lesions later turn brown and a white, felty proliferation of the endocarp occurs inside the pod. Oospores are formed in the endocarp. Mycelium of the pathogen invades the seed coat. The seeds ovules frequently abort or form seeds with sunken brown spots.[11]

Disease cycle — The pathogen overwinters as oospores in the soil, oospores in plant debris, and oospores on the seed. Oospores and mycelium of *P. viciae* have been found in the seed, but seed transmission of the disease has not been demonstrated. Seedlings are infected by oospore inoculum. The systemically diseased seedlings produce conidia which are disseminated by wind and rain to start the secondary cycle of disease.

Disease spread — The short-range spread of the disease is accomplished by dissemination of conidia by air currents and splashing rain drops. Long-range spread is by transport of soil, seed, or plant debris, and infected market produce.[11,27,39]

X. DOWNY MILDEW OF SUNFLOWER

Causal agent — *Plasmopara halstedii* (Farl.) Berl. and de Toni (*Peronospora halstedii* Farl.).

Host range — Thirty-five genera have been reported as hosts of *P. halstedii*. However, some researchers have suggested that *P. halstedii* is not a single species but a heterogeneous group containing several species, each virulent on only a portion of the reported host species. At present, there is insufficient experimental research available to resolve this controversy. Assuming that *P. halstedii* is a single species, the host of major economic importance is the cultivated sunflower (*Helianthus annus* L.) Other host generally reported include *Argeratum, Ambrosia, Aster, Bidens, Callistephus, Centaurea, Chrysanthemum, Cineraria, Clibadium, Coreopsis, Dimorphotheca, Elephanthopus, Emilia, Erigeron, Eupatorium, Franseria, Galatella, Gnaphalium, Heteropappus, Iva, Petasites, Ratiba, Rudbeckia, Saussurea, Scorzonera, Senecio, Silphium, Solidago, Tragopogon, Venidium, Verbisina, Vernonia, Xanthium, Ximinesia*, and *Zinnia*.[34]

Geographic range — Countries where *P. halstedii* has been reported on sunflower include Argentina, Bulgaria, Canada, Chile, Czechoslovakia, Ethiopia, France, Germany, Hungary, Iran, Italy, Mexico, Poland, Rumania, Dominican Republic, Spain, Turkey, Uganda, U.S., U.S.S.R., Yugoslavia, and Zimbabwe.[34]

Symptoms — Systemically infected sunflower plants may be stunted. The leaves display mosaic of green and chlorotic mottling, usually more intense on the younger leaves. Under moist conditions, downy fructification of the pathogen develops on the lower surface of the leaves. The heads of diseased plants may face upward. Severely diseased plants often die prematurely; those which survive produce smaller heads which are frequently barren. Secondary infection of leaves by zoospores of *P. halstedii* produces angular, pale green spots on the leaves. Under humid conditions sporulation occurs on the lower surface of these leaves.[34]

Disease cycle — *Plasmopara halstedii* survives over winter as mycelium in the rhizomes of plants of perennial *Helianthus* sp. and as oospores in soil, plant debris, and seed. The oospores germinate and produce zoospores which invade the roots of sunflower seedlings to initiate the primary disease cycle. Sporangia are produced on the leaves of the systemically diseased seedlings. The sporangia are disseminated by the wind to leaves of healthy plants to produce recurring cycles of infection.[34]

Disease spread — Wind-borne sporangia are the agents of disease spread over relatively short distances. Long-distance spread is dependent on the transport of oospores in infested soil or in seed.[30,34]

XI. DOWNY MILDEW OF GRAPE

Causal agent — *Plasmospora viticola* (Berk. and Curt.) Berl. and de Toni (*Peronospora viticola* (Berk. and Curt.) Casp.; *Botrytis viticola* Berk. and Curt.; *Botrytis cana* Lk.).

Host range — *Plasmopora viticola* attacks *Vitis vinifera* L., *Vitis sp. Parthenocissus quinquefolia* Planch. and *P. tricuspidata* Planch.[23]

Geographical range — The disease is endemic in grape-growing regions throughout the world.[23]

Symptoms — Greenish-yellow spots develop on the upper surfaces of leaves. A downy growth of sporangia and sporangiophores form on the undersurface of the leaves. The leaf spot becomes necrotic and infected leaves may fall from vines. The shoots and tendrils are stunted and brown spots occur on the berries. The downy sporulation develops on shoots, tendrils, and berries of diseased vines.[3,23]

Disease cycle — The pathogen overwinters as mycelium in buds, crowns, and shoots and as oospores in diseased canes and leaves. The oospores germinate and produce zoospores which invade stomata of leaves. Secondary spread is by means of sporangia produced on diseased plants.[3,23]

Disease spread — Sporangia disseminated by rain and wind account for the disease and spread over short distances. Long-distance disease spread occurs primarily though transport of infected canes.[3,23]

XII. DOWNY MILDEW OF CUCURBITS

Causal agent — *Pseudoperonospora cubensis* (Berk. et. Curt.) Rost. (*P. cubensis* var. *tweriensis* Rost.; *Peronoplasmopara cubensis* (Berk. et. Curt.) Clinton; *Peronospora cubensis* var. *atra* Zimm.; *Plasmospara (Peronoplasmopara) cubensis* (Berk. et. Curt.) Berl.; *Plasmopara cubensis* (Berk. et. Curt.) Humph.; *Peronospora cubensis* (Berk. et. Curt.).

Host range — *Pseudoperonospora cubensis* attacks a large number of species of the Cucurbitacae. The important economic hosts are *Citrullus lanatus* (Thumb.) Matsum. and Nokai L., *Cucumis melo* L., *C. sativus*, and *Cucurbita* spp. Other genera reported as host to *P. cubensis* are *Actinostemma, Adenopus, Benincasa, Bryonopsis, Mukia, Sechium, Sicyos, Trichosanthes,* and *Zehneria*.[8,31]

Geographic range — *Pseudoperonospora cubensis* occurs throughout the world. The pathogen has been reported on *Cucumis* sp. in 70 countries. Experimental trials indicated that different physiological races of the pathogen differing in virulence to host genotypes existed at different geographical locations.[9]

Symptoms — Yellow spots develop on the upper surface of infected leaves. A purplish mildew appears on the lower side of the leaf when humidity is high. On muskmelon the spots become reddish brown with age. On cucumber, infected leaves usually die. The plants may be killed or severely stunted. If the plants survive, the fruit may fail to ripen or if ripe, have an unpleasant taste.[8,9]

Disease cycle — *Pseudoperonospora cubensis* survives the fallow season as mycelium in wild cucurbits or in cucurbits grown in the greenhouse. These infected plants produce sporangia which serve as primary inoculum at the beginning of the growing season. The sporangia are disseminated by wind. The sporangia are deposited on leaves, germinate if water is present, and produce zoospores. The zoospores germinate and infect the host by invading the stomata. Sporangia produced on the infected leaves serve as inoculum for the secondary cycle of disease.[8,31]

Disease spread — The airborne sporangia of *P. cubensis* account for the spread of disease for short distances. Successive increments of sporangial inoculum progressing from new site of disease establishment would spread the disease for great distances in the absence of natural barriers. Although *P. cubensis* produces oospores, its ability to induce disease has not been demonstrated. If oospores are not involved in disease transmission, the wide geographical distribution of cucurbit downy mildew indicates that long-distance travel of viable airborne sporangia occurs.[9,31]

XIII. DOWNY MILDEW OF PEARL MILLET (GREEN EAR DOWNY MILDEW)

Causal agent — *Sclerospora graminicola* (Sacc.) Schroet. (*Sclerospora graminicola* (Sacc.) Schroet. var. *Setaria-italicae*; *Peronospora graminicola* (Sacc.) Sacc.; *Protomyces graminicola* Sacc.).

Host range — *S. graminicola* contains physiological forms virulent only to a single genus and in some cases only to certain species within a genus. The disease is of economic importance on *Pennisetum americanum* (L.) Leeke (pearl millet), and *Setaria italica* (Linn.)

P. Beauv. (foxtail millet). Other species attacked by *S. graminicola* are *Agrostis alba* Auctt. non Linn., *Echinochloa crusgalli* var. *frumentacea* (Roxb.) W. F. Wright, *Zea mexicana* (Schrad) O. Kuntze, *Panicum miliaceum* Linn., *Pennisetum leonis* Stapf and Hubb., *P. spicatum* (Linn.) Roem. and Schult., *Setaria lutescens* (Weig.) F. T. Hubbard, *S. magna* Griseb., *S. verticillata* (Linn.) P. Beauv., and *Zea mays* L. *Saccharum officinarum* Linn. was reported susceptible to artificial inoculation.[4,15,35]

Geographic range — The disease has been found in Canada, Chad, China, Czechoslovakia, Egypt, Fiji, France, Germany, Hungary, India, Iran, Israel, Italy, Japan, Korea, Malawi, Mozambique, Netherlands, Niger, Nigeria, Pakistan, Rumania, Senegal, South Africa, Spain, Sudan, Switzerland, Tanzania, Taiwan, Burkina Faso,Volta, Uganda, U.S., U.S.S.R., Yugoslavia, and Zimbabwe.[4,15]

Symptoms — Systemic infection of seedlings causes stunting and leaf chlorosis. The chlorosis extends from the base of the leaf upwards often appearing as a pattern of streak in the upper part of the leaf blade. Under cool, damp conditions, a downy whiter growth of spores and sporophores of the pathogen is produced on the infected leaves. As the disease progresses, oospores form in the leaf tissues and the leaves turn brown. The necrotic leaves may shred, leaf shredding is a characteristic of diseased *Setaria* and comparatively infrequent in *Pennisetum* and *Zea mays*. Systemically diseased seedlings frequently die; those that reach maturity often have abnormal inflorescence. The inflorescence may show various degrees of malformation and overgrowth. Phyllody of the infected inflorescence produces the leafy structure which gives the disease its name of green ear.[15,35]

Disease cycle — The disease survives the fallow season as oospores in the soil, in plant debris, and on infested seed. The oospores invade the underground parts of the plant and colonize the meristematic foliage. Sporangia are produced on the leaves of diseased plants and carried on air currents to the leaves of adjacent plants. Zoospores released from the sporangia invade the leaves of healthy plants to initiate the secondary cycle of the disease.[15,35]

Disease spread — The disease can be spread for short distances by sporangia carried by air currents or rain water. Transport of diseased plants, and oospores in soil, in plant debris, or on infested seeds provide the means of spread over long distances. Infection by mycelium in the seed has been reported but drying the seed presumably would inactivate the mycelium.[4,15,27,30]

XIV. JAVA DOWNY MILDEW OF MAIZE

Causal agent — *Peronosclerospora maydis* (Racib.) C. G. Shaw (*Sclerospora maydis* Racib.).

Host range — *Zea mays* L. is the only natural host known. However, *Z. mexicana* (Schrad.) O. Kunze and *Tripsacum* spp. were susceptible to artificial inoculation.[17,20]

Geographic range — The disease has been reported in Indonesia and Australia.[4,37]

Symptoms — Leaf invasion by conidia produces slight chlorosis. The fungus spreads downward through the invaded leaf and colonizes immature foliage in the leaf whorl and causes an acropetal chlorosis of the leaves. The discolored area of the leaf increases progressively in successively younger leaves until the emerging leaves are completely chlorotic. Maize seedlings infected by mycelium in the seed exhibit leaf chlorosis on the first leaf.[37]

Disease cycle — The pathogen does not produce oospores and survives between the regular maize cropping seasons as mycelium in undried maize seed or in maize plants growing out of season, usually in irrigated plots. Maize seedlings, if grown from undried infected seed, are invaded by mycelium of the pathogen present in the seed. These plants are systemically infected and conidia produced on their leaves are disseminated by air currents to leaves of healthy plants nearby. These conidia invade leaves, colonize the plant, and produce conidia which are disseminated to continue the disease cycle. In the absence of infected seed, maize seedlings can be infected by conidia from diseased maize plants grown under irrigation during the off season.[4,37]

Disease spread — The disease is spread for short distances by conidia. Spread for long distances would require transport of diseased plants or infected maize seed. Seed transmission of the disease is prevented by drying the seed.[4,37]

XV. PHILIPPINE DOWNY MILDEW OF MAIZE OR SUGARCANE

Causal agent — *Peronosclerospora philippenesis* (Weston) C. G. Shaw (*Sclerospora philippenesis* Weston).

Host range — *Zea mays* L. and *Saccharum officinarum* L. are the host species of economic importance. Other species attacked are *Avena sativa* L., *Z. mexicana* (Schrad.) O. Kunze, *S. spontaneum* L., *Sorghum bicolor* (L.) Moench, and *S. halepense* (L.) Pers.[4,18,20]

Geographical range — India, Indonesia, Nepal, Philippines, and Thailand.[4,18,20]

Symptoms — Initial invasion of leaves by conidia produce whitish yellow spots. Systemic infection of the plant produces yellow to white discoloration at base of early leaves; later leaves show progressively greater areas of discoloration until succeeding leaves become completely chlorotic. Some leaves may show chlorotic streaking. A white down composed of spores and sporophores of the pathogen develops on the leaves under humid conditions. Infection usually results in partial or complete floral sterility.[13,18]

Disease cycle — The pathogen survives between maize crops as oospores in soil and plant debris, and in diseased plants of other host species. Maize seedlings are infected by conidia from diseased plants nearby. The infected maize seedlings in turn produce conidia that continue the recurring cycles of inoculum production and infection characteristic of this disease.[13,18]

Disease spread — The disease spreads from plant to plant and to adjacent fields by dissemination of airborne conidia. Long-distance spread of the disease requires transport of diseased plants, or infected cuttings of sugarcane.[4,18]

XVI. DOWNY MILDEW OF SUGARCANE (SUGARCANE DOWNY MILDEW OF MAIZE)

Causal agent — *Peronosclerospora sacchari* (Miyake) Shirai and K. Hara (*Sclerospora sachari* Miyake).

Host range — The hosts of economic importance are maize (*Zea mays* L.) and sugarcane (*Saccharum officinarum* L.). In addition to other species of *Saccharum*, the pathogen attacks the genera *Echinochloa, Eleusine, Zea, Panicum, Polypogon, Setaria*, and *Sorghum*.[4,17,20]

Geographic range — The disease has been reported in Australia, Fiji, India, Japan, Nepal, New Guinea, Philippines, Taiwan, and Thailand.[20,28]

Symptoms — Diseased sugarcane plants originating from diseased buds are stunted and discolored. Later infections produce pale green to yellow stripes on the leaves, chlorotic mottling, leaf necrosis, and leaf shredding. Stems are often abnormally long, and weak, with more internodes than usual. In maize, infection causes pale yellow to white stripes extending upward from the base of the leaf. Diseased seedlings are stunted and often die prematurely. Those which survive produce small, poorly filled ears. The number of ears per plant may be abnormally large and ear shanks may be elongated. The tassel is often imperfect and may develop grains.[28,35,41]

Disease cycle — The pathogen survives between cropping seasons as oospores in plant debris, mycelium in perennial host species. Primary cycle starts most frequently with infection of maize or sugarcane seedling by conidia produced on adjacent perennial host plants. The infected seedlings produced conidia in turn which are disseminated by air currents to surrounding plants.[28,35]

Disease spread — The disease is spread for short distances by dissemination of conidia. The major means of spread over long distances are infected sugarcane sets. The pathogen can be transported as mycelium in infected seed but drying of the seed to a moisture content below 20% will prevent seed transmission of the disease.[4,17,28]

XVII. SORGHUM DOWNY MILDEW

Causal agent — *Peronosclerospora sorghi* (Weston and Uppal) C. G. Shaw (*Sclerospora sorghi* Weston and Uppal, *Sclerospora andropogonis-sorghi* (Kulk.) Mundkur, *Sclerospora sorghi-vulgaris* (Kulk.) Mundkur).

Host range — *Sorghum bicolor* (L.) Moench. and *Zea mays* L. are the economically important host species. Other species attached by *P. sorghi* are *Z. mexicana* (Schrad.) O. Kunze, *S. almum* Parodi, *S. arandinaceum* (Desv.) Stapf, and *S. halepense* (L.) Pers.[4,14,17]

Geographic range — The disease has been reported in Argentina, Brazil, China, Egypt, El Salvador, Ethiopia, Honduras, India, Iran, Israel, Italy, Kenya, Malawi, Mexico, Nepal, Nicaragua, Nigeria, Pakistan, Peru, Philippines, Somali, South Africa, Sudan, Tanzania, Thailand, Uganda, U.S., Venezuela, Zambia, Zaire, and Zimbabwe.[14,17]

Symptoms — Systemic infection of seedlings of corn and sorghum produces chlorotic areas at the base of leaf blades with successive leaves showing larger areas of chlorosis until entire leaf blades are chlorotic. A white downy growth of pathogen sporulation is produced on the infected leaves during periods of cool damp weather. In sorghum, leaves produced at later stages of growth are striped. Chlorotic interveinal areas alternate with green, veinal tissue. In sorghum, oospores are formed in the interveinal tissue. The leaves turn brown and shed. Conidial infection of the leaves produces localized lesions. Color varies with the host genotype ranging from yellow to purple in sorghum and form yellow to tan in maize. In sorghum the white downy sporulation of the fungus forms on these lesions when the weather is cool and damp. Systemically diseased seedlings often die prematurely. The inflorescence of plants which survive to maturity are partially or completely sterile.[14,16]

Disease cycle — The pathogen survives between cropping seasons as oospores in the soil, in plant debris and on infested seed, and as mycellium in wild hosts. Sorghum or corn seedling are infected by oospores of *P. sorghi* which invade the root and induce systemic infection by colonizing the meristematic foliage tissues. Conidia are produced on the leaves of diseased seedlings and disseminated to surrounding plants. The conidia germinate, and invade the leaves of the host to initiate the secondary cycle of the disease.[14,16]

Disease spread — The disease can be spread for short distances by airborne conidia. Long-range spread requires transport of diseased plants or movement of oospores of the pathogen. Oospores can be transported in infested soil, plant debris, or seed. Oospores are frequently found in the glumes of infected plants. Mycelium, but not oospores, have been found in infected seeds; however, drying the seed inactivates the mycelium.[4,14,16,17]

XVIII. CRAZY TOP OF MAIZE

Causal agent — *Sclerophthora macrospora* (Sacc.) Thirum., Shaw and Naras. (*Sclerospora macrospora* Sacc., *Sclerospora kriegeriana* Magnus, *Sclerospora oryzae* Brizi, *Nozemia macrospora* (Sacc.) Tasugi, *Phytophthora macrospora* L. (Sacc.) Tanaka).

Host range — The pathogen attacks several genera of the Gramineae. The more economically important host genera are *Elusine, Hordeum, Aryzae, Sorghum, Triticum*, and *Zea*. Other important host genera include *Alopecurus, Agrostis, Avena, Axonopus, Brachiania, Bromus, Digitaria, Echinochloa, Elymus, Elytrophorus, Eragrostis, Festuca, Glyceria, Holcus, Lolium, Panicum, Pennisetum, Phalaris, Phragmites, Saccharum, Sacciolepis, Schedonnardus, Secale*, and *Setaria*.[20,33,42]

Geographic range — The pathogen is found in many temperate or near-temperate climatic regions. Reported sites include Australia, Canada, China, Germany, India, Israel, Italy, Japan, Mexico, U.S., and U.S.S.R.[20,33,42]

Symptoms — The more striking symptoms are vivipary, proliferation of floral and vegetative parts into leafy structures, distortion, excessive tillering, and stunting. Leaves of infected plants may show chlorotic tripping or mottling. Leaves are often abnormally narrow, strap-like, and leathery in appearance. The white sporulation of the pathogen may occur on diseased plants particularly on the under surface of the leaves.[42]

Disease cycle — The pathogen overwinters as oospores in soil or in plant debris from diseased plants. The oospores germinate in the presence of water and produce zoospores which invade the foliage of host seedlings. Sporangia are produced on the leaves of the infected seedling. These sporangia release zoospores which initiate the secondary cycle of the disease. Oospores of *S. macrospora* are formed in the leaves and floral parts of systemically infected host plants.[42]

Disease spread — Disease spread over short distances is accomplished through dissemination of sporangia by run-off water from flooded disease sites and by transport of plant debris of soil infested by oospores. Long-distance spread results from transport of pathogen mycelium in undried, infected seeds, seed lots infested with oospores, movement of oospore-infested plant debris and soil by flood waters, and by transport of disease plants.[33,42]

XIX. BROWN STRIPE DOWNY MILDEW OF MAIZE

Causal agent — *Sclerophthora rayssiae* var. *zeae* Payak and Renfro.

Host range — *Zea mays* L., *Digitaria* sp.

Geographic range — India, Nepal, Pakistan, Sikkim, and Thailand.

Symptoms — The initial symptoms of disease on maize are parallel longitudinal stripes on the leaves of infected plants. Initially, these stripes are chlorotic or yellowish. Later, these stripes turn yellow or brown. Later, the stripes turn purple or brown.[32,38]

Disease cycle — The pathogen survives between cropping seasons in infected crab grass and as oospores in plant debris. Maize seedlings are infected by zoospores produced by germinated oospores or by zoospores form sporangia produced on crab grass. Sporangia are produced on the leaves of the diseased seedling and disseminated by air current, rain water, or leaf to leaf contact to healthy plants where the sporangia release zoospores which invade healthy plants to start the secondary cycle of disease.[32,38]

Disease spread — The disease is spread for short distances by airborne or waterborne sporangia and the movement of soil or plant debris containing oospores. Long-distance spread of the disease could occur by transport of diseased plants or plant debris containing oospores.[38]

XX. DOWNY MILDEWS IN THE FUTURE

The pathogens causing these diseases are very much a part of the agricultural environment of the world. The stunning and striking destructiveness of many of the diseases caused by these pathogens invoke alarm among plant protectionists universally. Nevertheless, we recognize great advances toward decreasing the threatening losses from these diseases because of advances in utilizing host resistance, management of host-resistance genes, development of highly effective systemic fungicides, and the integration of disease control strategies.[16] Equally important is the realization that carefully monitored regulatory programs can be utilized to reduce the introduction of these pathogens to new areas. For example, sorghum downy mildew has not been observed in Colombia even though the vast majority of the sorghums grown in the country are highly susceptible.

Biotechnology will be exploited as a tool for characterization of isolates and by inference of the taxonomic relations among species and pathotypes of the fungi causing these diseases. Other advances in technology and communication can be used to better exploit existing information for control of these diseases. In the past, without either an understanding of the pathogens and their biology, the diseases were not controlled, in fact some practices actually encouraged disease spread and subsequent damage. With the technology of today, the diseases can be modeled, optional and integrated disease control strategies can be implemented. Downy mildews, like most of the seed-borne smuts, are destined to become relics of the past.

REFERENCES

1. **Anon.**, Distribution Maps of Plant Diseases, No. 343, *Peronospora trifoliorum* DeBary, Commonw. Mycol. Inst., Kew, U.K., 1978.
2. **Armstrong, G. M. and Albert, W. B.**, Downy mildew of tobacco on pepper, tomato, and eggplant, *Phytopathology*, 23, 837, 1933.
3. **Barrett, J. T.**, Overwintering mycelium of *Plasmopara viticola* (B&C) Berl. & Det. in the California wild grape, *Vitis californica* Benth., (Abstr.), *Phytopathology*, 29 (Abstr.), 822, 1939.
4. **Bonde, M. R.**, Epidemiology of downy mildew of maize, sorghum and pearl millet, *Trop. Pest. Manage.*, 28, 49, 1982.
5. **Byford, W. J.**, Host specialization of *Peronospora farinosa* on *Beta, Spinacia,* and *Chemopodium, Trans. Br. Mycol. Soc.*, 50, 603, 1967.
6. **Channon, A. G.**, Downy mildew of brassicas, *The Downy Mildews,* Spencer, D. M., Ed., Academic Press, London, 1981, 321.
7. **Chilton, S. J. P., Henson, L., and Johnson, H. W.**, Fungi reported on species of *Medicago, Melilotus,* and *Trifolium,* U.S. Department of Agriculture, Misc. Publ. No. 499, 1943.
8. **Clinton, G. P.**, Downy mildew, on blight, *Peronoplasmopara cubensis* (B&C) Clint., of muskmelons and cucumbers, Conn. State Agric. Exp. Stn. Annu. Rep. for 1904, 28, 329, 1905.
9. **Cohen, Y.**, Downy mildew of cucurbits, in *The Downy Mildews,* Spencer, D. M., Ed., Academic Press, London, 1981, 341.
10. **Crute, I. R. and Dixon, G. R.**, Downy mildew disease caused by the genus *Bremia* Regel, in *The Downy Mildews,* Spencer, D. M., Ed., Academic Press, London, 1981, 421.
11. **Dixon, G. R.**, Downy mildews of peas and beans, in *The Downy Mildews,* Spencer, D. M., Ed., Academic Press, London, 1981, 487.
12. **Dunleavy, J. M.**, Downy mildew of soybean, in *The Downy Mildews,* Spencer, D. M., Ed., Academic Press, London, 1981, 515.
13. **Exconde, O. R.**, Philippine corn downy mildew, *Indian Phytopathol.*, 23, 275, 1970.
14. **Francis, S. M. and Williams, R. J.**, *Pennosclerospora sorgi, Commonw. Mycol. Inst. Descr. Pathog. Fungi Bact.*, 761, 1983.
15. **Francis, S. M. and Williams, R. J.**, *Sclerospora graminicola, Commonw. Mycol. Inst. Descr. Pathog. Fungi Bact.*, 770, 1983.
16. **Frederiksen, R. A.**, Sorghum downy mildew in the United States: overview and outlook, *Plant Dis.*, 64, 903, 1980.
17. **Frederiksen, R. A. and Renfro, B. L.**, Global status of maize downy mildew, *Annu. Rev. Phytopathol.*, 15, 249, 1977.
18. **Holliday, P.**, *Sclerospora philippinensis, Commonw. Mycol. Inst. Descr. Pathog. Fungi Bact.*, 454, 1975.
19. **Jones, F. R. and Torrie, J. H.**, Systemic infection of downy mildew in soybean and alfalfa, *Phytopathology,* 36, 1057, 1946.
20. **Kenneth, R. G.**, Downy mildew of graminaceous crops, in *The Downy Mildews,* Spencer, D. M., Ed., Academic Press, London, 1981, 367.
21. **Kenneth, R. G. and Kubicek, Q. B.**, *Peronosclerospora globosa* sp. *nov.* A downy mildew of prairie cup grass, *Eriochloa contracta,* in Texas, *Mycologia,* in press.
22. **Krober, H.**, Die Lebensdauer der Konidren einiger *Peronospora* arten unter Verschiedenen Umweltverhaltnissen, *Phytopathol. Z.*, 69, 64, 1970.
23. **Laflon, R. and Bulit, J.**, Downy mildew of the vine, in *The Downy Mildews,* Spencer, D. M., Ed., Academic Press, London, 1981, 602.

24. **Leach, L. D.,** Downy mildew of the beet caused by *Peronopsora schachtii* Fuckel., *Hilgardia,* 6, 203, 1931.
25. **Leach, L. D. and Borthwick, H. A.,** Distribution of downy mildew mycelium in spinach fruits, *Phytopathology,* 24, 1021, 1934.
26. **Ling, L. and Tai, M. C.,** On the specialization of *Bremia lactucae* on Compositae, *Trans. Br. Mycol. Soc.,* 28, 16, 1945.
27. **Melhus, I. E.,** The presence of mycelium and oospores of certain downy mildews in the seeds of their host, *Iowa State Coll. J. Sci.,* 5, 185, 1931.
28. **Mukerjii, K. G. and Holliday, P.,** *Sclerospora sacchari, Commonw. Mycol. Inst. Descr. Pathog. Fungi Bact.,* 453, 1975.
29. **Newhall, A. G.,** The spread of onion mildew by wind-borne conidia of *Peronospora destructor, Phytopathology,* 28, 257, 1938.
30. **Noble, M. and Richardson, M. J.,** An Annotated List of Seed-borne Diseases, 2nd ed., Commonw. Mycol. Inst., Kew, Surrey, England, 1968, 191 pp.
31. **Palti, J. and Cohen, Y.,** Downy mildew of cucurbits. *(Pseudoperonospora cubensis)*: the fungus and its host, distribution, epidemiology and control, *Phytoparasitica,* 8, 109, 1980.
32. **Payak, M. M. and Renfro, B. L.,** A new downy mildew of maize, *Phytopathology,* 57, 354, 1967.
33. **Payak, M. M., Renfro, B. L., and Lal, S.,** Downy mildew diseases incited by *Sclerophthora, Indian Phytopathol.,* 23, 185, 1970.
34. **Sackston, W. E.,** Downy mildew of sunflower, in *The Downy Mildews,* Spencer, D. M., Ed., Academic Press, London, 1981, 545.
35. **Safeeulla, K. M.,** *Biology and Control of the Downy Mildews of Pearl Millet, Sorghum, and Finger Millet,* Wesley Press, Mysore, India, 1976, 304 pp.
36. **Schiltz, P.,** Downy mildew of tobacco, in *The Downy Mildews,* Spencer, D. M., Ed., Academic Press, London, 1981, 578.
37. **Semangoen, H.,** Studies on downy mildew of maize in Indonesia with special reference to the perennation of the fungus, *Indian Phytopathol.,* 23, 307, 1970.
38. **Singh, J. P., Renfro, B. L., and Payak, M. M.,** Studies on the epidemiology and control of brown stripe downy mildew of maize (*Sclerophthora rayssiae* var. *zeae*), *Indian Phytopathol.,* 23, 194, 1970.
39. **Snyder, W. C.,** *Peronospora viciae* and internal proliferation in pea pods, *Phytopathology,,* 24, 1358, 1934.
40. **Stuteville, D. L.,** Downy mildew of forage legumes, in *The Downy Mildews,* Spencer, D. M., Ed., Academic Press, London, 1981, 355.
41. **Sun, M. H,** Sugar cane downy mildew of maize, *Indian Phytopathol.,* 23, 262, 1970.
42. **Ullstrup, A. J.,** Crazy top of maize, *Indian Phytopathol.,* 23, 250, 1970.
43. **Viranyi, F.,** Downy mildew of onion, in *The Downy Mildews,* Spencer, D. M., Ed., Academic Press, London, 1981, 461.
44. **Williams, R. J.,** Downy mildews of tropical cereals, in *Advances in Plant Pathology,* Vol. 2, Academic Press, London, 1984.
45. **Wolf, F. A. and Lehman, S. G.,** Diseases of soybeans which occur both in North Carolina and the Orient, *J. Agric. Res.,* 33, 391, 1926.
46. **Wolf, F. A., McLean, R. A., and Dixon, L. F.,** Further studies on the downy mildew of tobacco, *Phytopathology,* 26, 760, 1936.
47. **Yarwood, C. E.,** Onion downy mildew, *Hilgardia,* 14, 595, 1943.

Chapter 5

LEAF MINERS

Kenneth A. Spencer

TABLE OF CONTENTS

I. GENERAL CONSIDERATIONS

A. Definition

True leaf miners occur in four orders of insects — Diptera, Lepidoptera, Coleoptera, and Hymenoptera. The Agromyzidae are the best known group of dipterous leaf miners and it is this family which is of the greatest quarantine significance. However, in the Diptera, two *Scaptomyza* species (Drosophilidae) feed as leaf miners and empty mines may be confused with those of Agromyzidae. Some *Hydrellia* species (Ephydridae) are leaf miners on grasses and cultivated cereals and can be serious pests on rice. In the Anthomyiidae, 45 species in the genus *Pegomya* in North America are leaf miners, forming large blotch mines, the main families attacked being Caryophyllaceae, Chenopodiaceae, and Polygonaceae, but a few species feed on seven other families. The *Pegomya hyoscyami* (Rond.) group is well known in Europe and North America as "the beet or spinach leaf miners." This group is now believed to represent four species — one occurs in Europe and three in North America. *Pegomya terebrans* (Rond.) is also known in Europe as a pest of artichoke.

The most serious pests in this family are in the genus *Delia* feeding not as leaf miners but internally in roots and stems (Griffiths, personal communication). Many families of microlepidoptera feed predominantly or even exclusively as leaf miners, such as Elachistidae, Gracilariidae, Lyonetiidae, and Nepticulidae. In the Coleoptera the larvae of some genera in the families Buprestidae, Chrysomelidae, and Curculionidae are well known as leaf miners and some saw-flies in the Hymenoptera (Tenthredinidae) are also leaf miners. To the best of this author's knowledge no leaf mining species in these three orders have anywhere in recent years caused major concern as imported pests. The discussion below will be focusing primarily on the Agromyzidae, dealing in detail with 22 species in Section II.

Leaf miners, in the strict sense, can be defined as insects whose larvae feed internally within the leaf tissue, both the upper and lower epidermis being left intact. The term leaf miners is frequently used in a more generalized sense referring to all Agromyzidae, although of the 2500 species now known in the world, at most 75% are true leaf miners, the remainder feeding either as internal stem borers, external stem miners, or in seeds, the receptacle of the flower head, in pods, in the root or even in the wood of growing trees. Species of quarantine significance may thus be found in any part of the plant.

B. Literature on Agromyzidae

Extensive and up-to-date literature is now available on Agromyzidae in the U.S., Latin America, Europe, and elsewhere. An identification handbook of the 313 known British species was published by Spencer.[25] Species in Finland and Scandinavia were discussed by Spencer[28] and this work deals with a high proportion of the fauna known in central and southern Europe. Species in Africa were recently largely revised in a paper by Spencer specifically covering East African Agromyzidae.[33] Canadian species were discussed by Spencer and of the 290 species known in Canada many also occur in the U.S. Species in Florida[24] and in California[30] have been discussed in some detail and these works were followed by a manual dealing with the 540 species known in the U.S.[36] In Latin America (though not in Mexico) the Agromyzidae are now reasonably well known and have been discussed in four publications - Venezuela,[27] Chile,[31] Costa Rica,[32] and Colombia.[34] In addition, detailed accounts of research being carried out on *Liriomyza* pests, covering biology, effectiveness of different insecticides, development of resistance, and other aspects are contained in the proceedings of the three conferences on leaf miners held at Lake Buena Vista, Fla., 1981,[16] San Diego, Calif., 1982,[12] and Sarasota, Fla., 1984.[13]

Thus, with some experience, quarantine entomologists in Europe and North and South America may be able to identify the various stages of Agromyzidae which are found on imported plant material.

C. Biology

In the case of true leaf miners and some stem feeders females penetrate the leaf with their ovipositor and the egg is laid inside the leaf (or stem), in many species consistently at a particular point of the leaf, such as the tip, elsewhere on the margin, on the midrib or in one species, the Holly leaf miner, *Phytomyza ilicis* at the base of the midrib on the underside of the leaf. After hatching, the larva feeds away from the oviposition site, eating out a channel which widens as the larva grows to form a linear or serpentine mine, the form of which may be distinctive of the species concerned, or it may feed outwards in an irregular circle to form a blotch mine. In a number of species the larva forms a linear mine in the first instar and after the first molt the feeding instinct changes and it may then form a blotch; in such cases the early linear mine remains detectable from the arrangement of frass within the encircling blotch. Many species can be identified by the distinctive form of their leaf mine.[6]

The time taken by an agromyzid larva from hatching to pupation varies from species to species and is also considerably influenced by temperature. It may be as little as a week in the tropics but in most species it is from 2 to 3 weeks. There are exceptional species in which the larva may feed intermittently throughout the winter, pupating in spring, some 2 to 3 weeks prior to emergence of the fly. Frequently, inspection of plants with a leaf miner infestation will reveal that most, if not all, mines are already empty but caution is required here. On pupation the puparium normally falls to the ground, but in the case of *Liriomyza trifolii* and many other species in this genus the larvae fall to the ground prior to transforming and may wander around for up to 2 hr before pupating. Thus larvae or puparia may be present in boxes of packed flowers or vegetables, perhaps partially concealed among foliage or packing material. In a few species, such as *Chromatomyia syngenesiae* (see below), the puparium remains in the leaf at the end of the mine where it is readily visible, even with the naked eye.

D. Detection and Identification

Leaf miners are of particular quarantine significance in view of the large numbers of species involved and the wide range of cultivated plants which may serve as hosts. Fully developed leaf mines can be readily detected at ports of entry but young mines up to the first molt of the larva when about 3 days old are minute and insignificant and are frequently overlooked. A single egg will be even more difficult to detect and this presents plant quarantine inspectors with a virtually impossible detection problem. However, in the Agromyzidae the females of many species have the habit of making numerous "feeding punctures" in the leaf, penetrating the leaf with their ovipositor and then moving backwards to imbibe the exuding sap. Among such feeding punctures there may be a single egg or the eggs may be laid on another leaf. There is no consistent ratio between feeding punctures and eggs but there may be as many as 50 punctures in one leaf, with or without an egg. Leaves in which females have been feeding in this way are very conspicuous and the "stippling" effect caused by the punctures, characteristic of Agromyzidae will alert inspectors to the probability of a leaf miner infestation.

Plant quarantine officials should be familiar with the characteristic form of mine of any leaf miner species known to attack the various crops they are likely to encounter. Detection of puparia will be greatly facilitated if the inspector knows precisely what to look for, i.e., the color, shape, and size of the puparium. In the Agromyzidae these may be shining black, dark brown, or orange, yellowish, or white; they may be smooth and shining, rough and segmented, oval, or more elongate. In the case of stem borers and pod feeders in the genus *Melanagromyza*, the puparia remain concealed within the stem or pod, making detection impossible from superficial inspection, and stems or pods need to be opened up individually to establish whether larval feeding has been taking place.

Hering in Berlin made a lifelong study of leaf miners in all four orders. He described some hundreds of new species in the Agromyzidae and the leaf-mining microlepidoptera but his major contribution was the publication of a three-volume work on European leaf miners[6] with identification keys for all known species arranged by plant genera. This work made possible the identification of leaf miners, once the host plant had been established, from the characteristics of the leaf mines alone, without reference to the adults. Unfortunately, no such comparable work exists for North America or other regions of the world and the preparation of such keys will be a challenging task for future specialists on leaf miners.

E. Pest Status of Leaf Miners

A major problem with leaf miners — as indeed with all potential pests — is the relative rapidity with which their pest status can change from being insignificant to serious. This happened in 1964 to 1965 with *Agromyza megalopsis*, a barley leaf miner in Europe which had previously only been known in small isolated populations of little economic importance. Quite suddenly and for no obvious reasons the species developed enormous, largely uncontrollable populations extending from southern Germany through eastern Europe to Iraq. Cereal farmers were greatly concerned, major spray programs with helicopters were necessary, but after 3 years the infestation subsided and the species has not subsequently caused comparable damage, although well known as a pest on barley in the Paris area of France.

The most serious case of such a change of status is evident in *Liriomyza trifolii*, a highly polyphagous species which prior to the mid 1970s, although widespread in the eastern U.S., was not considered to represent a major threat to horticulture. The most important *Liriomyza* pest at this time was the closely related species, *L. sativae*. However, after 1975, following its introduction to the Netherlands, Kenya, and Colombia, *L. trifolii* has spread widely throughout much of the world. It is able to rapidly develop resistance to conventional organophosphorus and organochlorine insecticides and has caused losses running into millions of dollars. If the status of one species can change so dramatically and so rapidly, conceivably any other species might similarly change from a minor pest to one of great economic importance. Thus it is not only the known pests today among leaf miners which are cause for concern but any of the 150 or so species with hosts on cultivated plants,[26] even though the majority are currently of minor quarantine significance, might well develop into major pests comparable with *L. trifolii*. With this potential threat, it is clearly desirable to be able to identify any leaf miners detected on imported plant material and to have the maximum possible knowledge about their biology.

Conversely, the pest status of a leaf miner after introduction and successful establishment beyond its normal range may be greatly reduced by comparison with its customary environment, even though suitable or even identical hosts are available and climatic conditions are entirely comparable. *Chromatomyia syngenesiae* (known in earlier literature as *Phytomyza atricornis*) is a striking example of such a beneficial status change. In Europe, *C. syngenesiae* is a widespread, polyphagous species and a major pest on glasshouse chrysanthemums. It became established in the northeastern U.S. in the 1880s in glasshouses but appears not to have colonized wild hosts to any significant extent or spread south from the area of its original establishment. It is also now widespread in California but occurs mainly on the introduced weed, *Sonchus oleraceus*, and has never been reported as a serious pest attacking chrysanthemums in glasshouses. Similarly, in Colombia where it was recently discovered on *Sonchus* and on the ornamental bush, *Chrysanthemum coronarium*, introduced from southern Europe and widely grown in gardens in Bogotá, it has not yet succeeded in colonizing one of its favorite European hosts, the cultivated chrysanthemums grown on many large farms in the Bogotá area.

F. Prevention of the Export of Leaf Miners

There is no easy answer to the prevention of export of leaf mining pests and procedures for their detection at entry ports are necessarily largely limited to spot checks. Primary responsibility must be with growers themselves. The necessity for efficient control of leaf miners throughout the growing cycle is now appreciated by all growers, but the most important step is meticulous care in inspection when plants are being packed. In the author's experience this is not always given the necessary attention. Obviously, packing has to be rapid in order to meet deadlines to catch planes waiting on the tarmac, but it must be false economy to skimp inspection at this stage and allow infested plants to be shipped, only to risk fumigation and possible destruction following more careful inspection by quarantine officials at ports of entry.

Phytosanitary certificates are now a normal requirement for the export of plants which may harbor pests not present at their destination. However, experience has shown that these are far from foolproof and their value can only reflect the efficiency and availability of government inspectors certifying the shipment as pest free. The U.S. certainly has a system for issuing phytosanitary certificates which is as efficient as any country in the world and more efficient than most, but this did not prevent the export of infested chrysanthemum cuttings to Sardinia, the Netherlands, Kenya, and Colombia in the 1970s. At best, phytosanitary certificates can only be considered as a first line of defense against the import of leaf miners.

With the huge volume of international trade in flowers and vegetables today, only a small proportion of imported material can be properly inspected. This applies even at the Miami International Airport, Miami, Fla., which is the main port of entry for flowers in the U.S. where about 80 inspectors are available. If exporters are not conscientious in their own inspection procedures, the risk of leaf miners escaping detection at ports of entry will thus always be high.

If leaf mines are discovered, in many cases accurate identification, at least in advanced countries such as Canada, Great Britain, and the U.S. where experienced quarantine entomologists are available, is now possible and a rational decision regarding destruction, fumigation or return to the exporter of the infested shipment can be taken. For chrysanthemum cuttings effective procedures combining cold storage and fumigation with methyl bromide have now been developed, with little deleterious effect on the cuttings.[10] However, such fumigation applied to cut flowers can have disastrous consequences, causing wilting of flowers and foliage and shorter shelf life. In effect such a fumigated shipment may become largely unsaleable. The knowledge of this possible outcome should provide an overwhelming incentive for growers to ensure that no quarantine pests are present in their shipments. In 1983 at Bogotá this author saw a large shipment of carnation cuttings from Italy infested with *Liriomyza dianthicola*, a species not known anywhere in North or South America, and the Colombian quarantine authorities justifiably had the entire shipment destroyed.

G. Eradication after Establishment of a Leaf Miner Pest

If a leaf miner successfully establishes itself in an area where it was previously unknown, some efforts have been made at total eradication. However, only in England has this been achieved with success.

In Kenya, *L. trifolii* became established at the main production nursery at Masongaleni some 70 miles east of Nairobi in 1976. This large nursery employing 3000 people supplied chrysanthemum cuttings to most countries in western Europe, including 40 million annually to Britain. By 1979 this nursery was heavily infested with *L. trifolii* and on one occasion when the author was there early in December, 1979 the active adult population was estimated at 250,000. Early in 1979 the British Ministry of Agriculture imposed a 3-month suspension of the import license from Kenya and intensive efforts were made by the Kenyan nursery to control and eradicate the *L. trifolii* infestations. Four organochlorine and fifteen organophosphorus insecticides, five carbamates, and two pyrethroids were used in increasing quantities and at shorter and shorter intervals in a desperate effort to effect control, in view of the threat of the permanent loss of the license to the most important export market, Britain. The license had been reinstated at the beginning of August, 1979 for a temporary period until the end of December. However, overuse of insecticides proved to be largely counter productive and merely speeded up the resistance which *L. trifolii* has developed more effectively than in any other *Liriomyza* species. The British license was finally revoked at the end of December, 1979 and, as a consequence, production of chrysanthemums at the Kenyan nursery ceased.

In Kenya, *L. trifolii* rapidly spread to a number of weeds and vegetables in gardens near the main propagating nursery at Masongaleni. In April, 1983 it was found by the author on chrysanthemums at several small flower farms west of Nairobi and on the widespread weed, *Bidens pilosa*, and also in the gardens of the Blue Posts Hotel at Thika, also 30 miles north of Nairobi on *Bidens* and cultivated gerbera daisy. A small population was also present further north in uncultivated country on *Bidens* and on the introduced "sunflower" *Tithonia diversifolia*. Another farm cultivating chrysanthemums for the export of cuttings from Kenya has recently been established at Embu and the utmost vigilance will be required to prevent infestation of this farm with *L. trifolii*.

In the Mediterranean area, including France, Italy, Spain, and Israel, with a suitable climate permitting overwintering in the open and numerous suitable wild hosts, *Liriomyza trifolii* is now fully established and there is no possibility of eradication. *L. trifolii* reached the relatively small island Réunion in the Indian Ocean in 1978, either from Kenya or France and in such a restricted area eradication might conceivably have been possible but before the significance of the pest was appreciated it had spread to virtually all vegetable and flower crops and had colonized local weeds. By 1979 it was present on 79 cultivated and wild hosts.[41] Energetic control measures were taken using conventional insecticides but *L. trifolii* had by this time become strongly resistant to all available chemicals and eradication proved impossible.

Liriomyza trifolii first reached England in 1977 with chrysanthemum cuttings imported from Kenya and Malta and was discovered at 15 different nurseries. By 1979 infestations were more widespread and it was intercepted on 207 occasions, the sources including the Canary Islands, the Netherlands and the U.S.[18] Selective destruction of infested crops discovered at ports of entry was applied and insecticides were used against adults (fogs and smokes), against larvae (foliar sprays and persistent systemics incorporated in planting beds) and soil drenches and sterilants were used to kill pupae. By this energetic action *L. trifolii* was eradicated from all commercial nurseries where it was discovered.[1] In May, 1979 a more serious situation arose when *L. trifolii* spread from glasshouse chrysanthemums to glasshouse tomatoes and to cultivated and wild hosts in the open. The Ministry of Agriculture developed a total eradication campaign in the main infested glasshouses and monitored an area up to 4.5 km from the originally infested site where it had become established on 16 different genera of cultivated plants and 11 wild hosts. Known preferred hosts were thoroughly surveyed and inspected twice, in July and September, over a wide area and all potential hosts in a more limited area were inspected as often as practicable from July through to November. Inspection of 175 km of verges or hedges was carried out twice. No positive identification of *L. trifolii* was made from hosts more than 500 m from the original infestation. Treatments used outdoors included soil drenches and foliar sprays against actual infestations, and weeds were controlled by mowing and the precisely targeted use of selective and total herbicides. With cooler weather in the late autumn, populations declined and from November, 1979 to November, 1980, yellow sticky traps and regular monitoring revealed no *L. trifolii*.[14]

The establishment of *L. trifolii* in South Africa occurred in 1981, following importation of infested chrysanthemum cuttings from the Netherlands. It is now widespread in the Transvaal where it occurs on a number of vegetables, including tomatoes and is present on ornamentals in the Johannesburg area. It has been identified on 15 species of indigenous weeds and eradication is now impossible.

Ophiomyia phaseoli is a widespread and serious agromyzid pest on beans and other cultivated leguminous crops throughout the Old World tropics from the Cape Verde Islands off the west coast of Africa to northern Australia and most islands in the Pacific. *O. phaseoli* was first noted as an introduction to Hawaii in 1968 and damaging populations quickly built up on cultivated beans. The only native parasite recorded was *Haltictoptera patellana* (Pteromalidae). For possible biological control, field-collected puparia of *Ophiomyia phaseoli* were shipped to Hawaii from Uganda in April through May, 1969 where the braconid parasites, *Opius melanagromyzae* and *O. importatus* had previously been identified. Releases began on the island of Oahu in May, 1969 and the parasites had spread widely by October, when high levels of parasitism were achieved and the host population density began to fall. In July and August, 1969, releases began on other islands. On Maui, control was apparent within 6 months at lower altitudes below 300 to 500 m but at higher altitudes and on Kauai Island control was not effective for about 12 months. *O. phaseoli* thus began to be brought under control within 1 year of its detection and biological control has allowed the continuation of bean growing without the need for insecticides.

H. Losses Incurred and Costs of Control

The importance of effective quarantine control of imported leaf miners is apparent from the serious losses which they can cause and the high overall costs of control measures. Actual damage caused varies considerably from season to season and in varying degrees on different crops. In Florida, celery, lettuce, tomatoes, and chrysanthemums suffered severe losses in the 1970s and early 1980s. Losses of tomato production in Florida in the 1976 to 1977 growing seasons to leaf miners amounted to "hundreds of thousands of dollars"[5] and in 1982 many growers of celery were compelled to cease production owing to the losses suffered and the "astronomical populations" of *Liriomyza trifolii* which had built up.[38] The situation has now considerably improved with the use of the new "insect growth regulators", particularly cyromazine (Trigard®), developed initially for use against dipterous larvae breeding in the manure of livestock and poultry, and avid (Avermectin®), a natural product pesticide developed from the soil microorganism *Streptomyces avermitilis*.

The costs of leaf miner control or attempts at total eradication are inevitably considerable. A detailed analysis of fixed and variable costs incurred on two chrysanthemum farms in Florida in applying sprays and granules against *Liriomyza trifolii* was estimated by Prevatt et al.[15] at slightly over $2000 per acre on each of two farms when the pest was present, but when the pest was not present costs were reduced to $684 on farm A and $965 on farm B. In England the cost of the eradication campaign against *L. trifolii* in 1979, based on value of crops destroyed, lost production, pesticides, traps, and labor was established as £36,000.[14] These figures clearly indicate the damaging potential of leaf miners, although at the present time the number of species capable of causing critical economic losses is relatively few.

I. Natural Dispersal of Agromyzidae and Introduction by Human Agency

Although weak fliers, it is well documented that Agromyzidae have on numerous occasions successfully colonized areas far from their natural habitats. Following study of the fauna of New Zealand,[29] it was postulated that the endemic fauna of 34 species is derived from 17 ancestral immigrants which have reached New Zealand from eastern Australia by long-distance aerial dispersal during the past 5 million years. At the present time Australian insects which have crossed the 800 km water gap of the Tasman Sea with the help of the strong westerly winds are not infrequently found in New Zealand. Meteorological conditions suitable for airborne transport of insects from Australia to New Zealand occur at least 20 times a year and successful colonizations over a water gap of the size of the Tasman do in fact occasionally take place, although the chances are slight.

In southern Florida the presence of numerous insects of Caribbean or Central or South American origin is well known. The frequency of hurricanes sweeping across the Caribbean to Florida, carrying with them countless seeds and insects, obviously has greatly facilitated such wind-borne dispersal. Of the 86 species of Agromyzidae known in Florida,[35] it was established that 31 species, representing 36% of the total, are of southern origin, with the immediate source areas being Cuba and the Bahamas.

The long common frontiers between the U.S. and Canada in the north and Mexico in the south, although admittedly with some geographical barriers such as prairie or desert, nevertheless provide easy access to wind-borne insects.

Plant quarantine officials should be aware of this potential for adult leaf miners to reach the U.S. by natural long-distance dispersal, but this in no way detracts from the need for vigilance in their detection at ports of entry. The two main sources of such human transport are (1) commercial cargoes by ship and increasingly by air and (2) in passengers' baggage. In all, there are some 40,000 to 50,000 interceptions of pests annually at U.S. ports of entry and 4 to 5 new pests are reported annually.[7]

In view of the large populations of many species of leaf miners and the many species present on cultivated and imported crops, it is surprising that so few are intercepted or have become established in the U.S. or South America. Apart from the unique case represented by *Liriomyza trifolii*, there seems to be one-way traffic in leaf miner pests between Europe and America. No other American leaf miners have become established in Europe and only five European pests are now established in the U.S. The most recent introduction is *Ophiomyia phaseoli* in Hawaii (see above) and *Ophiomyia simplex* is host specific on cultivated asparagus, a plant of European origin, and is now present virtually wherever asparagus is cultivated, from the east coast to California. Introduction of this pest was certainly facilitated by the puparia remaining in the stems of the plant. *Agromyza frontella* is a minor pest on alfalfa (*Medicago sativa*) which is widespread in Europe and since the early 1970s has been causing some concern in view of its occurrence on alfalfa in Massachusetts and New York. *Phytomyza ilicis* is host specific on European holly (*Ilex aquifolium*) and was severely damaging commercial holly plantations in British Columbia by 1931. It is not known when this holly was first imported into the U.S., but *Phytomyza ilicis* is now well established wherever the plant is grown. Whether it reached the U.S. from Canada or England is not known. *Chromatomyia syngenesiae* (as *Phytomyza atricornis*) reached the northeastern U.S. in the 1880s. It became established on Long Island, N.Y., and in the Boston area, and is now present from Connecticut to Wisconsin, primarily on glasshouse-grown chrysanthemums and cineraria. At some unknown date it also became established in California. Fortunately these populations have not developed into the serious pest on chrysanthemums which the species represents in Europe and no serious attempt has ever been made at its total eradication. There is also a record of its interception on chrysanthemum in 1981 in British Columbia in the baggage of an air passenger arriving from California.[40] *Phytomyza rufipes*, a European pest of cruciferous crops is now established at several localities near Bogotá, Colombia.[34] Spain has recently been exporting several hundred tons annually of "German cabbage" to Colombia and *P. rufipes* almost certainly reached Colombia with these shipments. *Agromyza apfelbecki* is a very large Mediterranean species, with wing length up to 4.2 mm, which is host specific on thistles of the genus *Cynara* and is a minor pest on cultivated globe artichoke (*C. scolymus*). This plant was introduced to Chile, presumably in the 19th century or earlier, and *A. apfelbecki* is now widespread in Chile wherever globe artichoke is cultivated.

II. LEAF MINERS OF QUARANTINE SIGNIFICANCE

A. Liriomyza Group of Leaf Miners

The most serious pest in this group is without question *L. trifolii*. In the U.S. and the neotropical region, at the present time both *L. huidobrensis* and *L. sativae* appear to be of reduced significance compared to 20 years ago. In Europe a species related to *L. huidobrensis, L. bryoniae,* occurs naturally in the Mediterranean area and is widespread in glasshouses further north and is a considerable pest on a wide range of vegetables but does not attack chrysanthemums. *L. dianthicola*, a host-specific species on carnations, is widespread in Italy and has been spread by the trade in carnation cuttings to many countries in Europe and to Kenya. It has reached Colombia, but is not established there. The three onion miners *L. cepae, L. chinensis*, and *L. nietzkei* are present in continental Europe but not in Britain or in North or South America and represent pests of some significance.

1. Liriomyza trifolii (Burgess) — Chrysanthemum Leaf Miner (U.S.); American Serpentine Leaf Miner (Europe)

1. Synonyms — *Liriomyza alliovora* Frick (Iowa) and *Liriomyza phaseolunata* (Frost) (New Jersey).
2. Description — Small species, wing length 1.25 to 1.9 mm; head mostly yellow, hind margin of eye largely yellow, at most slightly darkened centrally; antennae yellow; mesonotum distinctively mat gray (contrast *L. huidobrensis* and *L. sativae*); legs largely yellow.
3. Host range — A highly polyphagous species, recorded on 47 genera in 10 families in the U.S.,[37] on 79 species in 17 families on the Indian Ocean island Réunion,[41] and on additional local hosts in Israel, Kenya, and South Africa. The total of known hosts is now in excess of 400 and a complete list is in preparation by P. Seymour, Harpenden, England. The most important crops attacked are beans, celery, chrysanthemum, cucumber, gerbera, gypsophila, lettuce, onion, potato, and tomato.
4. Biology — The egg is deposited randomly at any point on the upper leaf surface in the pallisa parenchyma and the larva normally hatches after about 3 days, forming an upper surface linear or serpentine mine. The mine is frequently rather broad and relatively short, with frass in irregular black strips at either side of the channel. After leaving the leaf it crawls around for several hours before transforming into a yellowish-brown puparium. The posterior spiracles of the larva (and puparium) each have three pores (contrast *L. huidobrensis* with six to eight). Up to 50 larvae have been found in a single leaf of gerbera daisy in France. The length of the life cycle is largely temperature dependent and can also vary considerably with the host plant. On Réunion, breeding on the weed *Spilanthes* from the time of egg to adult can be as little as 12 days, while on bean, 18 to 20 days in the cold season. In the Netherlands on chrysanthemum a complete generation takes 3 weeks in summer, 6 to 7 weeks in winter. Comparable variation has been noted in Florida on beans and chrysanthemum.
5. Distribution — Originally described from Washington, D.C., now certainly present in the U.S. in all eastern and southern states, also California and Hawaii; Ontario; Bahamas, Barbados, Colombia, Costa Rica, Dominican Republic, Guatemala, Guyana, Peru, Venezuela; by introduction to Kenya, Tanzania, South Africa, Nigeria; France, Italy, Malta, Netherlands, Spain (including the Canary Islands), and most countries in western Europe in glasshouses; Israel; Mauritius, Réunion; and Guam.

The explosive spread of this pest since 1975 from Florida to California, much of South America, Kenya, and then to several countries in Africa and to most in Europe, and also Israel, has primarily resulted from the export of infested chrysanthemum cuttings. Prior to 1977 *L. trifolii* was not known in California where it is now a major pest on celery and chrysanthemums, reaching this state with cuttings supplied from Florida. It apparently reached Hawaii in 1978.

Its natural distribution is not clear. Its original discovery in Washington, D.C. may represent the northernmost limit of its normal range and there is no reason to believe that it has not been present in other states south to the Gulf of Mexico at least for some thousands of years. Prior to the development of extensive horticulture, isolated populations were presumably effectively controlled by the numerous parasites which are known to attack the larvae. Only when this parasite complex was seriously disrupted by the increasing use of modern insecticides did *L. trifolii* proliferate and produce the large populations which are so difficult to control today.

The presence of *L. trifolii* on islands in the Caribbean, in Central America and in Guyana and Venezuela, is certainly in part due to recent introductions by commerce. However, as early as 1960 serious attacks on onions occurred in the State of Lara, Venezuela and there is no evidence to suggest that this was the result of introduction. Onion is a favored host and this author found a field of onions to be heavily attacked west of San José, Costa Rica in January, 1982.

Thus, although there may well have been extensive natural dispersal resulting in sparse populations in Guyana and Venezuela and throughout the Caribbean and Central America, the present large populations in Colombia, Costa Rica, and the Dominican Republic are certainly the result of the import of infested chrysanthemum cuttings from the U.S.

During the past 8 years *Liriomyza trifolii* has been detected on chrysanthemum cuttings or flowers imported into England from a number of countries, primarily Israel, Italy, the Netherlands, and Spain (including the Canary Islands), and on several occasions has established breeding colonies on local weeds. Energetic action undertaken by the Ministry of Agriculture has resulted in the rapid eradication of such populations and on present evidence it appears that *L. trifolii* cannot successfully overwinter in the wild owing to excessively damp rather than cold climatic conditions. However, in Israel, France (at least in the southern Mediterranean area), and in Spain, including the two main Canary Islands, Gran Canaria and Tenerife, the species is well established outside farms and eradication is not possible. The problem is particularly serious in these southern areas where host crops such as celery and tomatoes are grown in the open and there is thus no physical barrier against escape to wild plants once *L. trifolii* has become established on a crop, or conversely to the crop being reinfested from such wild populations.

During the past 6 years the detailed records kept by the British Ministry of Agriculture[17-23] show that there has been a steady decline in the annual number of interceptions of *L. trifolii* since the peak of 207 in 1979, the average now being about 70. The pest has been found at airports, docks, nurseries, markets, wholesalers, and in shops. Most identifications are based on leaf mines or dead larvae, but live larvae have been found on a number of occasions. The two most frequent hosts are chrysanthemum cuttings and celery but other hosts recorded include gerbera daisy, gypsophila, pepper, and tomato. On two occasions the species has been found on celery imported from the U.S.

With the justified determination of the British Ministry of Agriculture to prevent the introduction of *L. trifolii*, the frequency of its detection and the relatively large number of countries from which flowers and vegetables are imported where the pest is known to be well established, this pest now without question represents the most serious threat among leaf miners to horticulture in Britain.

2. *Liriomyza huidobrensis* (Blanchard) — Pea Leaf Miner (U.S.)

1. Synonyms — *Liriomyza cucumifoliae* Blanchard (Argentina), *Liriomyza langei* Frick (California), and *Liriomyza dianthi* Frick (California).
2. Description — Wing length 1.7 to 2.25 mm (significantly larger than *L. trifolii*); head with upper orbits and hind margin of eye partially darkened, third antennal segment normally at least somewhat brownish; mesonotum black but not fully shining; legs

variably yellowish-black; posterior spiracles of larva and puparium each with an ellipse of six to eight pores (contrast *L. sativae* and *L. trifolii* with only three).

3. Host range — A highly polyphagous species, with confirmed records on 11 families: Caryophyllaceae, Chenopodiaceae, Compositae, Cucurbitaceae, Leguminosae, Liliaceae, Linaceae, Solanaceae, Tropaeolaceae, Umbelliferae, and Violaceae. The most important crops attacked are beet, spinach, peas and beans, and potatoes and among flowers most commonly gypsophila, more rarely carnations and chrysanthemum.
4. Biology — The egg is normally laid towards the base of the leaf, on or near the midrib and the larva forms a distinctive leaf mine, largely following the midrib and branching along lateral veins. When many larvae are present in a single leaf, the exact form of the mine may be obscured and the larva will feed wherever leaf tissue is available. A further distinctive character of the mine is that the larva does not feed exclusively on the upper leaf surface but frequently changes to the lower surface.
5. Distribution — Originally described from Argentina and present in most of South America at higher elevations but not in lower, tropical areas; small populations present in Central America and the Dominican Republic; in the U.S. largely confined to California.

This species has caused concern to APHIS in view of its presence on mums in Colombia and its occasional detection on flowers exported to Florida. *L. huidobrensis* does not occur naturally in Florida or anywhere in the east. The supply of vegetables and flowers from California to Florida, doubtless occasionally with infestations of *L. huidobrensis*, has been far greater over the past 30 years than of chrysanthemums from Colombia but at no time has the pest succeeded in long-term establishment in Florida. On the the two occasions when it briefly became established in Florida and in southern Virginia recently following introduction from California on gypsophila, the wild population had died out in the year following its discovery. Thus, for climatic and other reasons which are not fully understood, in the opinion of the author *L. huidobrensis* poses no threat to horticulture in Florida or elsewhere in the east. Despite this, as discussed earlier, APHIS has adopted a cautious approach and has considered it necessary to treat *L. huidobrensis* as a quarantine pest. Fumigation is required for any shipment of flowers from Colombia or elsewhere outside the U.S. on which it is detected. However, similar quarantine measures are not applied to interstate shipments from California on which the species is found, in part for economic reasons, in view of the additional personnel this would entail.

It appears that *L. huidobrensis* is now of reduced economic importance in California compared to 30 years ago. It is rarely recorded today by the California Department of Agriculture on peas or carnations on which it was originally considered to be a major pest. The host on which it does still occur regularly is gypsophila, but if it is present in sufficient numbers to require chemical treatment effective control can be obtained by conventional insecticides such as dimethoate.[11]

L. huidobrensis has twice been intercepted in England, in 1980 as a live larva in peas from the U.S. and in 1981 as a dead pupa in chrysanthemums from Peru.[18,19] It could possibly become established in England, with many suitable wild hosts available but it is doubtful whether it would survive the winter except under protection. It is nevertheless treated as a pest of quarantine significance.

3. Liriomyza sativae (Blanchard) — Vegetable Leaf Miner

1. Synonyms — *Liriomyza pullata* Frick (Hawaii), *Liriomyza canomarginis* Frick (Hawaii), *Liriomyza multiseta* Frick (Hawaii), *Liriomyza munda* Frick (California), *Liriomyza guytona* Freeman (Alabama), *Liriomyza propepusilla* Frost (Kansas), and *Liriomyza pictella* (Thomson) — misidentification (California).

2. Description — Similar in size to *L. trifolii*, wing length about 2 mm, but differing in the black upper orbits and largely black hind margin of the eye; conspicuously differentiated from both *L. trifolii* and *L. huidobrensis* by the brilliantly shining black mesonotum; posterior spiracles of larva and puparium each with three pores.
3. Host range — A highly polyphagous species, attacking crops mainly in the three families, Cucurbitaceae, Leguminosae, and Solanaceae, including cantaloupe, cucumber, squash; lupin, alfalfa, beans; pepper, tomato, eggplant and potato. A further 20 genera in 10 families have been recorded as hosts.[26]
4. Biology — Essentially as in *L. trifolii*, larva forming a narrow upper surface linear mine, pupating externally. Populations can be very large and 80 larvae have been recorded in a single leaf of *Ricinus* (Euphorbiaceae).
5. Distribution — Widespread in southern states in the U.S. from Florida to California; Caribbean, Central America, and throughout much of South America to Argentina (original description) and Chile.

The pest status of this species appears to have changed during the past 30 years. In the 1950s it was the dominant leaf miner pest in Florida and California and was the only species reported attacking tomatoes. In the last few years it has been largely supplanted by *L. trifolii*, even on tomatoes. In Colombia and Venezuela the author has found it to be restricted to warmer, low elevation areas and at higher altitudes it is replaced by *L. huidobrensis*.

In view of its wide distribution in the U.S., it is of little quarantine significance to the U.S., although it almost certainly is occasionally present on shipments of vegetables from Central and South America. Leaf mines or dead larvae or puparia cannot be distinguished from those of *L. trifolii*.

In England it has only once been intercepted, on okra in 1978 from Mexico.[17] It certainly could not become established in the wild but populations could thrive in heated glasshouses. It is subject to the same quarantine regulations as *L. trifolii*.

4. Liriomyza brassicae (Riley)

1. Synonyms — *Liriomyza hawaiiensis* Frick (Hawaii).
2. Description — Indistinguishable on external characters from *L. sativae* and a positive identification may only be possible from the male genitalia;[26,36] larva and puparium with three pores on the posterior spiracles, also indistinguishable from *L. sativae*.
3. Host range — Primarily Cruciferae, most genera serving as hosts; also commonly on *Tropaeolum* (garden nasturtium), *Capparis, Cleome*, and *Gynandropsis* (Capparaceae) and occasionally on peas.
4. Biology — Similar to *L. sativae* and *L. trifolii*, larva forming narrow upper surface mine, pupating externally.
5. Distribution — Semicosmopolitan, widespread in the U.S., Old World tropics and Australia; uncommon in South America and in Europe largely restricted to botanical gardens, not in England.

L. brassicae represents only a minor pest on cultivated brassicas but its complete absence from England would justify stringent control measures should it be intercepted or become established. In recent years it has been intercepted only once, from Florida on "Chinese cabbage" in 1982.[22]

5. Liriomyza bryoniae (Kaltenbach)

1. Synonyms — *Liriomyza solani* Hering and *Liriomyza citrulli* Rohdendorf.

2. Description — Wing length 1.75 to 2.1 mm; head largely yellow but hind margin of eye normally narrowly black centrally; mesonotum black, with slight mat undertone; posterior spiracles of larva and puparium each with an ellipse of 7 to 12 pores (generally resembling *L. huidobrensis*).
3. Host range — A highly polyphagous species, recorded on 33 families but of economic importance on tomatoes; Cucurbitaceae; particularly melons, watermelon, and cucumber; and glasshouse-grown lettuce and beans.
4. Biology — The female has been recorded as laying up to 260 eggs on lettuce in France but the average in England on tomatoes was 70. Adult life is on average 8 days in the male and 14 in the female, exceptionally up to 4 weeks. The larva forms a narrow linear mine, avoiding main veins (in contrast to *L. strigata* with which the species can occur together which primarily follows the midrib). The entire life cycle at 21°C lasts 25 to 30 days and there can be numerous generations in a growing season.
5. Distribution — Essentially a Mediterranean species, occurring from Spain (including the Canary Islands) to Egypt and Israel; also present in the U.S.S.R., Ukraine, and Caucasus. Elsewhere in Europe frequently found in glasshouses and in botanical gardens.

This species causes considerable damage to tomatoes grown in the open in France and Spain and grown in glasshouses in England, and on cucurbits in France and Israel. It is regularly intercepted in England on produce from Mediterranean countries. *L. bryoniae* is not known in the U.S. and establishment there would represent a threat to horticulture. Leaf mines are not distinguishable from those of *L. trifolii* but the larvae can be identified by the more numerous posterior spiracular bulbs. In this character larvae of *L. bryoniae* could be confused with those of *L. huidobrensis* and even the male genitalia are similar. Positive identification in the U.S. on material imported from Europe is important but not easy and it is essential that adequate reference material should be available at the Systematic Entomology Laboratory, Washington, D.C.

6. Liriomyza strigata (Meigen)

1. Description — Generally similar to *L. bryoniae* but somewhat darker, particularly on the hind margin of the eye; posterior spiracles of larva each with 10 to 12 pores.
2. Host range — A highly polyphagous species, frequent on beet, lettuce, and peas, very common on the weed *Sonchus*.
3. Biology — Larva forming a distinctive mine, largely following the midrib, with short offshoots into the leaf blade.[26]
4. Distribution — Widespread in most of Europe, present in Israel, but eastern limit of range not known.

This is only a minor pest. On importation into the U.S. it could easily be confused with *L. huidobrensis*, with similar mines, larval spiracles, and male genitalia. Positive identification of dead larvae or puparia would be difficult but this is a species which should not be allowed to become established in the U.S. where conditions for its development are entirely suitable.

7. Liriomyza dianthicola (Venturi)

1. Synonym — *Phytomyza jannonei* Séguy.
2. Description — A delicate little yellow and gray species, with the third antennal segment angulate, the costa ending midway between veins R4 + 5 and M1 + 2 and the outer cross-vein lacking.

3. Host range — Host specific on carnations.
4. Biology — The whitish leaf mines are initially narrow and later greatly widen, becoming blotch like. Mines are frequently confined to the lower leaf surface, but can also occur on the stem.
5. *Distribution* — Italy, France, Greece, and certainly other Mediterranean countries, not confirmed in Israel. Introduced in glasshouses in Belgium and to Kenya where it is present on the main carnation farm near Naivasha.

This species regularly causes considerable damage to young carnations in the main growing areas in Italy in Liguria and Tuscany and severe damage has been reported in glasshouses in Belgium.

In England *L. dianthicola* has been intercepted on carnations from Italy, Greece (Crete), France, Spain, and once from Kenya.[22] In 1983 a shipment of carnation cuttings from Italy to Colombia was found to be heavily infested and was destroyed. In this author's opinion this species could become established in Colombia and in southern states in the U.S. and therefore must be treated as a potential pest of some significance.

8. Liriomyza quadrata (Malloch)

1. Description — Large species, wing length 2.3 to 3.2 mm, head largely bright yellow, mesonotum shining black but with yellow patch centrally before scutellum.
2. Host range — Restricted to Solanaceae, occurring on many wild genera and also on potato, tomato, and tobacco.
3. Biology — The larva forms a narrow linear mine in the first instar, later a large, irregular white blotch with frass in conspicuous black lumps.
4. Distribution — Widespread throughout much of South America from Argentina and Chile to Colombia and Venezuela, apparently not reaching Central America.

The author sees no reason why *L. quadrata* should not be able to establish itself on tomatoes in Florida or California and quarantine inspectors should be aware of this species when examining shipments of solanaceous plants from South America.

9. Liriomyza cepae Hering, *L. chinensis* Kato, *L. nietzkei* Spencer (Onion Leaf Miners)

These three very similar species are host specific on *Allium*, being pests on both *A. cepa* (onion) and *A. porrum* (leek).

1. Description — All are small, with wing length from 1.3 to 2 mm in *L. chinensis* and *L. nietzkei* but up to 2.5 mm in *L. cepae;* the head is largely yellow and the third antennal segment distinctly angulate; mesonotum grayish-black; scutellum entirely dark, blackish gray in *L. nietzkei*, partially yellow in *L. cepae and L. chinensis*; legs with femora yellow in *L. chinensis*, partially blackish in *L. cepae*. Positive identification of adults may be difficult on external characters but the male genitalia of the three species are entirely distinct.[8,26]
2. Host range — Exclusively *Allium*, pests on both onion and leek; in Asia *L. chinensis* has also been recorded on *A. bakeri, A. fistulosum* and *A. odorum*.
3. Biology — the larvae form a whitish linear mine and there are two generations annually. In *L. cepae* and *L. chinensis* the posterior spiracles of the larva each have 8 to 10 pores, while in *L. chinensis* there are consistently 14 to 15.
4. Distribution — *L. cepae* only confirmed in Germany; *L. nietzkei* is known in Germany, France, Italy, and Switzerland; *L. chinensis* is widespread in Asia from Japan to China, Malaysia, Singapore, and Thailand, also surprisingly in the south of France.[8]

Examination of male genitalia led to the diagnosis of these three species.[26] Serious damage has been caused to onions in France and Italy by *L. nietzkei*, and *L. chinensis* regularly causes damage in Asia, most recently in Thailand, and also in France.

One species in this group has recently been intercepted in England on a shipment of fresh onions from Cyprus. Live larvae were obtained and it is hoped to rear the species and make a positive identification in due course.[23]

In view of the considerable international trade in onions, and the absence of all three species from England and North America these three species must be treated as of considerable quarantine significance.

B. Species on Leguminous Crops

Many species are host specific on leguminous crops, feeding either as internal stem borers or in pods (*Melanagromyza*), external stem miners (*Ophiomyia*) and some as leaf miners (*Japanagromyza*). There can be some variation in the feeding habit of these genera. The most destructive species is *Ophiomyia phaseoli* (see below), feeding in succession in leaf, stem, and root. The internal stem borers for the most part cause little serious damage, as the number of larvae is normally restricted to one per plant. In India *Melanagromyza obtusa* feeding in pods of *Cajanus* spp. (tur, dhal, red gram, or pigeon pea) is considered as a serious pest and a related species in East Africa, *M. chalcosoma*, feeds in pods of pigeon pea but the damage caused is of less significance. Five species are discussed in more detail below.

1. Ophiomyia phaseoli (Tryon) — Bean Fly

1. Description — A small, conspicuously shining black species, wing length 1.8 to 2.2 mm. A distinctive character is the long, narrow, brilliantly shining ocellar triangle.
2. Host range — Many genera of Leguminosae, particularly *Glycine, Phaseolus*, and *Vigna*.
3. Biology — Oviposition takes place in young leaves, a female laying between 100 and 300 eggs over a 2-week period. The young larva forms a short leaf mine to the nearest vein which it enters and follows to the midrib, then feeding down the stem to the root. The larval stage occupies about 10 days and the pupal stage 9 to 10 days. The entire life cycle can be as short as 17 days but at higher, cooler elevations in Java it has been reported that the larval stage can last from 17 to 22 days and the pupal stage from 13 to 20 days. Numerous larvae can be present in a single plant and they pupate beneath the epidermis, normally near the base of the stem. The puparium is yellowish. The biology has been discussed in detail by Greathead.[3]
4. Distribution — Occurring throughout the Old World tropics from Africa, across Asia to northern Australia and islands in the Pacific; present in Egypt and the Jordan valley in Israel; also present in Hawaii by introduction.

This is an extremely serious pest on beans and other cultivated leguminous crops and chemical control is invariably necessary. In Hawaii biological control is apparently keeping populations at acceptable, nondestructive levels.[4]

O. phaseoli could certainly flourish in many areas of South and Central America, and in southern states in the U.S. It is a most important quarantine pest and every effort should be made to prevent its further establishment outside its normal range.

2. Ophiomyia spencerella Greathead

1. Description — Essentially as in *O. phaseoli* but ocellar triangle perhaps less narrow and shorter.

2. Host range — Most commonly found on beans (*Phaseolus vulgaris*), but also on other *Phaseolus spp., Lablab niger*, and *Vigna unguiculata*.
3. Biology — Oviposition takes place normally in the hypocotyl of emergent seedlings at ground level. The larvae feed mainly in the root, returning to ground level to pupate. The puparium is black (in contrast to the yellow puparium of *O. phaseoli*).
4. Distribution — Africa: Kenya, Uganda, and Nigeria, but probably more widespread and hitherto confused with *O. phaseoli*, apparently not present in Asia.

Greathead[3] considers that in East Africa *O. spencerella* causes more damage than *O. phaseoli*, as the larvae reach and destroy the root system first. The two species frequently occur together.

3. Melanagromyza phaseolivora Spencer

1. Description — Large species, wing length 2.85 to 3.2 mm; head largely black, mesonotum blackish green, abdomen shining greenish, squamal fringe white; male genitalia distinctive.[26]
2. Host range — Apparently restricted to beans; *Phaseolus vulgaris* (Ecuador), ''green beans'' (Colombia).
3. Biology — Pod feeders, almost certainly pupating in the pod; no detailed information recorded.
4. Distribution — Colombia, Ecuador, probably more widespread in South America.

The economic importance of this species has not been established but, in view of the concealed feeding habit of the larva within the pod where the pupa remains, detection of an infestation of imported beans would be difficult. There is no obvious reason why this species should not become established in southern states in the U.S.

4. Japanagromyza phaseoli Spencer

1. Description — Stout species, wing length 2.5 to 2.75 mm, head largely black but frons sometimes brownish, mesonotum predominantly mat grayish or with faint greenish tinge, abdomen shining green, halteres white.[32]
2. Host range — *Phaseolus* spp. but reported in Peru on *Brassica oleracea* (requiring confirmation).
3. Biology — Larva forming a large blotch mine, sometimes several in a single leaf, pupation on ground; puparium brownish orange.
4. Distribution — Peru, Venezuela, Costa Rica, probably more widespread in South and Central America.

Young plants can be completely destroyed when several leaves are attacked and this species is clearly a pest of some significance on beans. Establishment in southern states in the U.S. seems possible.

5. Chromatomyia horticola (Goureau)

1. Synonym — *pytomyza atricornis* Meigen, p.p.
2. Description — Wing length 2.2 to 2.6 mm, costa ending at vein R4 + 5, outer crossvein lacking; frons yellow, antennae black; mesonotum mat gray.
3. Host range — A highly polyphagous species, rarely found on Compositae but can be a serious pest on peas (India) and on lupins (South Africa).

4. Biology — The larva forms a narrow linear mine, pupating in the leaf at the end of the mine, with the anterior spiracles projecting through the leaf epidermis.
5. Distribution — Widespread in Europe, particularly in the Mediterranean area, extending eastward through Iran to India and Japan; small populations present in East and West Africa, more common in South Africa.

This species cannot be distinguished on external characteristics from *C. syngenesiae* (see Section II.A.9) but the male genitalia of the two are distinct.[26] With the puparia remaining in dead leaves, introduction to countries where it does not occur must be greatly facilitated and it is surprising that the species has not reached and become established in the U.S. Differentiation by quarantine inspectors of *C. syngenesiae* from *C. horticola* will not be possible from dead larvae or puparia and any male adults intercepted will require careful examination by experienced entomologists. This is a species which should not be allowed to become established in the New World where with its wide host range it could develop into a serious pest on any of the numerous acceptable hosts which are available.

C. Species on Miscellaneous Crops

Numerous species are known on a wide variety of crops, particularly in Europe and to a lesser extent in South America,[26] but relatively few are of major quarantine significance. Six species are briefly discussed below.

1. Melanagromyza tomaterae Steyskal

1. Description — Relatively large species, wing length 2.7 to 3.3 mm; head black, mesonotum blackish green, abdomen more conspicuously green; squamal fringe white; male genitalia: typical of the genus, illustrated by Spencer.[34]
2. Host range — Only known host, tomato, possibly feeding on other genera of Solanaceae.
3. Biology — The larva feeds and pupates internally in stems of tomato; puparium pale brownish to straw colored.
4. Distribution — Colombia, Ecuador, Venezuela, probably more widespread, range possibly extending northward to Central America.

Young plants have been severely damaged in Venezuela. Puparia might well be present in pieces of stem inadvertently included with tomatoes packed for export and this must be treated as a pest of some significance.

2. Ophiomyia pinguis (Fallén)

1. Description — Small, shining black species, wing length 1.9 to 2.3 mm; squamal fringe black.
2. Host range — *Cichorium endivia* (endive), *C. intybus* (chicory), more rarely on lettuce.
3. Biology — In cultivated chicory the larva feeds exclusively as a leaf miner and the mines frequently turn reddish in the white, forced leaves; the pale yellow puparium remains in the leaf, with the anterior spiracles projecting through the epidermis.
4. Distribution — Widespread in Europe, extending eastward to the Tadzhik and Uzbek Republics of the U.S.S.R.; also Egypt.

This species is a serious pest on cultivated endive and chicory, particularly in Belgium, the Netherlands, France, and Switzerland. This author has reared adults from chicory bought in a shop in north London, imported probably from Belgium. The salad crops serving as host for *O. pinguis* are considered to be a delicacy and, although commercial exports from

Europe to North America are not substantial, small amounts of infested material might easily be included in passengers' baggage.

3. Liriomyza braziliensis (Frost)

1. Synonyms — *Agromyza ecuadorensis* Frost and *Liriomyza subandina* Blanchard.
2. Description — Exceptionally large species, wing length up to 4.5 mm; head largely yellow but antennae frequently darkened; mesonotum black with conspicuous triangular yellow patch centrally.
3. Host range — Potato, including *Solanum andigenum.*
4. Biology — Oviposition takes place low on the stem, the larva then feeding down into the tuber where a mine is formed near the surface; pupation takes place at the end of the mine; larva up to 6 mm long, puparium dark brown.
5. Distribution — Widespread in South America at higher elevations from Argentina to Colombia.

In view of the restricted trade in potatoes, with the total prohibition of imports from South America into both Canada and the U.S., this species is of little quarantine significance.

4. Napomyza carotae Spencer

1. Description — Medium-sized species, wing length 2.7 to 3.1 mm, costa ending at vein R4 + 5; head largely yellow but antennae black; mesonotum mat gray, legs black with all knees bright yellow; squamae yellow, fringe black.
2. Host range — Carrot (*Daucus carota*), possibly other Umbelliferae.
3. Biology — Eggs may be laid in leaves but the stem is the preferred oviposition site. The average number of eggs per female is 173, with a maximum of 248 by a single female. The larva feeds down the stem into the main root of the carrot where a mine is formed, pupation taking place towards the top of the carrot, near the base of the leaves. At 23 to 24°C larval development lasts from 53 to 68 days, the pupal period, however, is short, only 7 to 8 days. As many as ten larvae have been found in a single carrot in Switzerland.
4. Distribution — England, Germany, Holland, Switzerland, probably more widespread in Western Europe.

This species has been overlooked in Europe, as it was first noted in Berlin in 1907 but not again recorded for more than 50 years. It could reach North America with commercial shipments of carrots from Europe and establishment in suitable localities would pose no problems. Quarantine inspectors should be aware of this pest.

5. Phytomyza ilicicola Loew

1. Description — Small species, wing length 1.6 to 2.3 mm; frons black below, brownish above; mesonotum silvery gray.
2. Host range — Holly, most common on *Ilex opaca* but also recorded on *I. aquifolium* and possibly *I. arenicola.*
3. Biology — The egg is laid randomly at any point in the leaf, the larva forming a narrow linear mine which later develops into an elongate blotch; pupation externally.
4. Distribution — U.S., widespread in northeast, extending south to South Carolina, Kentucky, and Florida.

Serious damage to cultivated holly is caused by the numerous leaf mines which can be found on a single plant. The species was intercepted in Canada in 1980 on *Ilex opaca* supplied from Massachusetts[39] and was once intercepted in England in 1982 on the same host, also from Massachusetts.[22] In England the only species occurring naturally on holly is *Phytomyza ilicis*. *P. iliciola* could well become established with the native British holly available as an acceptable host.

6. *Phytomyza rufipes* Meigen

1. Description — Large, yellow and gray species, wing length 2.5 to 3.5 mm; frons yellowish to orange brown, third antennal segment verging from almost black to largely yellow; mesonotum ash gray.
2. Host range — Host specific on Cruciferae, frequent on most cultivated brassicas, recorded on nine other genera.
3. Biology — Oviposition takes place in the leaf blade, the young larva then forming a short mine before entering the petiole and in young plants may continue feeding in the stem. It is not uncommon to find 10 larvae in a single stalk and as many as 25 have been observed.
4. Distribution — Widespread in western Europe; Canada: New Brunswick, Newfoundland; Colombia. References in U.S. literature to this species in Oregon are based on a misidentification.[36]

This species can be a serious pest on young plants and population levels can be very high. In eastern England in 1949 soil populations of puparia ranged from 50,000 to 1,650,000 per acre in fields of swedes (*Brassica napobrassica*).[26]

The restricted populations in northeastern Canada and in the Bogotá area of Colombia certainly represent introductions with cabbage or other brassica crops from Europe. Some 20 years ago this author identified larvae of this species intercepted in New York, on a consignment of cabbages from Holland.

D. Species on Cereals

Many species in the genera *Agromyza, Cerodontha,* and *Pseudonapomyza* attack all cultivated cereals and the larvae, when present in large numbers, can cause significant damage to young plants. There is little overlap in species between Europe and North America but the best known species in the U.S., *Cerodontha dorsalis*, which is probably present in all states and Puerto Rico, also occurs throughout much of South America as far south as Argentina and Chile. Species in the genus *Agromyza* have been recorded attacking barley, oats, rye, wheat, and corn (maize) in Europe, Africa, India, and the U.S.[26] *Cerodontha dorsalis* has been studied in some detail and the larvae have been known to cause minor damage to barley and corn (maize). However, it has not been established that leaf mining damage causes any reduction in the ultimate yield. In Europe a related species, *C. denticornis*, occurs commonly on cultivated cereals and numerous wild grasses. Slight damage has been recorded on barley in southern England, and also in Japan on the same host in the Tokyo area.

Agromyza parvicornis Loew is widespread in the U.S. and has been recorded in Canada, the most common cultivated host being corn (maize). Several larvae can be present in a single leaf and the species occasionally occurs in large numbers in restricted localities, retarding the growth of young plants.

In Europe a recent study has been made of four species of *Agromyza* attacking cereals in the Paris area.[2] *A. megalopsis* Hering has caused serious damage to young barley plants in spring, *A. nigrella* Rondani primarily attacks wheat, while *A. intermittens* (Becker) and *A.*

luteitarsis Rondani are found on both hosts. Damage has sometimes been sufficiently severe to require chemical control.

In the genus *Pseudonapomyza* four species have been recorded as pests: *P. atra* (Meigen) feeding on a number of cereals in Europe, *P. asiatica* damaging rice in Taiwan, *P. spicata* (Malloch) feeding on corn (maize) in Hawaii and Guam, and on corn and barley in Egypt and the Sudan, and *P. spinosa* Spencer, which had been confused with *P. spicata* until its description in 1973, which has been confirmed on barley in Egypt and on oats in Nigeria. This last species is now known from South Africa, India, northern Australia, and islands in the Pacific.

Although the quarantine significance of these species is not serious, with the extensive international trade in cereals there is a real possibility that one or another might reach and become established in a country outside its normal range. When visiting the APHIS entomologists at the plant quarantine inspection station at the Miami International Airport in January, 1983, this author observed that a young grass plant about 3 in. long was found growing in the soil at the base of a palm tree which had been impounded, imported from Saudi Arabia. Examination of this plant revealed a small leaf mine containing a puparium which the author was able to identify as a *Pseudonapomyza* sp., almost certainly either *spicata* or *spinosa*. Either of these species could become established in Florida and this is an interesting illustration of the way in which pests may succeed in reaching a country well beyond their normal range.

E. Species Feeding in the Wood of Growing Trees

The larvae of species in the genus *Phytobia* Lioy feed exclusively in growing wood in twigs or stems of trees or woody bushes. The egg is inserted into the bark and the larvae feed in the previous year's or even older wood. They may feed upwards or downwards and can form a channel up to 40 ft in length. They are long and elongated, due to restriction in the narrow, inflexible channel they bore in the wood, and in many species attain a length of 20 mm or more. The brownish larval tunnels later become infected with bacteria and occluded with callus cells and remain visible throughout the life of the tree. The value of commercial timber is greatly reduced by such larval feeding tracks, which have in the past been referred to as "pith flecks".[26] A detailed study of the biology of the European species feeding in poplar and willow in northern France has recently been made in Martinez et al.[9] The adults are large, with wing length normally of 3 to 5 mm, but in the largest species known in the Agromyzidae, *Phytobia gigas* Spencer from Burma, the wing length is 6.5 mm.

Over 50 species are known in this genus which occurs virtually throughout the world, with 8 species known in Europe, 15 in the U.S.[36] and 10 in Latin America. In an area of rain forest some 350 km northeast of Lima, Peru 20 different species were obtained using bait traps (von Tschirnhaus, personal communication). However, the hosts of relatively few species are known. In the U.S. species are known feeding on *Acer, Amelanchier, Betula, Crataegus*, and *Prunus*; in Europe known hosts are *Alnus, Betula, Crataegus, Malus, Prunus, Populus, Salix*, and *Sorbus*; in Canada *Fraxinus* is a known host and in Colombia larvae have been found in *Solanum quitoensis* ("lulo"), but the adult is not known.

The likelihood of introduction of species in this genus in young trees or freshly sawn timber is probably slight but quarantine inspectors should be aware of this risk.

ACKNOWLEDGMENTS

I am grateful to Mr. P. Aitkenhead, Plant Health, Entomology Branch, Harpenden Laboratory, Ministry of Agriculture, Fisheries and Food, Harpenden, England and to Charles E. Miller, APHIS, PPQ, USDA, Hyattsville, Md. for kindly reviewing the manuscript.

REFERENCES

1. **Bartlett, P. W. and Powell, D. F.,** Introduction of American serpentine leaf miner, *Liriomyza trifolii,* into England and Wales and its eradication from commercial nurseries, 1977-1981, *Plant Pathol.,* 30, 185, 1981.
2. **D'Aguilar, J., Chambon, J.-P., and Touber, F.,** Les *Agromyza* mineurs de feuilles de céréales (Diptères, Agromyzidae) dans la région parisienne, *Ann. Zool. Ecol. Anim.,* 8, 579, 1976.
3. **Greathead, D. J.,** A study in East Africa of the bean flies (Dipt., Agromyzidae) affecting *Phaseolus vulgaris* and of their natural enemies, with the description of a new species of *Melanagromyza* Hend., *Bull. Entomol. Res.,* 60, 541, 1969.
4. **Greathead, D. J.,** Biological control of the beanfly, *Ophiomyia phaseoli* (Dipt.: Agromyzidae) by *Opius* spp. (Hym.: Braconidae) in the Hawaiian Islands, *Entomophaga,* 20, 313, 1975.
5. **Hawkins, W.,** Industry Viewpoint, Tomatoes, in *Proc. IFAS Industry Conf. Biology and Control of Liriomyza Leafminers,* Schuster, D. J., Ed., Institute of Food and Agricultural Science, University of Florida, Gainesville, 1982, 8.
6. **Hering, E. M.,** *Bestimmungstabellen der Blattminen von Europa,* Vols. 1 to 3, Junk, Netherlands, 1957.
7. **Husnik, D.,** Quarantine aspects of importing cut flowers into the United States, *Proc. 4th Annu. Industry Conference on the Leafminer,* Poe, S. L., Ed., Society of American Florists, Alexandria, Va., 1984, 1165.
8. **Martinez, M.,** Contribution à l'étude des Agromyzidae de France, *Bull. Soc. Entomol. France,* 87, 302, 1982.
9. **Martinez, M., Gumez, J. L., and Munnier, P.,** Un ravageur mal connu: la mouche mineuse du cambium des peupliers, *Phytoma,* 372, 51, 1985.
10. **Mortimer, E. A. and Powell, D. P.,** Development of a combined cold storage and methyl bromide treatment to control the American serpentine leaf miner (*Liriomyza trifolii,* Diptera: Agromyzidae) in imported chrysanthemum cuttings, *Ann. Appl. Biol.,* 105, 443, 1984.
11. **Parella, M. P. and Bethke, J. A.,** Biological studies of *Liriomyza huidobrensis* (Diptera: Agromyzidae) on chrysanthemum and aster, in *Proc. 3rd Annu. Industry Conf. Leafminer,* Poe, S. L., Ed., Society of American Florists, Alexandria, Va., 1983, 110.
12. **Poe, S. L., Ed.,** *Proc. 3rd Annu. Industry Conf. Leafminer,* Society of American Florists, Alexandria, Va., 1983.
13. **Poe, S. L., Ed.,** *Proc. 4th Annu. Industry Conf. Leafminer,* Society of American Florists, Alexandria, Va., 1984.
14. **Powell, D. F.,** The eradication campaign against American serpentine leaf miner, *Liriomyza trifolii,* at Efford Experimental Horticultural Station, *Plant Pathol.,* 30, 195, 1981.
15. **Prevatt, J. W. and Price, J. F.,** An economic analysis of costs for leafminer control on two pompon chrysanthemum farms in Florida, in *Proc. 3rd Annu. Industry Conf. Leafminer,* Poe, S. L., Ed., Society of American Florists, 1983, 125.
16. **Schuster, D. J., Ed.,** *Proc. IFAS Industry Conf. Biology and Control of Liriomyza Leafminers,* Florida Institute of Food and Agricultural Sciences, University of Florida, Gainesville, 1982.
17. **Seymour, P. R. and Kilby, L. J.,** Insects and Other Invertebrates Found in Plant Material imported into England and Wales, 1978, Ministry of Agriculture, Fisheries and Food, Harpenden, U.K., Ref. Book 442, 1978.
18. **Seymour, P. R. and Kilby, L. J.,** Insects and Other Invertebrates Found in Plant Material imported into England and Wales, 1980, Ministry of Agriculture, Fisheries and Food, Harpenden, U.K., Ref. Book 442, 1980.
19. **Seymour, P. R. and Kilby, L. J.,** Insects and Other Invertebrates Found in Plant Material Imported into England and Wales, 1981, Ministry of Agriculture, Fisheries and Food, Harpenden, U.K., Ref. Book 442/83, 1983.
20. **Seymour, P. R., Roberts, H., and Kilby, L. F.,** Insects and Other Invertebrates Found in Plant Material Imported into England and Wales, 1983, Ministry of Agriculture, Fisheries and Food, Harpenden, U.K., Ref. Book 442/83, 1985.
21. **Seymour, P. R., Roberts, H., and Davis, M. E.,** Insects and Other Invertebrates Found in Plant Material Imported into England and Wales, 1984, Ministry of Agriculture, Fisheries and Food, Harpenden, U.K., Ref. Book 442/84, 1985.
22. **Seymour, P. R., Roberts, H., and Davis, M. E.,** Insects and Other Invertebrates Found in Plant Material imported into England and Wales, 1982, Ministry of Agriculture, Fisheries and Food, Harpenden, U.K., Ref. Book 442/82, 1986.
23. **Seymour, P. R., Roberts, H., and Davis, M. E.,** Insects and Other Invertebrates Found in Plant Material imported into England and Wales, 1985, Ministry of Agriculture, Fisheries and Food, Harpenden, U.K., Ref. Book 442/85, 1986.
24. **Spencer, K. A.,** The Agromyzidae of Canada and Alaska (Diptera), *Mem. Entomol. Soc. Can.,* 64, 1, 1969.

25. **Spencer, K. A.,** *Diptera, Agromyzidae,* Handbooks for the identification of British insects. I, Part 5 (g), 1972.
26. **Spencer, K. A.,** *Agromyzidae (Diptera) of Economic Importance,* Series Entomologica 9, Junk, The Hague, 1973.
27. **Spencer, K. A.,** The Agromyzidae (Diptera) of Venezuela, *Rev. Fac. Agron. Maracay,* 7, 5, 1973.
28. **Spencer, K. A.,** The Agromyzidae (Diptera) of Fennoscandia and Denmark, *Fauna Entomologica Scandinavica 5,* Vols. 1 and 2, Klampenborg, 1976.
29. **Spencer, K. A.,** The Agromyzidae of New Zealand (Insecta: Diptera), *J. R. Soc. N.Z.,* 6, 153, 1976.
30. **Spencer, K. A.,** *A Revisionary Study of the Leaf-mining Flies (Agromyzidae) of California,* Spec. Publ. 3273, Division of Agricultural Sciences, University of California, Berkeley, 1981.
31. **Spencer, K. A.,** Agromyzidae (Diptera) in Chile, *Stuttg. Beitr. Natur.,* Series A. 357, 1, 1982.
32. **Spencer, K. A.,** Leaf Mining Agromyzidae (Diptera) in Costa Rica, *Rev. Biol. Trop.,* 31, 41, 1983.
33. **Spencer, K. A.,** East African Agromyzidae (Diptera): further descriptions, revisionary notes and new records, *J. Nat. Hist.,* 19, 969, 1985.
34. **Spencer, K. A.,** The Agromyzidae (Diptera) of Colombia, including a new species attacking potato in Bolivia, *Rev. Colomb. Entomol.,* 10, 3, 1985.
35. **Spencer, K. A. and Stegmaier, C. E.,** The Agromyzidae of Florida, with a supplement on species from the Caribbean, Arthropods of Florida and Neighboring Land Areas, Florida Department of Agriculture and Consumer Services, Gainesville, 7, 1, 1973.
36. **Spencer, K. A. and Steyskal, G. C.,** Manual of the Agromyzidae (Diptera) of the United States, Agricultural Research Service, U.S. Department of Agriculture, Washington, D.C., Agric. Handb. 638, 1986.
37. **Stegmaier, C. E.,** Host plants and parasites of *Liriomyza trifolii* in Florida (Diptera: Agromyzidae), *Fla. Entomol.,* 49, 75, 1966.
38. **Talbott, G. M.,** Activities by the Florida celery and lettuce industries seeking a control for leafminers, in *Proc. 3rd Annu. Industry Conf. Leafminer,* Poe, S. L., Ed., Society of American Florists, Alexandria, Va., 1983, 8.
39. **Tymchak, A. E.,** Entomology, Intercepted Plant Pests, 1980-81, Plant Health Division, Agriculture Canada, Ottawa, 1983.
40. **Tymchak, A. E.,** Entomology, Intercepted Plant Pests, 1981-82, Plant Health Division, Agriculture Canada, Ottawa, 1984.
41. **Vercambre, B.,** Etudes réalisées à la Réunion sur la mouche marachère: *Liriomyza trifolii* Burgess, *Rev. Agric. Sucriere Ile Maurice,* 59, 147, 1980.

Chapter 6

WOOD-BORING INSECTS

Thomas Wallenmaier

TABLE OF CONTENTS

I. INTRODUCTION

In spite of the economic importance of wood-boring insect pests, plant quarantine activities related to such pests are often relegated to a relatively low level of enforcement in competition with a wide spectrum of plant quarantine (PQ) activities. Many countries, underscoring the importance of forest resources and resulting wood products, have established forestry agencies to deal with production and utilization problems. However, the exclusion of wood-boring pests is usually under the PQ service. Exclusion is a difficult task because of the wide range of pathways for wood-borer entry. Quarantine practices are not always effective or practical in blocking the pathways — yet, the threat still remains. The objective of this chapter is to describe the biological characteristics and quarantine threats of four groups of wood-boring pests and the regulatory approaches that might be taken to meet these threats.

II. PATHWAYS

A. Natural Pathways

While this chapter focuses on man-made pathways over which exotic pests may travel to other countries, natural pathways should be acknowledged. Where two countries are contiguous and not separated by a natural barrier, there will be many possibilities for natural spread. Since endemic pest populations have been in existence for very long periods, the probabilities are high that these pests will have already expanded along natural pathways to the limits of their environmental niches. Consequently, any natural extension of their geographical range should already have occurred. It is likely that there will not be any additional natural range extension unless natural environmental barriers are changed as a result of human or natural activity. Although it is possible that wood-boring pests could be carried through upper air currents or other natural agencies over natural barriers to other countries, the probability for these events occuring appears to be quite low.

B. Man-Made Pathways

Wood-boring pests may move on imported articles along man-made or artificial pathways. Such articles consist of four types of commodities: (1) logs, (2) lumber, (3) wood packing materials, and (4) living plants or plant propagative material.

Logs usually represent the cut or sawn trunks or boles of trees and are sometimes shipped commercially with the bark attached. These logs often represent high volumes of pulp wood or low volumes of exotic hardwoods. The risk of transporting immature stages of insects, i.e., eggs, larvae, pupae, is high, especially if the logs are allowed to lie in the forest where they certainly attract many pests.

Lumber would encompass any form of wood trimmed or sawn into a regular shape. Lumber is usually made from high grades of wood. Wood borers are less likely to survive in lumber than logs because with lumber the bark is usually removed and the wood cut into smaller pieces. These milling operations produce wood products in which borers are less likely to survive.

Table 1
EXAMPLES OF COMMODITIES SHIPPED WITH WOODEN MATERIALS THAT HARBORED WOOD-BORING PESTS INTERCEPTED ENTERING THE U.S. AT PORTS OF ENTRY

Aluminum	Ironware
Artware	Loose crating[a]
Automobile parts	Loose dunnage
Brassware	Loose pallets[a]
Chemicals	Machinery
Electrical parts	Marble
Glass	Steel
Granite	Tiles
Hardware	Tubing

[a] Wooden packing and shipping materials shipped with cargo.

From Anon., List of Intercepted Plant Pests, Fiscal Year 1983, Animal and Plant Health Inspection Service, U.S. Department of Agriculture, Washington, D.C. APHIS 82-10 (undated).

Wood packing materials including crates, boxes, pallets, bracing, and dunnage serve as important pathways because they are used throughout the world in the international transportation of goods. Risk, represented by the survival of wood borers in transit, increases as the size of the pieces of wood and amount of bark present increases. Quite often, packing materials consist of lower grade wood — and sometimes, bark is attached. The sizes of packing materials vary widely. Since the wood used in pallets consists of very small pieces, pallets are not good sites for wood-borer survival. However, other types of wood-packing materials such as large beams used for bracing and thick wood used for crates containing heavy machinery are more likely to carry viable stages of wood-boring pests. The list of commodities that are imported or exported is almost endless, and much of these goods are accompanied by the wooden packing materials. It is not feasible to provide meaningful data on the volume of commodities accompanied by wood-packing material. However, Table 1 gives the major types of commodities whose wood-packing material was found, at U.S. ports of entry, to carry viable stages of wood-boring insects.[4]

Living trees in various stages of growth, moved from one country to another, may harbor wood-boring insects. Such insects are almost always represented by immature stages since adults would not remain very long on plants when they are handled during international shipment. The movement of living trees is very closely regulated by many countries. Often many tree species are prohibited entry into a country because of the threat of specific pests or pathogens. Trees that are enterable are inspected very carefully at ports of entry.

Probably the most dangerous article, as far as the introduction of wood-borers is concerned, is ship cargo. The fact that cargo takes longer to reach its destination than does planting material is not a drawback for most wood-boring insects that often have long life cycles with extended periods spent in the protection of tunnels in wood. Also, the heaviest and,

Table 2
EXAMPLES OF COUNTRIES INVOLVED IN THE EXPORT OF WOOD FROM CHILE AND THE IMPORT OF WOOD INTO THE U.S.

Exports from Chile, 1981[a] to		Imports to the U.S., 1982[b] from
Argentina	Libya	Australia
Austria	Malaysia	Belize
Barbados	Mexico	Bolivia
Belgium	Morocco	Brazil
Bolivia	Netherlands	Burma
Brazil	Panama	Canada
Canada	Paraguay	Chile
Colombia	Peoples Rep. China	Cameroon
Costa Rica	Peru	Costa Rica
Denmark	Philippines	Ecuador
Dominican Republic	Poland	Fed. Rep. Germany
Ecuador	Portugal	Ghana
Egypt	Puerto Rico	Guatemala
El Salvador	Rep. South Africa	Guyana
Ethiopia	Saudi Arabia	Honduras
Fed. Rep. Germany	Somalia	Hong Kong
France	Spain	Indonesia
Greece	Sweden	Ivory Coast
Guatemala	Switzerland	Malaysia
Haiti	Taiwan	Mexico
Honduras	Thailand	New Zealand
Hong Kong	Tunisia	Paraguay
Iran	United Arab Emirates	Peru
Israel	U.K.	Peoples Rep. China
Italy	U.S.	Rep. South Africa
Japan	Uruguay	Singapore
Jordon	Venezuela	Surinam
Korea	Yugoslavia	Taiwan
Kuwait		U.K.
		Zaire

[a] Source: Anon., *Chilean Forestry News,* No. 56, 1982, 10.

[b] Sources: Anon., U.S. Imports for Consumption, U.S. Department of Commerce, Washington, D.C., FT-246, 1982.

therefore, largest pieces of wood are moved by ship. Air cargo represents a lower level of risk because heavy pieces are not usually shipped by air. Because of the higher cost of shipping by air cargo, wood-packing material is often of a higher grade. Passenger baggage and mail represent a relatively low risk for the introduction of wood-boring pests. Dry wood or wood products carried by passengers or mailed could not be very large.

Large volumes of wood or wood products are moved along various pathways leading to and from many countries. The general directions of pathways can be described as follows: (1) logs and lumber from tropical countries of Asia, Africa, and the Americas are imported by temperate countries in Europe and North America, and (2) temperate hardwoods are exported by countries of Europe and North America to other countries. Countries that Chile[2] exports wood products to are shown in Table 2 along with countries that ship wood products to the U.S.[3] Wood markets change from year to year so countries, in addition to those listed in Table 2, may be brought into the export/import market — thus, providing variation in the man-made pathway system. In 1982, the volume of importations of logs and rough lumber into the U.S. from all countries except Canada was 169,906,000 board feet.[3]

The extent of the pathway involving wood-packing material almost defies description since the commodities shipped are so varied (e.g., machinery, furniture, chemicals, rubber, building stones, etc.). Almost any country that exports represents a threat to the importing country for the entry of wood-borers. Any two countries that trade regularly with each other have probably established inadvertently a pathway for the entry of wood-boring pests on packing materials. However, the threat of this pathway has been reduced by countries that use higher quality wood as packing material as discussed later in this chapter.

III. BARK BEETLES, AMBROSIA BEETLES, AND PIN-HOLE BORERS

A. Introduction

Bark beetles make up part of the 6000 species, worldwide, in the family Scolytidae while ambrosia beetles or pin-hole borers are represented by some species in the family Scolytidae and all of the 1000 species in the family Platypodidae.

B. Feeding Habits

Members of these two families can be divided into five groups based on insect feeding habits, as follows: (1) *phloeophagy,* species that feed in the phloem or cambium layer; (2) *spermatophagy,* species that feed on seeds or fruits; (3) *xylomycetophagy,* species that construct galleries in the wood or twigs and then feed on fungal growth on the walls of the cavities; (4) *herbiphagy,* species feeding on tissue of soft herbs, including twigs and seedlings, and (5) *xylophagy,* species feeding on the wood. Ambrosia beetles feed on wood and their larva feed on fungi that grow in the tunnels (xylomycetophagous). Bark beetles feed on phloem tissues (phloeophagous). Both wood- and bark-feeding beetles present a risk for entry on wood and many wood products moving in international commerce.

C. Life Cycles

Members of the Scolytidae and Platypodidae have one of the following three life cycles: (1) *monogamous;* the Platypodidae many phloeophagous species of Scolytidae, and a small number of xylomycetophagous species are members of this group that is characterized by association of one male and one female in a gallery system; (2) *polygamous,* some phloeophagous species belong to this group that is characterized by the association of one male with several females in a gallery system; and (3) *spanandrous,* usually a group of female siblings raises the brood in a gallery system; low numbers of haploid males are produced from unfertilized eggs (arrhenotocic parthenogenesis). Species of the family Scolytidae produce pheromones which are chemicals used to attract other members of the same species. Many species produce a secondary attractant which serves to draw more individuals to a tree. Frequently, the attracted mass of beetles is able to overcome the tree often leading to its death.

D. Hosts

The hosts of bark beetles are usually conifers (softwoods) such as species of *Pinus*, (pine); *Abies*, (fir); *Larix*, (larch); and, *Pseudotsuga*, (Douglas fir). Bark beetles, in general, are more host specific than most ambrosia beetles which have a fairly large host range. Hosts of ambrosia beetles include: *Eucalyptus* (eucalyptus); *Nothofagus*, (beech); *Juglans* (walnut); *Shorea* (sal tree); *Triplochiton*, (obeche); *Terminalia*, (myrobalan); *Swietenia*, (mahogony); *Abies*, (fir); *Acer*, (maple); *Alnus*, (alder); *Castanea*, chestnut); *Morus*, (mulberry); *Pinus*, (pine); *Populus*, (poplar); *Quercus*, (oak); *Psuedotsuga*, (Douglas fir), and many other genera. Ambrosia beetle larva do not feed directly on these trees but rather on a fungus that grows in the gallery system produced by the beetles in the host.

E. Damage

Bark beetles damage trees particularly in temperate areas. In the U.S., bark beetles cause 90% of all tree mortality due to insects and 63% of growth reduction.[5] For example, 4 billion board feet of Englemann spruce, *Picea engelmannii*, were destroyed in Colorado by a pine beetle, *Dendroctonus engelmanni*, in an epidemic that occurred from 1942 to 1948. In tropical areas, the bark beetles are not as often ranked as major pests but much damage is caused by ambrosia beetles. Much of the damage takes place on fallen trees resulting in lower quality of lumber derived from these trees. For example, in 1929, 25% of Philippine wood graded as "wormy" was sold at a loss. Also in the same year, lumber from West Africa lost 50% of its market value due to ambrosia beetle damage.[7]

F. Control

One of the strongest arguments in favor of quarantine restrictions against wood-boring insects is that once a borer has become established in a new area, the pest is difficult to control. Most control methods are simply good management practices that attempt to keep insect populations below economic levels. Trap trees, chemical controls, and rapid removal of felled timber are all recommended methods that are often impractical to implement. Debarking and rapid drying of logs or timber are two more promising approaches to population control.

IV. POWDER POST BEETLES

A. Introduction

Powder post beetles are classified in the families Anobiidae, Bostrichidae, and Lyctidae. Species in these families have the similar habit of attacking dry or seasoned wood. Some members of these families are also scavengers or stored-product pests.

B. Life Cycles

Adult females of the family Anobiidae lay eggs in cracks and in old emergence holes in the wood surfaces. The emerging larvae feed in the sapwood. After pupation, the adults emerge and are active for several weeks. Females are strong fliers although often a female will oviposit in the same wood from which she emerged. The life cycle takes 2 to 3 years in temperate climates.

Most species of the Bostrichidae feed in dead branches of trees. Adult females lay eggs in tunnels bored in the wood. The larvae tunnel in the sapwood and adults emerge in the midsummer. The life cycle takes about 1 year in the temperate climates.

Members of family Lyctidae attack hardwoods. Adult females oviposit in exposed pores of the sapwood, or, sometimes in cracks in the wood. The larvae bore throughout the wood. The life cycle lasts about 1 year in temperate climates. Some research has been conducted to determine the conditions favorable for infestation. It was shown that the beetles require a moisture level of 7 to 30% and there must be a favorable level of starch in the wood.[6]

C. Hosts

Powder post beetles (Lyctidae) prefer hardwood trees as hosts (e.g., temperate hardwoods such as oaks, or various tropical hardwoods). Soft woods such as pine seem to be more resistant to attack perhaps either because of their small vessel size or the lack of certain nutrients. Members of the Anobiidae and Bostrichidae will attack both hardwoods and softwoods. Common hosts of Bostrichidae in Africa include species of *Khaya*; *Terminalia*, (myrobalan); *Strombosia*; *Azadirachta*, (margosa); *Afzelia* (African mahogany); *Brachystegia; Entandrophragma; Berlinia;* and *Chlorophora*,[1] (African teak). A few species of Bostrichidae attack fruit trees, e.g., *Apate monachus* and *Schistocerus bimaculatus*. Asian species of the genera *Heterobostrychus* and *Dinoderus* attack bamboo (e.g., *Bambusa* and other genera) and mimosa (*Mimosa* spp.).

D. Damage

Powder post beetles appear to cause more damage in tropical than in temperate climates. The damage usually occurs in seasoned wood and finished wood products in the temperate areas. Consequently, these insects are easier to control and they do not rank very high in pest damage. Although a large number of treatments are performed each year, the number of pests is less than for the other categories. In the tropical areas, damage is much greater. The warmer climates and the low level of pest control given to wood products are major reasons for a greater importance in these areas.

E. Control

Control methods for powder post beetles generally consist of either applications of an insecticide to the exposed wood surfaces or fumigation of infested structures. The latter treatment is more costly, but if done properly should eliminate the infestation.

V. TERMITES

A. Introduction

Termites, which comprise another group containing wood-boring pests, are classified in the order Isoptera. The appoximately 1800 species of termites that have been described can be grouped into one of the following four categories: (1) damp wood termites, (2) subterranean termites, (3) dry wood termites, and (4) fungus-feeding termites. About 80% of the termite species are fungus feeding and are classified in the family Termitidae.

B. Life Cycles

The reproductive forms that are winged swarm at certain times of the year. A male and female will pair together and start a colony. Most of the colony developed by a pair will consist of workers and soldiers. Termite species occurring in the tropics usually build exposed nests that may be quite large. Many species in the temperate climates build nests in the ground or in wood.

C. Hosts

Many species of termites have a wide host range. Some species consume wood, books, or other organic matter. Some species will feed on dead parts of living plants or in living roots. Protozoa and bacteria in the gut of the termite digest the cellulose in the wood eaten by the termites. Thus, the species of a tree is not an important consideration in the diet of termites.

D. Damage

Termites are responsible for a large amount of destruction to wood and wood products and some termites damage living plants. The dry wood termites (Kalotermitidae) and subterranean termites (Rhinotermitidae) are the most destructive. However, these two families contain only a small number of all known termite species. Some of the fungus-feeding termites (Termitidae) produce very large mounds, but usually do not pose a threat to wood.

E. Control

Termites account for a significant portion of the cost of insect control in many countries. Because of the threat of damage to man-made structures, insecticide application to structures and to the surrounding soil is a common control practice. Because of the internal nature of insect boring, which is usually hidden, treatment must be applied frequently after a nest has been found and some damage has occurred.

VI. TIMBER BORERS

A. Introduction

Timber borers comprise another category of wood pests. Timber borers consist of certain species in the beetle families Buprestidae, Cerambycidae, and Curculionidae; moth families Sesiidae and Cossidae; and, the wasp family Siricidae. Most of the species bore into dead or dying trees in their larval stages. They prefer fresh or unseasoned timber that has been recently cut. Timber borers attack both hardwoods and softwoods and their damage is usually confined to sapwood.

B. Life Cycles

Timber borers lay their eggs in or on the wood. The larvae then hatch and bore into the wood where they feed — often for many years. Pupation occurs in the boring gallery near the surface where the adults emerge.

C. Hosts

Most timber borers are fairly specific in the tree species they infest but others are not so host specific. Both coniferous and deciduous trees are attacked by one species or another and both temperate and tropical climates have a fauna of timber borers.

D. Damage

The timber borers as a group do not cause important damage but some species can cause economic losses. Most species attack fallen logs, and thus, they participate in the recycling of biomass in the forest ecosystem. However, if these fallen logs are harvested, the damage caused by the timber borers can reduce yields, quality, and profits for the forest industry.

E. Control

Control measures are available but difficult to apply when needed. When fallen or freshly cut trees lay in the forest, they are attacked. The principal control is to avoid these infestations by harvesting the logs as soon as possible but such procedures may not be compatible with industry practices. If trees on the forest floor are valuable, they should be treated with insecticidal applications. Unfortunately, practical and economically justified treatments are very difficult to find.

VII. CASE STUDIES OF INTRODUCTIONS

Much can be learned by examining actual case studies of introductions of wood-boring insects. In an analysis of introductions of Scolytidae, Wood[9] reviewed 62 species that were either introduced into the U.S. or were exported from the U.S. to other countries. Of the 62 species, 4 were bark beetles and 58 were ambrosia beetles. Most of the ambrosia beetles reproduce parthenogenetically, i.e., one female can establish an entire breeding population. This mode of reproduction contributes to the success of the establishment of species in new locations. Host material among the 62 species was quite variable and included seeds, twigs, branches, and logs. Marchant and Borden[8] discussed 54 species of Scolytidae and 3 of Platypodidae that became established outside of their natural geographic range. These scientists also reviewed various factors related to the establishment of exotic species. There is still a continuing need for more detailed investigations of the exact pathway involved in various introductions.

Additional examples of introductions are two species of termites — *Cryptotermes brevis* (a dry wood termite) and *Coptotermes formosanus* (a subterranean termite). *C. brevis* is a termite of neotropical origin that became established in Hawaii, Florida, and Louisiana as well as Queensland, Australia and in various parts of Africa. As a dry wood termite, this species commonly infests furniture. Consequently, the species is easily moved in commerce and its range is increasing. In the U.S., the species is now found in small colonies in many states in addition to the three listed above where the species is well established.

C. formosanus occurred originally in Japan and China but has become established in Hawaii, Louisiana, and South Africa. This subterranean termite was first detected in the U.S. in 1965 in a shipyard warehouse in Houston, Tex. In 1966, it was detected in a shipyard in the port area of New Orleans, La. and in 1967 in Charleston, S.C. In Louisiana, the major infestations were found at Camp Leroy Johnson, a World War II transportation facility, and at other port locations. Consequently, it is believed that returning military equipment from the Pacific area after World War II served as a pathway for introduction into the U.S.

VIII. REGULATORY ACTIONS

Government regulatory agencies can take regulatory actions that will protect the forest industry from the introduction of exotic wood-boring pests.

A. Regulations

Many countries have established import quarantine regulations that govern the entry status of imported wood products. For example, Australia has a strict quarantine law in place directed against infested wood or wood products. Wood products must be fumigated prior to shipment to Australia or on arrival if pests are found. The principal drawback to the use of quarantine laws is difficult enforcement. Some countries will be very stringent about infested wood products, but unless they have equipment and manpower available, they will not be able to enforce such regulations.

B. Phytosanitary Certificates

Phytosanitary export certification is conducted in accordance with the International Plant Protection Convention of 1951, as amended (also known as the Rome Convention). Many countries require that imports of logs and lumber as well as specified wood products be accompanied by a phytosanitary certificate stating that the consignment is free of pests. In addition, countries may require a specific declaration about freedom from certain pest conditions. The most commonly specified pest group is the Scolytidae or bark beetles. Other requirements include the removal of bark before shipment. In the U.S., for example, logs and lumber, depending on their source, are certified as free of the oak wilt disease or agent.

C. National Policy or Guidelines

In addition to regulations, countries may have policies or guidelines of a general nature that are based on a national quarantine act. Such policy might instruct inspectors to be alert to and inspect for certain pest situations such as dunnage from ships discharging nonwood cargo.

D. Inspection

Still another quarantine action to meet the wood-borer threat is inspection. Inspection requires access to the incoming cargo. Cargo must be arranged so that inspection for various types of pests can be accomplished. The written records relating to the cargo, e.g., a ship's manifest, or other custom documents, are important in the inspection process. Also of importance is availability of trained officers who can physically inspect the cargo. In addition, it is useful to be able to conduct random inspections of pier areas, docks, warehouse, etc. to look for wood products. A major problem for inspections is the increasing use of containerized shipping that often makes access to wood products difficult and time consuming.

IX. THE IDENTIFICATION PROBLEM

A major problem in meeting the threat posed by the entry of exotic wood-boring pests on imported articles is obtaining a timely identification of exotic pests found in association with imported articles. Identification is important in assessing the risk and prescribing safeguards. The inspecting officer must be reasonably sure that the detected specimens are, in fact, wood-boring pests. Some inspectors are able to identify wood-boring pests based on training or on-the-job experience. A system of insect identifiers to back up inspecting officers is quite useful in handling the wood-borer problem.

Identification is also a problem in the detection of and regulatory response to early infestations. For example, in the early 1920s, a wood wasp of the family Siricidae was identified in New Zealand as *Sirex juvencus* Linnaeus. However, after several years, it was determined that this pest was misidentified and was actually *S. noctilio* Fabricius, and represented a new introduction. The pest in New Zealand was exhibiting quite different behavior than it did in Europe where it is native. In New Zealand, the pest eventually became widespread and caused economic damage to the commercial stands of Monterey pine, *Pinus radiata*.

REFERENCES

1. **Akanbi, M. O.,** Trend in entomology of wood in use and in storage in Nigeria, in *International Research Group on Wood Preservation,* 10th Annu. Meet., Peebles, Scotland, September, 1978.
2. **Anon.**, Chilean Farmers News, No. 56, 1982, 10.
3. **Anon.**, U.S. Imports for Consumption, U.S. Department of Commerce, Washington, D.C., FT-246, 1982.
4. **Anon.**, List of Intercepted Plant Pests, Fiscal Year 1983, Animal and Plant Health Inspection Service, U.S. Department of Agriculture, APHIS 82-10, (undated).
5. **Baker, W. L.,** Eastern Forest Insects, Misc. Publ. No. 1175, Forestry Service, U.S. Department of Agriculture, Washington, D.C., 1972.
6. **Cymorek, S.,** Experiments with *Lyctus* (transl.), Experimente mit *Lyctus, Beihefte zu Material und Organismen,* No. 1, 1966, 391.
7. **Fisher, R. C., Thompson, G. H., and Webb, W. E.,** Ambrosia beetles in forest and sawmill. I, *For. Abstr.,* 14, 388, 1953.
8. **Marchant, K. R. and Borden, J. H.,** *Worldwide Introduction and Establishment of Bark and Timber Beetles (Coleoptera: Scolytidae and Platypodidae),* Pest Management Papers No. 6, Simon Fraser University, British Columbia, Canada, June, 1976.
9. **Wood, S. L.,** Introduced and exported American Scolytidae (Coleoptera), *Great Basin Nat.,* 37, 67, 1977.

Chapter 7

KHAPRA BEETLE

Michael J. Shannon

TABLE OF CONTENTS

I. COMMON AND SCIENTIFIC NAMES

Trogoderma granarium Everts.

1. Synonym — *Trogoderma affrum* Priesner
2. Common Names — Khapra beetle, Trogoderma (dermeste) du grain, Khaprahafer, and Escarabajo khapra.

II. TAXONOMIC CONSIDERATIONS

The larvae of this species may be easily confused with other members of the genus and other genera of Dermestidae. Taxonomomic separation from these species requires the services of a skilled taxonomist. The reader is referred to Beal[3] for further information on the separation of *Trogoderma* species.

III. HOST RANGE

Although there appears to be little in the way of dried vegetable and animal products, both processed and unprocessed, that the insect will not attack, its most common food sources are grain and cereal products. It is also dispersed by commerce in various materials not known to act as a food source and as infestations in commercial carriers.[5,24]

IV. SYMPTOMS AND DAMAGE

As with most species of dermestids, feeding takes place in the larval stage. Because of its high reproductive rate under favorable environmental conditions, the telltale signs of a khapra beetle infestation are the larvae and their cast skins. The larvae are yellowish or reddish brown. Clothed with long barbed brown hairs, the larva has a tuft of longer hairs which gives it the typical carpet beetle larva look. Adults are brown to blackish in color with indistinct red-brown markings on the wing covers. Although elytra are hairy on the dorsal surface, they may have a slick appearance when the hairs are rubbed off. Mature larvae and adult females are about one eighth inch long, and males and immature larvae somewhat smaller.[12,24]

Damage caused by larval feeding eventually reduces infested host material to a useless powder. Early-stage and low-level infestations may be less obvious or masked by the activity of other stored product pests. Low populations of larvae are also transported in apparent nonhost situations, such as in the ears and seams of used burlap bagging, used grain sacks, host material residues, or under paint scales in ship holds or containers.[15]

Distribution of the population within an infested commodity will be affected by biotic factors, the nature of the commodity, the manner of packaging, and the effect of the presence of other stored product insects.[21] In bulk grain, infestations were observed to be almost restricted to the top 12 in. of grain, with some larvae present at depths of 12 ft or more.[15] In another case described by Smith,[20] distribution of *Trogoderma granarium, Tribolium castaneum*, and *Cryptolestes ferrugineus* in a bag of Nigerian ground nuts were compared to variable moisture and temperature conditions within the bag. *T. granarium* numbers were greatest in the center of the sack, which was hotter and drier than the outer layers. Obviously, the longer a host material is in storage under conditions favorable to population buildup, the more apparent the infestation will become.

V. LIFE CYCLE

Each female usually deposits about 50 eggs in its lifetime. At temperatures above 90°F (32°C), up to 100 eggs per female may be produced. Eggs are deposited loosely and usually singly in host material. Hatching usually takes place in 1 to 2 weeks. The larvae pass through five to nine molts in a period ranging from several weeks to several years. The larvae are capable of survival for more than 7 years without food. They pupate inside the last larval skin. The pupal stage ranges from 2 to 23 days. The adult usually reaches sexual maturity 2 days after emergence. The female begins to deposit eggs 1 to 3 days after mating, with the largest number being on the first day. Adults are usually short-lived, but have been known to survive several months or even years at temperature below 60°F (16°C). Activity occurs primarily at dusk. Adults are winged but do not fly. Completion of a life cycle generally takes 4 to 6 weeks, but can extend several years depending upon the temperature and available food supply. There are usually 4 to 5 generations per year but can be as many as 12 under optimum conditions.[25] The khapra beetle can survive cold winter months in heated warehouses, grain storage tanks, or any establishment offering protection from freezing temperatures.[16]

The strategies for excluding or eradicating this insect are complicated by the following problems resulting from its biology and behavior: (1) detection of life stages as low-level contaminants on imported cargo, in structures, and commercial carriers is difficult; (2) transportation of the pest occurs not only in host material, but also in a wide variety of "nonhost" situations; and (3) the range of eradicative treatments is limited to those that will reach and cause mortality to diapausing larvae, the most difficult life stage to kill and the one least accessible to direct contact by pesticides.

VI. GEOGRAPHICAL DISTRIBUTION

Stored-product insects are in general highly successful colonizers readily dispersed by commerce as demonstrated by the wide distribution of a number of species. Champ[9] documents this phenomenon well. Khapra beetle is no exception in this regard, but is of primary importance as a pest to countries with hot, dry climates. In a review of climatic factors favoring this pest, Banks[2] correlates areas having at least 4 months with a mean temperature greater than 20°C and a relative humidity below 50% with areas where khapra beetle either is or could be a significant pest problem. Quarantine regulations[22] imposed to guard against introduction of the pest into the U.S. list the following countries as considered to have endemic infestations at the time of this report:

Afghanistan	Israel	Senegal
Algeria	Libya	Sri Lanka
Bangladesh	Mali	Sudan
Burma	Mauritania	Syria
Cyprus	Morocco	Turkey
Egypt	Niger	Tunisia
India	Nigeria	Upper Volta
Iran	Pakistan	
Iraq	Saudi Arabia	

In addition, the pest has been reported to occur in specialized environments in more temperate areas such as maltings in Great Britain.[13,14] The pest may also occur in other areas as continuing infestations or periodic reinfestations from imports. In dealing with imported shipment, it is important to be alert to situations where the origin of a consignment may be misleading where risk from this pest is concerned. Such situations include transshipment of infested goods through noninfested countries and infestation or contamination occurring in transit in infested vessels or sea containers.

Because the pest is commonly moved in commerce, reports of its occurrence in other countries may be found in the literature which are actually quarantine interceptions, i.e., pests found on imported products. In other countries, it has been introduced and eradicated. In the U.S., for example, introductions of this insect have been found a number of times and eradicated in each case. An intensive ongoing detection program is in place that provides for early detection and elimination of reintroductions should they occur in that country.

VII. QUARANTINE SIGNIFICANCE

Once an organism that damages plants or plant products has been targeted for exclusion, some judgment is made on the significance of the threat posed. This provides a basis for choosing the severity and nature of the restrictions that will be imposed on commerce to reduce the chances that the pest will enter and become established.

''Quarantine significance'' is a term denoting the relative importance of a pest that either does not occur or occurs in only a limited area in a country. The degree of significance of individual organisms relates directly to the severity of exclusion measures applied and the amount of resources a quarantine agency devotes to a given pest or disease problems. The degree of quarantine significance of individual organisms is established based on three primary factors: (1) the value placed on the production of products impacted by the pest, (2) the nature and anticipated damage and economic losses caused, and (3) the availability of technology to control or eradicate the organism to migrate its impact.

Exotic stored-product pests have at least some importance to most countries as a target for quarantine exclusion. This importance is a reflection of their role in causing economic loss to a wide range of products, resistance to control measures, and impact on the export of goods to receiving countries where the pest does not occur.

Champ[9] reviewed the pest status of 15 major stored-product pests on five commodities in representative countries, and the comparative importance of khapra beetle, along with *Rhizopertha dominica* and *Tribolium castaneum*, was greater in dry, very hot conditions.

Although the khapra beetle appears to do its greatest damage in hot, dry conditions, its mere presence in a country can have economic impact through exclusion restrictions placed by other countries on imports of agricultural commodities from infested sources.

Khapra beetle is a pest of high quarantine significance to the U.S. The production of grain crops is a significant portion of the agricultural economy. Export of these crops is a significant aspect of the country's international trade. Establishment of this pest would disrupt both domestic and international uses of these and other products. Past introductions of the pest have shown two important things. First, the pest can establish itself over a wide area of the U.S. and attain pest status. Second, federal and state plant protection services have the resources and technical capability to rapidly detect and eradicate infestations at an acceptable economic and environmental cost.

The first infestation of this pest in North America occurred in 1953 in California, and infestations were found in Arizona, New Mexico, Texas, and also in parts of Mexico.[23] This infestation of grain storage and transport systems was eradicated in a 12-year cooperative effort. Since the 1966 eradication, a number of infestations have been detected and rapidly eliminated from isolated situations such as spice warehouses, bagging companies, and animal-hide processors,[1] but the pest has been successfully excluded from reestablishing itself in grain storage situations.

In the case of the 1953 infestation, a joint concerted effort by both Mexico and the U.S. eliminated the pest from the continent. Had either party undertaken eradication on its own and the pest allowed to spread in its neighbor's environment, lasting eradication would be a difficult objective to maintain. When khapra beetle is endemic in a neighboring country, overwhelming introduction pressure is exerted through extensive movement of people, agricultural products, and commercial carriers, especially across land borders. Such a situation can well make lasting eradication a noneconomic or impractical course of action.

The only eradicative treatment accepted by the USDA is precisely monitored and conducted fumigation under tarpaulin or in chambers using methyl bromide gas. The literature reports signs of resistance to certain commonly used control chemicals such as phosphine[4] and malathion.[30]

VIII. PATHWAY ANALYSIS FOR KHAPRA BEETLE ENTRY

The greater the quarantine significance of a pest, the more intensive the measures applied to prevent its introduction. However, the use of resources at a quarantine facility are dictated not only by what pests are of significance but, more importantly, how those pests are moved by commerce and the steps necessary to reduce the potential for establishment to occur.

Having made a judgment on the relative importance of khapra beetle should it become established, one may then analyze potential routes of introduction and the establishment threat inherent in each to delineate the problem into areas that can be considered for imposition of quarantine barriers or safeguard actions. In this problem analysis process, it is important to also identify nonprogram barriers to establishment. Utilization and handling of potentially infested materials or carriers, unsuitable climatic conditions, and other factors may greatly diminish probabilities for colonization to occur from introductions.

Physical inspection of international commerce provides intelligence information to identify introduction pathways. The dispersal pathways used by this pest are so diverse and sometimes unpredictable that such information is important in focusing quarantine restrictions on those that are high risk.

A. Review of U.S. Pest Interceptions

As discussed earlier, the U.S. places a high priority on the exclusion of khapra beetle, and significant resources are invested in inspections to detect and eliminate infestations. The results of inspections are computerized[29] and analyzed to assist in focusing efforts on higher-risk materials, make decisions on the need for more restrictive action, and to guide detection surveys. Khapra beetle interceptions at U.S. ports of entry can be utilized to illustrate dispersal pathways for this pest in international commerce (Table 1).

The data in Table 1 show that imported cargo is the most common means of entry. The biology of the pest suggests that entry by cargo is the most likely situation to result in colonization in susceptible environments. A more specific picture of the introduction threat to the U.S. is provided by examining the origin and identity of contaminated cargoes (Table 2). As can be seen in Table 2, a large portion of the threat to the U.S. emanates from four countries. The specific problem from each source is demonstrated by the following interception data.

B. Cargo Analysis

Air analysis of the data in Tables 1 through 6 and reports from inspecting officers support the following operations:

Table 1
INTRODUCTION PATHWAYS OF KHAPRA BEETLE INTO THE U.S. 1982—1984

	Number of interceptions		
	FY 82	Fy 83	FY 84
Baggage	26	55	48
Mail	1	—	1
Cargo	101	82	62
Ship's stores	14	14	13
Ship's quarters	—	1	
Holds	8	1	5
Total	150	153	129

Table 2
HAZARDOUS SOURCES FOR CARGO 1971—1984

	Number of	
Origins	Interceptions	Percentages
India	586	44
Sudan	317	24
Pakistan	198	15
Iran	63	5
Total		88

Note: 32 countries with 24 or fewer = 2.

Table 3
NUMBER OF INCIDENCES WHERE KHAPRA BEETLE WAS DETECTED WITH VARIOUS CARGOES 1971—1984

Gum	132	Brassware	15
Sheepskin	121	Hibiscus	10
Artware	100	Peanut	8
Capsicum	81	Rice	8
Cotton piece goods	47	Pistachio	8
Cucurbit	41	Screens	7
Cumin	40	Coffee	6
Carpets	34	Niger	5
Foodstuff	27	Goatskins	5
Automobile	24	Container	5
Celery seed	17		
Cassia	16	Total	757

Note: Number of other identified hosts with less than five = 44; total identified hosts = 801 (60 %); total unidentifiable hosts = 535 (40 %).

Table 4
CARGO INTERCEPTIONS FROM INDIA

	FY 82	FY 83	FY 84
Nonregulated articles			
Crating (artware, handcrafts, housewares, books, household goods, sheeting, rubber, tools, auto parts, metalware, and glass	20	9	3
Bagging (with cargo)			
Niger seed (bagging)	1		1
Celery seed (bagging)	1	8	2
Cloth (bagging)	2		
Coffee (bagging)	1		
Foodstuffs (bagging)	4	3	
Rice (bagging)	1	2	
Cumin seed (bagging)	4		1
Lentil (bagging)	1		
Dill seed		2	
Coriander (bagging)		2	
Carpets		1	
Beans (bagging)		1	
Psyllium sd.			1
Gum (bagging)	5	1	4
Gum		10	
Cargo not identified	7	9	6
Regulated articles[a]			
Brassware (crating); India	4	3	—
Sheepskins	1		
Bagging as cargo	1		

[a] Regulated articles are fumigated as a condition of entry from these sources.

Table 5
CARGO INTERCEPTIONS FROM PAKISTAN

	FY 82	FY 83	FY 84
Nonregulated articles			
Bagging (with cargo)			4
Rice (bagging)	1		
Rosa sp.	1		
Chickpea, *Cicer arientinum* (seed)		1	
Foodstuffs		2	2
Fennel (bagging)		1	
Sheepskins		1	
Cargo not identified		1	
Regulated articles[a]			
Cucurbit seed	7		
Cumin seed	14	2	

[a] Regulated articles are fumigated as a condition of entry from these sources.

Table 6
CARGO INTERCEPTIONS FROM SUDAN

	FY 82	FY 83	FY 84
Nonregulated articles			
Bagging (with cargo)			
Cassia senna (bagging)	1	2	2
Pepper (bagging)	1		
Hibiscus (bagging)	1	2	6
Gum (bagging)		2	2
Cargo not identified		3	8
Regulated articles[a]			
Sheepskins	10	6	
Cucurbit seed	9	4	4

[a] Regulated articles are fumigated as a condition of entry from these sources.

1. There is an extreme diversity of material which have been found contaminated.
2. Most materials are accompanied by burlap.
3. There is a variation from year to year in the frequency of problems on given cargoes, but a relatively constant proportion of the problem (88%) is from four countries — India, Pakistan, Iran, and Sudan.
4. The Port of New York accounts for 65% of all cargo interceptions, and eight ports account for 95% of the total interceptions.
5. Interceptions, usually single specimens, associated with crating only involved India for the last 2 years, but inspections and findings are not well enough documented at this point to delimit the problem for additional regulatory action.
6. A recent upsurge in finds from celery seed has occurred.
7. The presence of infested material in passenger baggage remains high and appears to be increasing.

Each of the interceptions cited represents a detection in a single lot or consignment of goods. In most cases, infestation detected consist of small numbers of larvae. The U.S. does not import large volumes of raw agricultural products such as grains, pulses in either bagged or bulk form from heavily infested countries. If this were the case, it is likely that this would be shown to be a high-risk entry pathway. At the time the analysis was done, a series of isolated infestations was detected at a number of locations in the U.S. which included spice processing firms and warehouses, used bagging companies, and a glove manufacturer.

All infestations were subsequently eliminated. Treatment as a condition of entry was subsequently required for certain items implicated as the source of these infestations.

C. Carrier Problem

Infested carriers provide a means of entry for khapra beetle. Seagoing vessels may harbor the pest as residual infestations in cargo holds, barges, or sea containers.[8] Such contamination can provide a continual source of infestations of cargoes carried. Infestations of food stores in vessels may also pose a threat in materials carried ashore. For the U.S., the threat posed by carriers has diminished in importance in recent years with the advent of sea containers and improved sanitation on arriving vessels. Ten years ago, one third to one half of annual interception totals were from this entry pathway. In recent years, they constitute only a small percentage (9.5% over the last 3 years).

Surface carriers such as trucks and railcars are not currently a pathway of concern to the U.S. since no infestations of the pest exist in North America. However, once khapra beetle has established itself on a continent, countries considering khapra beetle to be a quarantine pest face a formidable problem because of the extensive efforts required to prevent its introduction and spread through the use of truck and rail transport systems, the movement of seed and feed grain related products, and contaminated used bagging. Experience with the first khapra beetle introduction in the U.S. and Mexico in the 1950s and 1960s showed that the infestations had been spread widely and rapidly by this mechanism prior to detection.

D. Baggage and Mail Problem

Movement of infested materials in travelers' baggage and mail provide another means for introductions of khapra beetle. These pathways pose a comparatively lesser threat of establishment than those in commercial importations of seed and grain due to restricted destinations and utilization of infested materials. As with other entry pathways, degree of threat will vary with the nature, volume, and point of origin of materials received. Cultural ties to another country may increase risk through the extensive movement by people of high-risk host materials from areas with heavy infestations.

IX. SAFEGUARDS

A. Pest Risk Analysis Process

Pest risk analysis is a decision-making process which determines the level and types of safeguard measures required for specific situations where a quarantine-significant pest is known to exist or have a likelihood of movement. It defines whether movement should be allowed and, if so, what conditions will be imposed to reduce the risk of pest establishment to an acceptable level.

Based on the biology of the pest, its relative quarantine importance to a country, the ways in which it is being transported into that country, and the estimated probability of establishment resulting from those introductions, actions can be selected to mitigate introduction of the pest.

This problem analysis must be recognized as an extremely subjective process. Factors required for establishment of introduced insects are not well understood. Even purposeful introductions of beneficial insects where it would be expected that optimum conditions for establishment are selected, succeeded only an estimated 25% of the time.[10]

Based on the biology of the target pest or pests and the situation in which it is being moved, an evaluation of the relative potential of establishment occurring considers the following key points:

1. Probability that the pest will be present
2. Life stage and number of life forms present at the time of movement
3. Introduction history of the organism or related organisms
4. Probability of the organism reaching a favorable environment
5. Probability that port inspection will detect presence of the organism
6. Predicted frequency that infestations will be present
7. Suitability of the carrier to sustain the organism
8. Actions available which mitigate risk
9. Quality of mitigating actions

The plant protection and quarantine (PPQ) decision process is a subjective balance between imposition of measures to mitigate risk and interference with commerce. The safeguards imposed against pest establishment vary with each pest/pest carrier situation and range in stringency from prohibition to enterability at any entry point subject to inspection.

B. Nonprogram Barriers to Colonization

Although the khapra beetle is widely distributed in Europe, Africa, and Asia, and is one of the insects most frequently encountered in international commerce by U.S. quarantine inspectors, it is significant that it is not known to have become established in Australia or Latin America. Also, establishment of the now eradicated population in the U.S. did not occur until the mid 20th century. The failure of the pest to gain a foothold in these areas may be attributed to certain characteristics that these countries have in common. In general, the countries are exporters rather than importers of commodities likely to be infested. An important additional deterrent to colonization are control measures routinely applied in the U.S. and other economically developed countries against such cosmopolitan stored grain pests as the granary weevil, rice weevil, angoumois grain moth, Indian meal moth, and many others. For these reasons, it is not surprising to find that the khapra beetle is a poor colonizer in North America despite the frequency with which it is encountered or associated with imported commodities as well as its surprising longevity in the absence of food.[18]

C. Program Barriers

Where nonprogram barriers do not reduce the threat of establishment occurring from introductions to an acceptable level, the quarantine agency may impose additional entry requirements on international commerce. Because the imposition of these safeguard actions requires the restrictions which necessarily interfere with international commerce, the actions must have a defensible biological justification. To have a reasonable chance of success, the quarantine agency must have the resources necessary to carry out the action in an efficient manner, and the legal authority including, presumably, the necessary public and political support to carry out restrictive actions on the entry pathways of khapra beetle.

Khapra beetle is moved in commerce in a wide range of animal and vegetable products as well as various material not known to act as a food source and as infestations in commercial carriers.[24] This diversity of potential entry pathways poses a difficult problem for quarantine exclusion compared to exclusion of other types of crop pests and disease agents that enter as an infection or infestation of a well-defined range of botanical entities whose entry may be practically controlled through the imposition of regulatory controls on specific types of imported consignments. Imposing quarantine restrictions on all possible entry pathways may be difficult to justify economically, biologically, or socially. The nature and volumes of trade are peculiar to each country, and each country must assess its own situation. Since trade patterns may vary over time, and pest situations at infested sources change, this assessment is an ongoing process.

Assuming that regulatory controls on arriving international commerce are in place, two basic exclusion methodologies can be examined in more detail: inspection and treatment.

2. Inspection

While the importance and likelihood of detecting heavy infestations in host material may be rather obvious in quarantine inspections and detection surveys, detection of less obvious low-level infestations and larval forms hidden in crevices, etc., including contamination of nonhosts, is also important to preclude dissemination and colonization of new environments. The secretive habits of the insect, the small size and neutral coloration of the larvae, the most common life stage encountered, and the great diversity of situations in which this insect can be moved by commerce, makes inspection the least effective of potential deterrents to introduction.

Detection of the presence of the pest requires a knowledge of the biology, behavior, and appearance of the insect and its entry pathways. Where detection of low-level contamination is an objective, hands-on training and experience in performing inspections is also critical.

Detectable signs of infestation are obvious where materials have been in storage for a long period of time under the optimal conditions, and extensive product damage and buildup of cast skins has occurred. This situation is infrequently observed in imports of higher-risk products or in carriers arriving at U.S. ports of entry. This is undoubtedly a reflection of the rapid movement of products, sanitation measures used by industry, and the low amount of bulk or bagged cereals, grains, legumes, and other preferred khapra beetle host materials imported. Because of this, inspections concentrate on detection of low-level contamination that is often associated with marginal, poor, or nonhost situations. The interception analysis provided earlier illustrates the nature of these introduction situations.

Incoming consignments of cargo are detained at U.S. ports of entry and visually inspected. Emphasis is placed on materials packaged in used burlap because of the greater risk of exposure to the pest during the prior use of that packaging material. This phenomenon is reflected by inspection findings and experiences with the pest in the U.S. If the shipment is double bagged, a random inspection of bags is made, the outer bag is slit and the inner seam and corners (i.e., "ears") examined for life forms, usually larvae. If the imported material is a host, contents of the bags are also examined. Random observations, largely at the discretion of the inspector, are made throughout the consignment. If the consignment is in a sea container, partial devanning of a portion of it is required. In some cases, inspections may take place at the importer's premises after delivery.[28] Inspections are a time-consuming, manpower-intensive task that are far from foolproof. This approach has, however, been a viable one for the introduction problem confronting U.S. quarantine. Where inspections reveal a history of infestation on a commodity from a country, treatment as a condition of entry may be required in lieu of inspection unless the problem can be cleaned up at the source.

The analysis of pest interceptions at U.S. ports of entry cited earlier in this chapter revealed a recurrent pattern of khapra beetle introductions that, given the limitations of visual inspection and incidences of pest introduction, mandated more restrictive exclusion measures. After reviewing the problem with the affected and involved groups, PPQ implemented a risk-reduction strategy to require fumigation under USDA supervision as a condition of entry for products posing unacceptable hazards. The reader is referred to USDA regulations for currently regulated articles.

Although infestations in carriers pose a lesser hazard than those in cargo, they are still of concern. At the U.S. ports, as is true for cargo, heavy infestations of vessels are no longer seen. Bulk carriers which have conveyed agricultural products and older, break bulk vessels receive cargo-hold inspections. Officers concentrate inspections on agricultural cargo residues, and the undersides of paint scales and other crevices in the cargo compartments. Sea containers also pose a hazard in that cargo residues and subsequent infestations may build up; however, the U.S. has not observed this as a significant problem to date.

Ship stores are also inspected during the agricultural clearance which is made on every vessel entering the U.S. from a foreign source. Inspectors generally examine dry storage areas for signs of infestation. More emphasis is placed on carriers whose past itineraries have included countries with endemic khapra beetle infestations.

3. Treatment

Total elimination of infestations is the objective of quarantine treatments of infested cargoes and carriers. The only method currently approved by U.S. quarantine for such treatments is methyl bromide (MB) fumigation. Such fumigation may be supplemented by surface sprays using malathion. No data are available showing efficacy for other chemicals where eradication of all life forms is a treatment objective. Aluminum phosphide has been proposed for quarantine treatments in the past, but data showing efficacy on diapausing larvae have been deficient to date.

Effective eradicative treatments with methyl bromide require strict attention to gas tightness of the enclosure, maintenance of appropriate gas concentration throughout the specified duration of exposure, and circulation of the gas within the enclosure. The concentration, time, and temperature schedules authorized for regulatory treatments for khapra beetle by the U.S. Quarantine Service are as follows:

- MB at NAP (chamber or tarpaulin)
- 40 g/m^3 (2$^1/_2$ lb/1000 ft^3) for 12 hr at 3°C (90°F or above — (20 g (oz) minimum gas concentration at 2 to 4 hr, 15 g (oz) minimum gas concentration at 12 hr)
- 56 g/m^3 (3$^1/_2$ lb/1000 ft^3) for 12 hr at 26.5 to 31.5°C (80 to 89°F) — (30 g (oz) minimum gas concentration at 2 to 4 hr, 20 g (oz) minimum gas concentration at 4 to 12 hr)
- 72 g/m^3 (4$^1/_2$ lb/1000 ft^3) for 12 hr at 21 to 26°C (70 to 79°F) — (40 g (oz) minimum gas concentration at 2 to 4 hr, 25 g (oz) minimum gas concentration at 12 hr)
- 96 g/m^3 (6 lb/1000 ft^3) for 12 hr at 15.5 to 20.5°C (60 to 69°F) — (50 g (oz) minimum gas concentration at 2 to 4 hr, 30 g (oz) minimum gas concentration at 12 hr)
- 120 g/m^3 (7$^1/_2$ lb/1000 ft^3) for 12 hr at 10 to 15°C (50 to 59°F) — (60 g (oz) minimum gas concentration at 2 to 4 hr, 35 g (oz) minimum gas concentration at 12 hr)
- 144 g/m^3 (9 lb/1000 ft^3) for 12 hr at 4.5 to 9.5°C (40 to 49°F) — (70 g (oz) minimum gas concentration at 2 to 4 hr, 40 g (oz) minimum gas concentration at 12 hr)[26]

These are rates that kill the pest; however, where certain sorbitive commodities are involved, e.g., flour, the schedule is modified to provide for an extended exposure.

Fumigation of bulk commodities such as grain requires special measures to ensure penetration and circulation of the fumigant throughout the load and effective aeration. For example, U.S. plant protection service procedures[26] require three pipes to be driven into the cargo to approximately 1 ft from the bottom. The pipes should be located in a line 10 ft from the side and positioned approximately 14 ft apart. They must be connected together with 6-mil, heavy-duty plastic tubing as ducts, and this series connected to one of the blowers. The same procedure is followed on the opposite side of the barge to form two parallel lines. This is to assist with both gas introduction and aeration of the commodity to safe levels after treatment. A number of other requirements are outlined in these procedures. Methyl bromide fumigations must be precisely set up, monitored, and aerated both to avoid potential hazards from the chemical and to ensure complete eradication of the pest from the consignment.

3. Implementing Program Barriers

The extent and nature of actions applied to entry pathways is a subjective judgment based on many factors which vary according to the resources and priorities of a country. The objective of their imposition should ideally be to reduce the introduction threat to an acceptable level, whatever that level may be determined to be.

a. Actions at Origin

Obviously, the most desirable answer to the threat from an exotic pest is that it never arrive at a country which has concern for it. Some of the measures that could accomplish this have been discussed earlier. The limited effectiveness of inspection for exclusion of this pest is usually only amplified if some party at point of origin is relied upon to perform the function. Treatment at point of origin has the potential to be an effective approach provided there is assurance that acceptable fumigations are carried out in a manner which assures uniform gas distribution at prescribed temperatures, gas dosages, and time intervals. This is no simple matter to be accepted. Involvement of the knowledgeable counterpart personnel in the plant protection service of the exporting country is generally beneficial. In some case, commercial exporters are willing and able to undertake preexport fumigation as a normal trade practice. The U.S. has had some success with this approach with a major importer of plant gums. Port-of-entry inspection of incoming materials or foreign site visits can be a part of monitoring such preexport treatment efforts. It is essential, because of the nature of this pest, that fumigated materials not be exposed to pest recontamination during storage or land and sea transport.

Interaction with foreign and domestic import industries can do much to reduce introductions of this pest. Sanitation measures at origin will greatly reduce importer's problems and, more significantly, introduction potential.

Requirements that designated imported products be accompanied by certification of freedom from khapra beetle can be a useful method for controlling actions taken prior to export. Such certification may be based on inspection, methyl bromide fumigation, or acceptable evidence that the product has not been exposed to infestation. The degree of reliance that can be placed on such documentation is dependent upon the adequacy of the measures for providing the certification and the confidence that can be placed in those measures. Because of the limitations of inspection discussed earlier, it is difficult to place reliance upon it as the sole basis for certification. Properly conducted treatments will provide a pest-free commodity. Nevertheless, because of the technical requirements for effective treatments, care should be exercised in accepting such certificates without verification. In general, oversight by plant protection service personnel who have a commitment to the task is desirable. In addition, where treatments are performed prior to export, a risk of recontamination by the pest unless appropriate precautions are taken is ever present.

In some cases, khapra beetle may occur in a country, but its distribution may be limited. Khapra-beetle-free certification can be based on the certainty that the material has not been exposed to infestation. Because of the biology and wide host range of this pest and its tendency to contaminate transportation systems, acceptance of pest-free certification on this basis carries some risk unless there is assurance that intensive survey and eradication measures are underway in the country of origin.

b. Actions at Port of Entry

Inspection of cargo can be applied at port of entry as the sole means of excluding khapra beetle or to verify the quality of actions taken prior to export to prevent its occurrence in the exported product. The limitations of this strategy have been detailed under the discussion of this safeguard strategy.

Given evidence that an introduction pathway exists that cannot be satisfactorily addressed through other means, fumigation under supervision of the plant protection service of the importing country prior to release into national commerce may be necessary. Imposition of this restriction should be triggered by evidence that risk is present and limited only to those items where a threat can be demonstrated.

There may be situations where the above actions still have not reduced risk to an acceptable level. In such cases, the most drastic action, i.e., prohibition, must be considered, but only as a last resort when (1) an article with a high risk of khapra beetle introduction is subject to damage, (2) an accumulation of undesirable residues from the fumigation process may develop, or (3) where efficacious treatments are either not available or physically practical.

X. RESPONSES

A. Detection Survey

If significant resources and restrictions on commerce are justified to counter the introduction threat posed by khapra beetle, then an appropriate investment in survey efforts to assure detection in time to contain and eradicate infestations should be made. Survey and early detection serve as a second line of defense against the pest. Because of its high reproductive rate and ready dispersal by commerce, there is high potential for rapid spread of infestations from the point of introduction that make timely detection important.

Detection surveys should be designed to detect the pest as soon as possible after introduction. Therefore, emphasis should first be directed at premises with a relationship to introduction pathways which will vary from country to country according to the nature and sources of foreign commerce received.

Visual inspection of premises can be an effective method, but the difficulty of such inspections has already been reviewed under safeguard discussions. Exit and entry areas are among the obvious points to check, as are the storage portions of the premises. Movement of products and containers as well as people handling such products, which could be subject to khapra beetle exposure, should be observed. The inspector should be alert for cartons, sacks, debris, woodwork, cracks, loose plaster, paint, and other hiding places. Rat bait stations with grain or cereal should be carefully inspected. Visual inspection may be carried on as part of other collecting and trapping activities.

A review of survey procedures used by U.S. quarantine inspectors provides proven methods for the conduct of surveys. Prior to inspection, a plan should be developed that divides a large structure into smaller units for survey purposes. The following guidelines on effective inspectional procedures are suggested from inspection procedures used in the U.S.[27]

1. Use a portable hand- or bench-type vacuum cleaner equipped with a shoulder strap and smooth-walled hose to remove suspect material from corners, cracks, and crevices. The vacuum cleaner and hose must be cleaned and a separate paper filter vacuum bag used for each location or warehouse. It is desirable to change vacuum bags before sampling different portions of the same building, depending on its size and nature of its use. Each used vacuum bag should be placed in a plastic bag. Debris samples are to be placed in separate bags labeled with the location and collector. This procedure should be followed throughout the inspection unless suspect skins or specimens are collected directly.
2. Use a brush to sweep out any corners, cracks, under containers, etc. A sturdy probe or knife may be used to break away areas near cracks (as in concrete walls) to facilitate examination.
3. Inspect every room of the premises, emphasizing those used for storage of host material, shipping and packing materials, coffee rooms, and other places where food might be kept. Untidy areas or areas where refuse is kept for lengthy periods should also be carefully inspected.
4. Inspect the perimeter of the premises, particularly areas in or near exit points which are likely pathways for movement of suspect material, and any outside storage or trash areas.
5. Bulk grain storage areas should be thoroughly inspected. Larvae tend to congregate in the surface grain and on or near walls, especially when heavy infestations occur. Stored material such as grains should be sifted carefully.
6. During inspection, any suspect cast skins, larvae, pupae, or adult beetles should be placed in vials. Specimens from different portions of a building must be kept in separate vials and identified as to the part where found and whether they are dead or alive. It may be necessary to warm the specimens to determine if they are alive.

Debris may be examined using a sieve set and a small artist's brush. All levels of the sieved material are examined independently by placing debris on white paper and removing any insects present with tweezers. Examination under a lamp equipped with a magnifier is recommended. Larvae are not readily apparent in screened samples as they tend to become covered with fine dust and may be sluggish or immobile. Therefore, careful examination of appropriately sized and shaped pieces of debris with a brush is quite important.

Traps may also be used in premises at risk of khapra beetle exposure as a supplement to inspections.[6,7] Traps are placed randomly, approximately 50 ft apart in locations where they will not be disturbed, but still can be easily retrieved. Experience indicates that locations near walls, edges, and corners are productive trapping sites. Traps may be secured, if necessary, to surfaces with rubber or similar cement that will allow easy removal. They should be checked at between 1- and 3-month intervals. Checking at shorter intervals is probably not efficient, while longer intervals will increase the risk of lost or misplaced traps or damage to insects within the traps. Sampling is accomplished by holding the trap over a plastic or paper sheet, bag, or other sample container to check contents for any insects. Traps may be replaced without rebaiting when checked at less than 3-month intervals.

Dermestid larvae suspected to be khapra beetle should be identified by a qualified taxonomist because larvae may be easily confused with related species.[3]

B. Actions After Detection

1. PPQ Plan for New Introductions

Once an infestation is confirmed at a premise, regulatory action to contain further spread, additional survey of the property, and investigation to locate further spread will be required. The following procedures are detailed in the USDA, Animal and Plant Health Inspection Service, PPQ action plan for new introductions.[25]

When an area, establishment, or premise has been determined to be infested with khapra beetle, it will be necessary to expand the inspection to determine the extent of infestation as an aid in determining the type of control measures to apply. If surrounding buildings pose a hazard because of their proximity or their use they will be inspected to delimit the extent of the infestation in the area.

Secondary inspection lists should be developed to include distribution outlets supplied from the infested property and other similar or related businesses that may be infested. State cooperators will be notified of high-risk khapra beetle premises in their respective states, based on the site inspection lists that have been prepared.

All inspections are to be carried out in accordance with the methods outlined in the survey procedure section of the Khapra Beetle Program Manual.

When notification is received from any source that a premise or area is infested with khapra beetle, the following actions will be carried out in an attempt to identify any additional infestations.

First, when the presence of khapra beetle is confirmed, a survey will be carried out in the local area. The "local area" is usually defined as the immediate practical political division, such as a town, city, or county. Specific targets in order of priority are (1) all establishments and premises within the local area considered to be a high hazard risk (inspections should be carried out in the order of the potential risk of the property); (2) other likely establishments, premises, or areas as may be identified; and (3) retail establishments that handle, store, distribute, or market the potentially infested material.

Second, any location which can be determined to have served as a possible source of khapra beetle infested material or to which infested material may have been sent will be inspected.

In addition to the movement of khapra beetle in an infested commercial commodity, there are a number of less obvious ways in which the insect may be spread. Any locations that may have become infested through the following pathways will be inspected: (1) movement of raw materials/equipment into and out of the area; (2) trash movement and dump and fill sites. Inspectors should be aware that people may carry khapra beetle infested trash away from such sites; (3) employee movement (if applicable). Khapra beetle can be carried out by employees in any number of ways, such as vehicles, clothing, and/or containers of any kind; and (4) adjacent structures sharing a common wall. Khapra beetle can move in and through walls constructed of porous materials.

Regulatory actions are imposed on the infested premises until eradication of infestations are complete. Articles that may move the pest are subject to treatment prior to movement being allowed.

Eradication measures applied to the infestation include fumigation, surface pesticide treatment, and various supplemental actions described below.

2. Fumigation

a. Structures and Equipment

Fumigations of infested structures can be accomplished with MB. Treatments are made in accordance with provisions of Section III of the PPQ Treatment Manual. Fumigation will be carried out after all openings in a structure are sealed with masking tape, putty, polyethylene, or other materials as needs indicate. Shipping or storage containers may be tarped and fumigated separately when required. Stationary equipment will be tarped and fumigated separately.

b. Commodities

Commodities (foam rubber, electrical equipment, etc.) that are not tolerant to MB may be removed and subjected to an alternate treatment.

3. Surface Pesticide Treatment

a. Crack and Crevice Treatment

The application of a malathion treatment or other approved pesticide will be required on the interior and/or exterior surfaces of an infested structure, giving particular attention to areas where cracks, crevices, and other khapra beetle habitats exist or may exist. This treatment is to be applied only after sanitation measures have been carried out.

b. Broadcast Pesticide Application

The application of a full-coverage pesticide treatment (spray or fog) will be applied throughout the inside and/or outside of an infested structure. This treatment is to be applied only after sanitation measures have been carried out.

4. Supplemental Actions

a. Extensive Trapping

In addition to regular detection trapping, extensive trapping with adult and/or larval traps baited with attractant and approved pesticide may be carried out. Trap density and frequency of servicing will depend on individual judgment. See the Khapra Beetle Program Manual for general trapping information.

b. Cleaning and Sanitation

The structure will be thoroughly cleaned out prior to fumigation and all rubbish, dirt, debris, etc. will be fumigated with MB and dumped or otherwise destroyed. The surrounding grounds of the premises will be likewise cleaned and all trash disposed of in an approved manner.

c. Outside Surface Treatment

Upon removal of any debris present, the grounds surrounding the structure and applicable building surfaces will be given two or three pesticide applications of malathion at 3- to 7-day intervals. The last application is to be made after the structure is covered for fumigation or after the crack and crevice or broadcast applications are completed.

The following parameters or criteria will determine the minimum treatments to be used in achieving eradication. Expanded or additional treatment actions can be applied if mutually agreed upon with cooperating agencies. Control measures will be conducted only as long as necessary to achieve eradication. Posttreatment inspections will be conducted.

1. If khapra beetle is confirmed in a structure that can be made gastight, MB fumigation, outside surface treatment, commodity fumigation or treatment, and sanitation measures will be carried out.
2. If some structures on the property are not confirmed infested, these may be excluded from treatment unless postinspection indicates fumigation is necessary.
3. Structures that cannot be fumigated will be subject to eradication measures utilizing full-coverage pesticide application and a sanitation program.

C. Criteria for Eradication

In general, eradication may be declared when complete fumigation of exposed portions of an infested property has been accomplished. Where other less intensive treatment approaches are used, at least three inspections are made of the entire property over a 1-year period prior to a declaration of eradication.[26] It is emphasized that methyl bromide fumigation of infested structures has a number of attendant hazards because of the toxic properties of the gas. Careful attention should be paid to potential leakage of the gas from above and underground openings such as sewer systems, around doorways and vents, etc.

XI. CONCLUSION AND SUMMARY

As illustrated through this discussion of the khapra beetle problem, stored product pests pose a unique and difficult problem to quarantine agencies. In the case of the U.S. a successful strategy has been implemented which relies heavily on the professional skills of inspectors in visual examination of entry pathways of international commerce and detection surveys. Analysis of inspection results allows for profiling of entry pathways for resource allocation, and the identification of entry pathways where inspection must be augmented by more restrictive quarantine measures. This hazard profile is also utilized to direct detection efforts to backstop the exclusion system through timely detection and eradication measures. Such measures are economically and environmentally feasible with currently available technologies.

The U.S. situation was used in this chapter for illustration purposes. Although the risk-analysis process described has general applicability, the conclusions reached will vary depending on both biological and nonbiological variables unique to each country.

REFERENCES

1. **Anon.,** Khapra beetle eradicated from the United States, *FAO Plant Prot. Bull.*, 32, 150, 1984.
2. **Banks, H. J.,** Distribution and establishment of *Trogoderma granarium* Everts: climatic and other influences, *J. Stored Prod. Res.*, 13 (19), 183m, 1977.
3. **Beal, R. S.,** Descriptions, Biology and Notes on the Identification of Some *Trogoderma* Larvae (Coleoptera: Dermestidae), Technical Bulletin No. 1228, U.S. Department of Agriculture, Washington, D.C., 1960.
4. **Borah, B. and Chahal, B. S.,** Development of resistance in *Trogoderma granarium* Everts to phosphine in the panjab, *FAO Plant Prot. Bull.*, 27, 77, 1979.
5. **Burges, D. H.,** Diapause, pest status and control of the khapra beetle, *Trogoderma granarium* Everts, *Ann. Appl. Biol.*, 50, 614, 1962.
6. **Burkholder, W. E.,** *Biomonitoring for Stored-Product Insects: Management in Insect Pests with Semiochemicals,* Plenum Press, New York, 1981.

7. **Burkholder, W. E. and Hammack, L.,** Calling behavior in female *Trogoderma granarium* Everts (Coleoptera: Dermestidae), *J. Stored Prod. Res.*, 17, 25, 1981.
8. California Department of Food and Agriculture (CFDA), Division of Plant Industry, Khapra beetle, *Trogoderma granarium* Everts, *Plant Pest Detection Manual,* D. T. 3, 47, 1977.
9. **Champ, B. R.,** FAO global survey of pesticide susceptibility of stored grain pest, *FAO Plant Prot. Bull.* 25, 192, 1977.
10. **DeBach, P.,** Colonizing entomophagous insects, in *The Genetics of Colonizing Species,* Baker, H. G. and Stebbins, L. G., Eds., Academic Press, New York, 1965, 14.
11. European and Mediterranean Plant Health Organization, *Data Sheets on Quarantine Organisms, Trogoderma granarium,* 1979.
12. **Hinton, H. E.,** *A Monograph of the Beetles Associated with Stored Products,* Vol. 1, British Museum, London, 1945, 443.
13. **Howe, R. W.,** A theoretical evaluation of the potential range and importance of *Trogoderma granarium* Everts (Coleoptera: Dermestidae) in North America, *Proc. 10th Int. Congr. Entomol.*, 4, 23, 1958.
14. **Howe, R. W. and Lindgren, D. L.,** How much can the khapra beetle spread in the U.S.A.?, *J. Econ. Entomol.*, 50, 374, 1957.
15. **Lindgren, D. L. and Vincent, L. E.,** Biology and control of *Trogoderma granarium, Hilgardia,* 24, 1, 1955.
16. **Lindgren, D. L. and Vincent, L. E.,** Biology and control of *Trogoderma granarium* Everts, *J. Econ. Entomol.*, 52, 312, 1959.
17. **Lindgren, D. L., Khrone, H. E., and Vincent, L. E.,** The khapra beetle, *Trogoderma granarium* Everts, *Hilgardia,* 1, 1, 1962.
18. **McGregor, R. C.,** The Emigrant Pests, A Report to the Administration of the Animal and Plant Health Protection Service, U.S. Department of Agriculture, Washington, D.C., 1973
19. **Shepherd, D. R.,** Khapra beetle eradication, *FAO Plant Prot. Bull.*, 5, 75, 1957.
20. **Smith, K. G.,** The study of an insect population living on bagged groundnuts stored in Southern Nigeria with particular reference to the behavious of *Trogoderma granarium, J. West Afr. Sci. Assoc.*, 8, 44, 1963.
21. Trop. Dev. Res. Inst., Insects and Arachnids of Tropical Stored Products: Their Biology and Identification, Training Manual, 1984.
22. U.S. Code of Federal Regulations No. 40, 319.75, 1985.
23. U.S. Department of Agriculture, Khapra Beetle Program, Annual Report, Calendar Year 1955, Agricultural Research Service, Plant Pest Control Branch, 1955.
24. U.S. Department of Agriculture Agricultural Research Service, Khapra Beetle Eradication, Plant Pest Control Programs, Western Region, Plant Pest Control Division, Yearly reports, 1956—1969.
25. Action Plan for Elimination of Khapra Beetle Infestations, Animal and Plant Health Inspection Service, U.S. Department of Agriculture, Washington, D.C., 1984.
26. U.S. Department of Agriculture, Animal and Plant Health Inspection Service, Washington, D.C., Commodity and Warehouse Treatment Manuals, updated.
27. U.S. Department of Agriculture, Animal and Plant Health Inspection Service, Washington, D.C., A Domestic Program manual, 1981.
28. U.S. Department of Agriculture, Animal and Plant Health Inspection Service, Washington, D.C., Cargo Entry Procedures, a Program Manual, 1984.
29. U.S. Department of Agriculture, Animal and Plant Health Inspection Service, Washington, D.C., Automated Pest Interception Data, 1971—1984.
30. **Yana, A.** Note sur l'efficacite d'un insecticide de la serie das organophosphores: le sumition (fenitrothion sur guelgue, souches tuisiennesd'insectes des grains, *Tunisira Doc. Tech. Inst. Nat. Rech. Agron,* 25, 1, 1979.

Chapter 8

PARASITIC WEEDS

Lytton J. Musselman

TABLE OF CONTENTS

I. INTRODUCTION

The purpose of this chapter is to give a short introduction to parasitic weeds, briefly discussing aspects of their biology, hosts, and distribution, which may be of value in quarantine measures. For a more complete treatment of the biology and control of these serious agricultural pests see References 1 through 3. Here the treatment of the subject is restricted to four groups: mistletoes, witchweeds, broomrapes, and dodders.

II. MISTLETOES

Mistletoes are parasitic, epiphytic shrubs in the families Loranthaceae and Viscaceae. Very little attention has been paid to this group from the standpoint of spread and plant quarantine, perhaps because the greatest diversity of mistletoes and mistletoe hosts are in the tropics. Logs for timber would seem to be a possible vehicle for the spread of mistletoe seeds. Since the seeds are very sticky, they could easily remain on logs. The short-lived nature of the seeds, however, may limit their potential for spread. Another source of spread of mistletoe is by infected nursery stock. The only example of this known to this author is that of the European mistletoe *Viscum album* being introduced to California in apple (*Malus sylvestris*) stock by Luther Burbank.[4,5] Considerably more effort is needed to understand how mistletoes are spread. certainly, they should be considered to be a potential threat in several tree crops such as rubber (*Hevea brasiliensis*), cola (*Cola nitida*), citrus (*Citrus* spp.), and teak (*Tectona grandis*), and are also one of the most difficult pests to control in these crops.

III. WITCHWEEDS

Witchweed is the common name usually applied to species of the genus *Striga* of the family Scrophulariaceae. These pathogens should be considered the most serious of all parasitic plants because they attack grains. A great deal has been written on the successful eradication and control program of the American witchweed *Striga asiatica* and will not be repeated here.[6] This section presents a brief discussion of *Striga* biology and possible means of introduction of witchweeds to North America as well as the potential of other species to become pests.

The life cycle of *Striga* species is complex and highly specialized not only for the parasitic habit, but also in terms of adaptation to the semiarid grassland communities where most species are found. The following is a brief summary of the life cycle of most of the agronomically important species. The seeds are shed at the end of the growing season and will not germinate without an after-ripening period which usually corresponds to the time of the next rain. When after-ripening occurs, the seed will respond to moisture if the water is of sufficient duration and at the right temperature. Water and temperature requirements of the parasite are, therefore, closely linked with those of the host. Only then will the seed respond to a signal from a host root which will enable it to germinate and attach to and penetrate that host root.

All species of *Striga* are native to the Old World although two, *S. gesnerioides* and *S. asiatica*, have been introduced to the U.S. In some regions of the world they are the single most important factor in yield reduction and therefore have a great impact on food production and human welfare. Much of the food loss during the recent disastrous droughts in Ethiopia and other African countries is due to the effects of witchweed on sorghum and millet.

Witchweeds have tremendous potential for spread via seeds. The seeds are among the smallest of all dicots belonging to a category known as "dust seeds". A single *Striga* seed is smaller than the period at the end of this sentence. One plant may produce thousands of seeds and the seeds remain viable for many years.

It is not clear how witchweeds have been spread to the New World. The first and most serious infestation was *S. asiatica* in North and South Carolina on maize which was discovered in the mid 1950s. A variety of theories and speculations have surfaced regarding the introduction of *S. asiatica*, but none of them have much scientific merit. Two theories which one often hears repeated are what this author calls the boots and bah-bah theories. The boots theory states that the seeds came on the footwear of U.S. servicemen returning to this country through one of the ports in North or South Carolina. The difficulty with this theory is that this author knows of no area in the *S. asiatica* belt of Africa where there was U.S. military activity in the second World War, and thus favors the bah-bah or wool-carding theory which states that seeds arrived in wool from South African sheep. This has been disputed by at least one eminent South African scientist, J. H. Visser, who provides the information that few sheep are raised in the *S. asiatica* region of South Africa (personal communication). However, it would take only *one* sheep to accumulate enough seed to cause a serious infestation. That only one introduction occurred seems evident from recent research on the genetic diversity of American *S. asiatica*.[7]

To me, a far more intriguing question is why *S. asiatica* has not spread further in this country. There is a vigorous quarantine and eradication program, but the biopotential of this pest seems far greater then the efforts against it. Several factors may mitigate against spread, including the prevailing winds from the west which would hinder movement of seeds eastward; the same is true of the flow of rivers which all flow into the Atlantic Ocean. The witchweed quarantine area is in one of the busiest highway routes in the country, being the main thoroughfare between the cities of the northeast and Florida. Just one flat tire on one rainy day picking up only a few seeds on an automobile headed for the corn belt of the U.S. does not seem totally improbable in light of the millions of vehicles which pass through the area each year.

In short, we are not only ignorant of how *S. asiatica* arrived in this country but also of why it has not spread. To answer such questions would require a study of the movement of seeds within fields. While such studies may be of value in an understanding of the spread of the parasite, they have not been seriously considered.

S. asiatica is the most widespread of all witchweeds.[8] Patterson[9] has studied the ability of the American strain to survive under conditions prevailing in the American corn belt and found that in environmental chamber tests the parasite was able to complete its life cycle under conditions prevailing in Iowa. His observations provide valuable insight into understanding the potential of the species and are supported to some degree by field observations. For example, in regions of Botswana where frosts are of annual occurrence, *S. asiastica* is a frequent problem in grains.[10] On the other hand *S. hermonthica*, the most serious member of the genus which is a widespread problem in Africa, is not found in Botswana. It is a question in the mind of this author if this is due to the climatic regime. Establishing the environmental parameters for survival of witchweeds has a valuable practical application in being able to predict where an infestation is possible.

The situation with *S. gesnerioides* seems clearer. This widespread witchweed attacks dicots and is a serious problem in some parts of Africa on food legumes. It was discovered in the U.S. in January 1979 on old phosphate mine sites in central Florida.[11] Here it attacked only *Indigofera hirsuta* and *Alysicarpus vaginalis*, both native African legumes grown for erosion control and soil enrichment in mine-reclamation projects in Florida. Like *S. asiatica*, *S. gesnerioides* has developed various ''strains'' which are usually very host specific.[12] A series of host tests showed that the Florida strain of *S. gesnerioides* posed little problem to American agriculture.[12] Since the two hosts are known to support the parasite in Africa, it may not be too far-fetched to assume they were introduced with their hosts.

It seems reasonable that the seeds of witchweeds could be looked for as contaminants of grain which they infect. However, in the experience of this author, this seldom happens, simply because the grain is so far removed from the parasite that it is not harvested with the crop. Also, the surfaces of grains are usually smooth so that seeds do not stick to them.

Unfortunately, there are about 28 other species of *Striga* that may be serious problems if introduced into areas where they do not at present occur. Only a few of these are known to cause damage to crops,[8] but the native and presently nonvirulent species must be considered regarding potential damage to agriculture. There is evidence of species once thought benign becoming serious pathogens, as in the case of *S. latericea* on sugarcane in Somalia (C. Parker, personal communication).

IV. BROOMRAPES

In terms of their biology, broomrapes (genus *Orobanche* of the family Orobanchaceae) share many similarities with witchweeds — tiny seeds, obligate parasites, immense fecundity, etc. Differences between the two include the chiefly temperate distribution of *Orobanche* species and the preference for dicot hosts, especially of the families Solanaceae, Asteraceae, Fabaceae, and many others.

There are many native broomrapes in the U.S. and elsewhere which only occasionally are problems. One example would be *O. ludoviciana*, which has been of local importance in tomato in California and tobacco in Ohio.[13]

Broomrapes have repeatedly been introduced into different parts of the world. Perhaps the most widely spread is *O. ramosa*. In the U.S. this species was a serious hindrance to the success of a hemp (*Cannabis sativa*) growing industry in Kentucky from about 1880 to 1905.[14] It survives in this state only in burley tobacco grown for seed,[15] as regular tobacco culture involves soil sterilization. In California, *O. ramosa* became a serious problem in tomatoes. At present, it is extant in only three very limited areas in the U.S. — California, Texas, and Kentucky. In these states the infestations are very small and are carefully monitored. In Sudan, *O. ramosa* was introduced in about 1910, [13] and within the last 25 years has exploded into a major threat to the tomato industry. All commercially produced tomatoes are grown under irrigation which spreads the seeds. A recent volume[17] summarizes our current understanding of *Orobanche* and its control.

V. DODDERS

Unlike *Striga* and *Orobanche* species, dodders (species of *Cuscuta* of the Convolvulaceae) are not root parasites. They are orange or yellow vines which twine about the host and penetrate into the vascular system through modified roots termed *haustoria*. Dodders are found on every continent, and of many of the over 100 described species, 16 are native to North America. In fact, the only parasitic weed native to North America which has been widely spread is *C. campestris* (= *C.* pentagora), now found all over the world.

The life history of dodders is simple, and although seed germination is not host dependent, dodders are obligate parasites which cannot mature without hosts. The seeds of dodders are large compared to those of witchweeds and broomrapes and are produced in less quantity. A dodder seed may be more than 1.0 mm long; usually four are produced in a single capsule. In many species the seeds will germinate immediately after ripening if adequate moisture is present. If not, the seeds become refractory and can remain viable for many years. We do not know how these seeds are scarified in nature. Upon germination, the green seedling elongates and through nastic movements "searches" for a host. Once a suitable host is contacted, the seedling wraps around the stem and penetrates via haustoria.

We know very little regarding host selectivity in dodders, although monocots are rarely, if ever, seriously attacked. Dodders have been widely spread through contaminated crop seed. Unlike witchweeds and broomrapes, dodders produce their seeds intermingled with the seeds of the host. *Cuscuta campestris* is a serious problem on legumes, especially alfalfa (*Medicago sativa*). The seeds of host and parasite are of similiar size and shape. The surface of the dodder seed, however, is minutely rough. When alfalfa seed is cleaned commercially by passing it through felt rollers, the smooth seeds of the alfalfa pass through while those of the dodder stick and are thrown out. There are other examples of the spread of dodders through seeds of other crops.[16]

REFERENCES

1. **Anon.,** *Symposium on Parasitic Weeds,* European Weed Research Council, Wageningen, 1973.
2. **Musselman, L. J. and Worsham, A. D., Eds.,** *Proc. 2nd Int. Symp. Parasitic Weeds,* North Carolina State University, Raleigh, 1979.
3. **Parker, C., Musselman, L. J., Polhill, R. M., and Wilson, A. K.,** *Proc. 3rd Int. Symp. Parasitic Weeds,* International Center for Agricultural Research in the Dry Areas, Aleppo, Syria, 1984.
4. **Scharpf, R. F. and Hawksworth, F. G.,** Luther Burbank introduced European mistletoe into California, *Plant Dis. Rep.,* 60, 740, 1976.
5. **Hawksworth, F. G. and Scharpf, R. F.,** Spread of European mistletoe *(Viscum album)* in California, U.S.A., *Eur. J. Forest Pathol.,* in press.
6. **Eplee, R. E.,** Striga's status as a plant parasite in the United States, *Plant Dis.,* 65, 951, 1981.
7. **Werth, C. R., Riopel, J. L., and Gillespie, N. W.,** Genetic uniformity in an introduced population of witchweed (*Striga asiatica*) in the United States, *Weed Sci.,* 645, 32 1984.
8. **Musselman, L. J., Ed.,** Identification of witchweeds, in *Parasitic Plants in World Agriculture,* Vol. I, CRC Press, Boca Raton, Fla., 1987.
9. **Patterson, D. T.,** Environmental factors affecting witchweed growth and development, in *Parasitic Plants in World Agriculture,* Vol. I, Musselman, L. J., Ed., CRC Press, Boca Raton, Fla., 1987.
10. **Musselman, L. J. and Riches, C. R.,** Witchweeds of Botswana, *Bull. Agric. Res. Botswana,* 3, 1985.
11. **Wunderlin, R. P., Musselman, L. J., and Shuey, A. G.,** *Striga gesnerioides* (Scrophulariaceae) first record of the species in the New World, *Plant Dis. Rep.,* 63, 251, 1979.
12. **Musselman, L. J. and Parker, C.,** Studies on indigo witchweed, the American strain of *Striga gesnerioides* (Scrophulariaceae), *Weed Sci.,* 29, 594, 1981.
13. **Musselman, L. J.,** The biology of *Striga, Orobanche,* and other root parasitic weeds, *Annu. Rev. Phytopathol.,* 18, 463, 1980.
14. **Garman, H.,** The broomrapes, *Ky. Agric. Exp. Stn. Bull.* 105, 1903.
15. **Musselman, L. J.,** *Orobanche ramosa* in Kentucky, *Castanea,* 50, 57, 1985.
16. **Musselman, L. J.,** Parasitic weeds of arable land, in *Biology and Ecology of Weeds,* Holzner, W. and Numata, M., Eds., W. Junk, The Hague, 1982, 169.
17. **Ter Borg, S. J., Ed.,** *Proc. Orobanche Workshop,* Agricultural University, Wageningen, 1986.

Chapter 9

SOUTH AMERICAN LEAF BLIGHT OF RUBBER

S. M. Kang

TABLE OF CONTENTS

I. BACKGROUND

The commercial potential of *Hevea* rubber was recognized when the vulcanization process was developed by Charles Goodyear in 1838.[1] The demand for rubber was accelerated with the invention of the pneumatic tire in 1888 followed by the advent of the automobile age in the 1920s.[1] The Amazon basin,* the natural home of *Hevea* species, became the focus of world's attention in the search for wild *Hevea* rubber trees. *Hevea brasiliensis* Muell.-Arg., the most productive of all the *Hevea* species was selected for eventual commercial cultivation. The cultivated rubber in the Far East is derived from the 70,000 seeds collected in 1876 by Henry A. Wickham from the banks of the Tapojos River in the Amazon.[14,19,24] Seedlings obtained from this collection were established in Sri Lanka (1919 plants), Singapore (22 plants), and Bogor, Indonesia (18 plants), from which the cultivated rubber in Asia and Africa originated.[16,23]

Prior to 1912, Brazil produced 98% of the world's rubber which came entirely from the wild trees.[16] Today the situation has reversed. Asia and Africa produce 99.5% of the world's production which is derived from about 7.5 million ha of cultivated plants (Figure 1, Table 1).[13] The obliteration of the Amazon as the world's leading producer of rubber was the result of its inability to cultivate the crop on a commercial scale. The failure of all early attempts to domesticate *Hevea* and to cultivate the oriental clones has been attributed to the devastating effects of South American Leaf Blight (SALB), a most deadly disease of rubber. SALB, almost a century after its appearance in the Amazon, is confined only to tropical America. It has been, and still is, the biggest constraint to the development of viable rubber industries in that region. It has been predicted since the beginning of the century that the rubber industry in Asia and Africa would suffer a similar fate should SALB make its way to these regions.

II. DESCRIPTION OF SALB

A. The Pathogen

SALB is caused by a fungus, *Microcyclus ulei* (P. Henn.) Arx. It belongs to the family Dothideales, order Ascomycete, also known as *Melanopsammopsis heveae* Stahel, *M. ulei* (P. Henn.) Stahel, *Dothidella ulei* P. Henn., *Fusicladium macrosporum* Kuyper = conididial stage, *Aposphaeria ulei* P. Henn. = pycnidial stage.

The conidia, abundantly produced and longer surviving than the other spore types, are important in the distant spread of SALB.[5] The ascospores play an important role in the survival of the fungus from one season to the next.[5] Unlike the conidia and the ascospores, the pycnospores do not initiate an infection.[5]

M. ulei is known to exist in nine physiologic races and two strains which give rise to variability in the susceptibility of *Hevea* clones to SALB.[3,8]

The viability of the conidia and ascospores is affected by temperature and moisture.[5] The survival period of conidia shortens with increasing levels of relative humidity. At 85 to 95% relative humidity (r.h.) conidia attached to leaf lesions or detached conidia survive up to 3 weeks, which increases to 4 weeks at 65% r.h.[5,10] Under normal ambient temperature and r.h., the ascospores within the perithecia survive for about 3 weeks and for about 2 weeks after release from the perithecia.[10] The optimum temperature for growth, sporulation, and infection of *M. ulei* is 24°C.[8]

In infected rubber holdings ascospores are present in the air throughout the year with peak

* The Amazon basin refers to "Amazonia" which is defined in Webster's Geographical Dictionary as "the regions about the Amazon river in South America, including the greater part of Brazil, and parts of bordering countries, especially Colombia, Ecuador, Peru and Bolivia." It shall be referred to as "the Amazon" in the rest of the text.

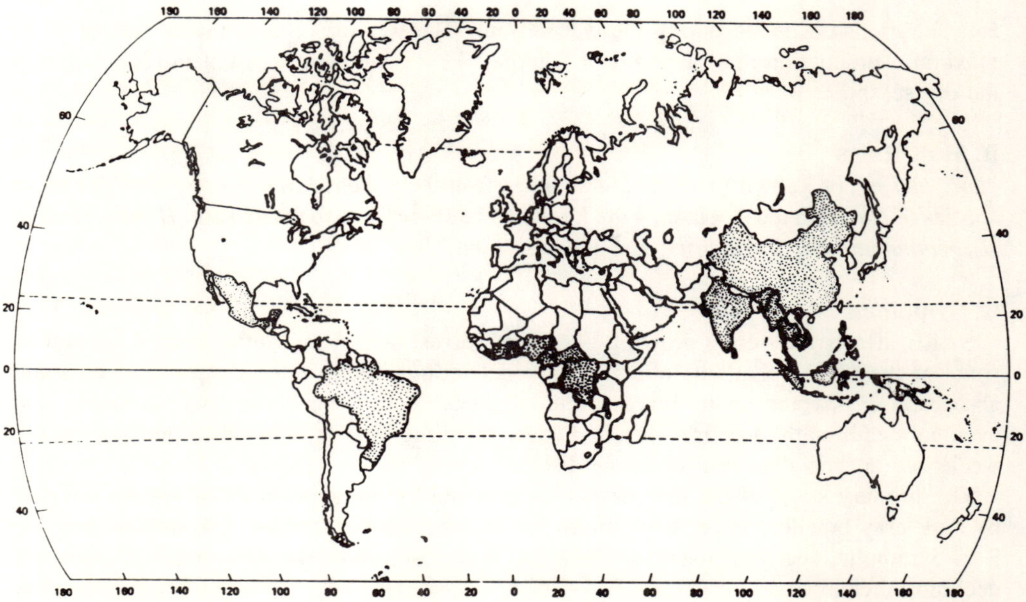

FIGURE 1. Countries with commercial rubber plantations (shown as dotted areas). Asia: Burma, India, Indonesia, Kampuchea, Malaysia, Papua New Guinea, Philippines, Peoples' Republic of China, Sri Lanka, Thailand, Vietnam. Tropical America: Brazil, Mexico. Africa: Central African Republic, Ghana, Ivory Coast, Liberia, Nigeria, United Republic of Cameroon, Zaire.

Table 1
AREA UNDER PLANTATION RUBBER
(1000 ha)[a]

Country/region	Total area	Year
Asia	7101.2	—
Indonesia	2327.5	1977
Malaysia	2006.9	1982
Thailand	518.0	1979
China	453.0	1982
India	241.5	1980
Sri Lanka	227.6	1975
Vietnam	85.6	1981
Burma	62.5	1964
Philippines	53.9	1981
Papua New Guinea	13.7	1970
Kampuchea	12.0	1982
Africa	488.1	—
Nigeria	185.0	
Liberia	119.8	1973
Zaire	93.0	1959
Ivory Coast	41.5	1982
Cameroon U.R.	36.0	1982
Ghana	11.6	1965
Central African Republic	1.2	1965
Tropical America	28.2	—
Brazil	20.0	1965
Mexico	8.2	1982

[a] Data obtained from the IRSG, *Rubber Stat. Bull.*, 39(2), 1984, 44. With permission.

concentrations occurring during the wet season.[6] The wet season also marks the period of maximum production and dispersal of conidia.[6] Few conidia are present in the air during the dry season.

B. Host Range

M. ulei is not known to affect plants other than the genus *Hevea*. Out of the 11 known species of *Hevea* in the Amazon, 4 are known to be susceptible to the disease, *H. brasiliensis, H. benthamina*, *H. guianensis*, and *H. spruceana*.[9,16]

C. Symptoms

SALB affects the young growing tissues of leaves, stems, and inflorescence as well as fruits of the *Hevea* plant. Its effect is most deleterious during refoliation when maximum abscission of affected young leaves occur. Repeated defoliation causes twig die-back, and eventual death of the tree may result. Leaves from the time of emergence until they are 2 weeks old are most susceptible to the disease.[8]

The primary stage of the disease is characterized by the appearance of lesions covered by dark gray powdery masses of conidia on the abaxial leaf surface. Sporulation lasts for 2 to 3 months, reaching its peak when the leaves are about 14 days old and gradually declining thereafter.

If the diseased leaves do not absciss, the conidial lesions develop into the secondary stage, which is characterized by the presence of spherical black bodies, the pycnidia, on the adaxial surface. The third stage of the disease is the replacement of pycnidia by raised black masses of perithecia on leaves 3 months old and onwards. Although much less common, infected petioles, inflorescence, young stems, and green fruits exhibit similar succession of spore stages.[9]

III. GEOGRAPHICAL DISTRIBUTION

The disease is confined to tropical mainland America and the adjacent islands covering an area approximately bounded by the Tropic of Capricorn, the Tropic of Cancer, and the meridians of longitude 30 and 120°W (Figure 2).[11] Countries within this region where the disease has been reported are Brazil, Bolivia, Peru, Colombia, Costa Rica, Panama, French Guiana, Guatemala, Guyana, Honduras, Mexico, Nicaragua, Trinidad, Venezuela, Haiti, and Surinam.[14,17]

IV. POSSIBLE CARRIERS OF *M. ulei* AND THEIR SIGNIFICANCE

SALB was endemic to the Amazon Basin, the area where species of *Hevea* are indigenous. The disease was disseminated through the *Hevea* planting materials to other parts of tropical America during the early part of the century. Similar occurrence has not taken place with planting materials transferred to the Far East and Africa.

Besides the contaminated or infected planting materials of *Hevea*, it has been speculated that the dissemination of SALB within the Americas could have been brought about through long-distance transport of spore-laden clouds and their deposition by rain.[17] The uredospores of coffee rust (*Hemileia vastatrix* Berk and Br.) was believed to have been blown by air currents across the 1000-km span of the Atlantic ocean from West Africa to South America.[19] However, the situation in reverse in the movement of the *M. ulei* conidia or the ascospores is unlikely because of the opposite wind direction.[17,19] Any dissemination of SALB across the Atlantic or the Pacific Ocean would require man to act as agent of transfer.

The role of man in bringing about the dispersal of disease-causing microorganisms and other pests became significant only after the introduction of the aircraft and other modes of

FIGURE 2. Countries (shown as dotted areas) in tropical America where South American leaf blight of rubber (*Hevea brasiliensis*) caused by the fungus *Microcyclus ulei*, is known to occur: Bolivia, Brazil, Colombia, Costa Rica, Ecuador, French Guiana, Guatamala, Guyana, Haiti, Honduras, Mexico, Nicaragua, Peru, Surinam, Trinidad and Tobago, and Venezuela.

long-distance transport. The long-distance transfer of pests is brought about through materials moved by man or by man being the vector. The *Hevea* plant, other plant species, plant products, nonplant articles, and insects originating from rubber growing areas in tropical America if moved by man may result in the transfer of the SALB pathogen. Man becomes the vector when he visits a diseased nursery, as is evident from the recovery of conidia from the nails and clothes of a person after such a visit in Trinidad.[10]

The *Hevea* plant where *M. ulei* can be present as contaminants or as systemic infections on the young leaves, twigs, pods, seeds, or budwood is undoubtedly a carrier of the highest risk. The chances of other carriers acquiring contamination by the conidia or ascospores of *M. ulei* is dependent on several factors, such as the severity of the disease, time of the year, weather conditions, proximity to infected plants, and the manner of handling of the materials. The chances of contamination would be greatest during the wet months of the year, especially at the time of refoliation when both the conidia and the ascospores in the atmosphere are at the highest level.

In the absence of information on the possible presence of airborne conidia or ascospores in urban areas or other localities away from cultivated or wild rubber in any of the countries in tropical America, the significance of non-*Hevea* plants and other materials originating from these areas as possible carriers of the SALB pathogen is difficult to establish. However, if it can be assumed that the aerial spores are not present, or if present their densities are negligible, it can be concluded that all non-*Hevea* materials originating from these areas are of little or no quarantine significance in relation to SALB dissemination.

The importance of man and other materials as carriers of SALB is dependent upon the rapidity upon which they can be conveyed to rubber-growing areas presently free from SALB. A journey exceeding 4 weeks under normal circumstances would be fatal to both the conidia and ascospores. With the exception of an infected *Hevea* plant, all other carriers would not constitute a risk if they are transported by sea route to the Far East, a journey which takes about 4 weeks. However, transportation by sea from tropical America to rubber-growing countries in Africa do not remove the risk of pathogen transfer since the journey could be accomplished within a much shorter time. Viable conidia and ascospores can be transferred by air to Africa and Asia within a matter of hours and few days, respectively.

V. CHANCES OF ESTABLISHMENT

It has been speculated since the beginning of the century that SALB, if introduced into the Far East, would establish and assume epidemic proportions. Factors such as high rainfall without marked dry season, continuous presence of young leaves, and the susceptibility of all clonal materials to the disease have been cited as factors favoring the development of the disease.[10]

Before infection could commence, the pathogen from the carriers must be deposited on the host tissue at its susceptible stage. What are the chances that such an event would occur? The pathogen on the *Hevea* seed or the budwood stands the likeliest chance of encountering a ready host, that is, the emerging seedlings or the stock plants. The possibility of such infection occurring increases if germination or budding of the imported material is carried out within the *Hevea* nursery.

The chances of spores borne on other carriers landing on a susceptible host is dependent on the proximity of the rubber plantation to the ports of entry as well as the final destination of the carriers. Due to the lethal effects of atmospheric ultraviolet (UV) irradiation on the conidia and ascospores, the chances of these spores being carried alive decreases the further away the host plants are from the point of introduction. The disease would stand a greater chance of being established if the pathogen is introduced during the period of refoliation.

VI. SUGGESTED SOLUTIONS

SALB, a disease whose spread to Asia and Africa would only be possible through the activities of man, is best excluded from these regions by controlling those activities that are likely to bring about its introduction. The first step towards this control would be the promulgation of plant quarantine legislations providing the government the powers to regulate the entry of all categories of carriers originating from the rubber-growing areas of tropical America. Regulations currently enforced in most of the Asian countries planted with rubber restrict the entry of all plant species from all the countries in tropical America irrespective of its specific place of origin. If it is established that the chances of spore contamination on non-*Hevea* carriers is negligible in urban areas, or areas away from wild or cultivated rubber, imposition of restrictions on the movement of these carriers would be unnecessary.

Besides preventing introductions of the disease, the quarantine regulations should provide for remedial measures for eradicating it in the event of its inadvertant introduction.

Sole dependability on legislative measures to protect the rubber industry against a disease as virulent as SALB would be risky. Research programs geared towards the development of clones resistant to the disease as well as development of economical and effective methods of control should be vigorously pursued. Similar attention should be given towards educating target groups, especially those connected with the rubber industry, on the importance of the disease, its possible mode of transfer, and the need for preventive measures.

A. Prevention of Entry

Regulatory measures and safeguards that may be prescribed for each category of carriers are described below. Unless indicated otherwise, the mode of transfer of these carriers is by air.

1. All Carriers except Hevea sp. by Sea

As the spores of *M. ulei* would not survive by the sea journey to Asia, quarantine measures on non-*Hevea* materials on entry would be unnecessary. In as far as exclusion of SALB is concerned, this category of materials should be freely enterable into Asia, but subject to normal control when entering rubber-producing countries in Africa.

2. Man

Quarantine actions in the form of interception, questioning, and possible decontamination should be imposed on all incoming travelers from tropical America. This requirement would apply also to travelers entering West Africa by sea. Interception of target persons is most effectively achieved by imposing a requirement for mandatory declaration by all incoming travelers of their presence in tropical America. If upon questioning there is reasonable cause to believe that the traveler may be carrying the SALB pathogen, he should be subjected to decontamination procedures at the port of entry, such as taking a complete shower. All used clothing should either be soaked in soap solution or subjected to heat treatment (75°C for 30 min) in a commercial clothes dryer. Suitable facilities must be provided at the ports of entry for this purpose.

An alternative procedure, which is less likely to give rise to unpleasantness to travelers, would be to impose a mandatory stopover in a third, non-rubber-producing country for decontamination. However, effective enforcement of this requirement depends very much on the sincerity of the traveler.

3. Nonplant Materials

Included in this category of carriers are mens personal belongings such as luggage, clothings, shoes, cameras, and parcels. Treatments which may be applied to these materials include 15 min exposure to UV irradiation (253.7 nm), 30 min exposure to moist heat at 55°C or dry heat at 75°C, soaking in soap solution at 40 mg/ℓ, or fumigation with formalin at the rate of 100 mℓ/m^3 for 30 min.[5,8] In cases where existing treatments are not suitable, an acceptable alternative measure would be to quarantine the articles in a spore-proof enclosure at the first port of entry for an adequate period in order to render the conidia inviable. The quarantine period estimated from the day the material leaves the country of origin in the American tropics should not be less than 4 weeks.

4. Insects

In the absence of an established method for disinfecting live adults, field-collected adults of beneficial organisms such as pollinators, honey bees, and biocontrol agents should be prohibited from entry. Any insect introductions from rubber-growing areas of tropical America should be made from laboratory-bred cultures, preferably at the pupal stage. Prior to dispatch, the pupae should be surface sterilized by dipping in 1% chlorothalonil or 0.1% benomyl. To eliminate possible conidial contaminants on the packaging materials, the cultures should be dispatched to an intermediate quarantine station for repacking in fresh packing materials.

5. Non-Hevea Plants

Disease risk can be reduced by subjecting the plants or parts of the plants to fungicidal treatments in the country of origin, and at the intermediate quarantine station followed by change of packing materials. On importation, the materials should be quarantined for at least 1 month in a postentry quarantine station situated away from rubber plantations. Treatment methods developed for the transfer of rubber seeds and budwood should be used for non-*Hevea* plants.

6. Hevea sp.

Risk of SALB introduction through the movement of *Hevea* planting materials may be minimized through the application of one or more of the following measures:

First, restrict introductions of *Hevea* planting materials for research purposes only.

Second, prohibit direct introductions of seeds or budwood materials. Seeds to be introduced as seedling should be germinated in an intermediate quarantine station situated outside tropical America, Africa, or Asia. Similarly, budwood should be budded and established in an intermediate quarantine station. Seedlings or budwood materials obtained from these stations should only be permitted for import after they are certified free from SALB by a competent plant pathologist.

Third, permit direct introductions of budwood or seeds on conditions that the materials are treated in the country of origin, the intermediate quarantine station, and on entry in the country of import. After treatment, at an intermediate quarantine station, the materials must be repacked in fresh packing materials before dispatch to the country of destination. The budwood or seeds are to be germinated or grafted and kept under quarantine for a period of 2 to 6 months in a spore-proof quarantine glasshouse. The following treatment procedures are suggested:

1. At the country of origin: dip budwood in 0.2% alcoholic mercuric chloride for 30 sec and seeds in formalin (1:200) for 15 min.
2. At the intermediate quarantine station: dip both budwood and seeds in 1% chlorothalonil for 15 min.
3. On entry in the country of import: dip budwood in 0.2% alcoholic mercuric chloride for 30 sec and seeds in 0.1% mercuric chloride followed by dusting with benomyl-TMTD.

B. Eradication Measures

If there is to be any hope at all of eradicating a disease as virulent as SALB, there must be an immediate notification of the outbreak, a readily-prepared method of eradication, with all equipment and materials instantly available, and an effective legislation to enforce eradication wherever the disease may appear.[12] To be in the state of preparedness, each country should (1) establish a system of detection and notification for SALB; (2) ensure the availability of expertise in providing rapid diagnosis of the disease; (3) identify and assign duties and responsibilities of persons and agencies responsible for commencing the eradication measure; (4) establish an emergency fund for SALB eradication; (5) keep stock of defoliants and fungicides recommended for SALB eradication; (6) ensure the availability of aircraft, spray attachments, and fogging machines for aerial or ground application of defoliants and fungicides; (7) prepare contingency plans incorporating all aspects of the detection and eradication procedure for the information of all concerned; and (8) carry out mock exercise to test the soundness of the contingency plan with respect to logistics, spray techniques, and effectiveness of the defoliants.

Based on the scheme developed by the Rubber Research Institute of Malaysia, the eradication procedure would involve (1) the establishment of a control zone consisting of the infected area and a guard belt of 500 m around it, (2) isolation of the control zone, (3) application of defoliants from the aircraft or mobile fogging machines, (4) spraying with fungicides following defoliant application, (5) inspection of fallen leaves in the guard belt for presence of the leaf blight lesions, (6) inspection of young leaves on trees outside the control zone over a distance of 1 to 2 km for SALB symptoms, and (g) repeat application of defoliants and fungicides at monthly intervals for a period of 6 months.

Suitable defoliants that have been developed for aerial and ground application are *n*-butyl ester of 2,4,5-T in diesoline or the less hazardous alternative, trichlorpyr.[12,23]

Fungicides recommended for supplementary application are thiophanate methyl, benomyl, chlorothalonil, triadimefon, and mancozeb.[7,12,23]

C. Regional Cooperation

SALB, restricted in its distribution to tropical America and separated from the *Hevea* rubber in Africa and Asia by the Atlantic and the Pacific Ocean, is a good example of a disease where legislative measure is necessary to prevent its long-distance transfer by man. Judging by the distribution of cultivated rubber in Asia and Africa (Figure 1) and the highly virulent nature of SALB, it is a certainty that the disease once introduced would spread to the whole region in a matter of time. Singular action by any one country in excluding the disease would be in vain. The threat of SALB is a clear case where international cooperation is required.

The government of Malaysia, concerned over the threat of the disease on its rubber industry, initiated the move in 1948 to secure cooperation from neighboring countries to keep the region free from the disease.[12] Malaysia's initiative led to the promulgation of the FAO Plant Protection Agreement for the Asia and Pacific Region which came into force in 1956.[11] All of the 11 rubber-producing countries in the Asia and Pacific Region, in addition to 13 other non-rubber-growing countries, are signatory to the agreement.

Over the years as more non-rubber-producing countries became signatory to the agreement, lesser emphasis was placed on SALB.[21] The decreased effectiveness of this agreement led to the promulgation of yet another agreement, called the Association of Natural Rubber Producing Countries (ANRPC) Agreement on South American Leaf Blight which was signed in 1984 by six of the seven member countries.[2] Intercountry cooperation among the rubber-producing countries in Africa comes under the auspice of the Inter-African Phytosanitary Council (IAPSC) established in 1956.[15]

Securing the cooperation of countries within the same geographical area to adopt and to implement common quarantine legislations to prevent the introduction into the region of an exotic pest is not an easy task to achieve, especially so when dissimilar economic implications of such an introduction exists. However, in the Asian region, the first hurdle of obtaining the agreements of the governments of all the 11 rubber-producing countries to adopt a common legislation pertaining to the movement of planting materials from tropical America was overcome when all the countries ratified the Plant Protection Agreement for the Asia and Pacific Region. The next important step is the determination of modalities for effective implementation of the Agreements by all parties concerned. Taking into consideration the existence of established procedures in the implementation of the Asia and Pacific Plant Protection Agreement and the ANRPC Agreement on SALB, further measures should be explored through these two agreements to secure more effective regional cooperation in protecting the rubber industry in the region from SALB. It is suggested that the executive bodies of the agreements, that is the Asia and Pacific Plant Protection Commission and the ANRPC Technical Committee on SALB, work hand in hand to (1) streamline the agreements with respect to prevention of entry and eradication of SALB, (2) formulate effective and practical quarantine measures for implementation by all the 11 rubber-producing countries as well as other signatory countries strategically located within the region, (3) identify problems and requirements of member countries in implementing the agreements as well as the quarantine measures and to propose solutions to the problems, and (4) develop modalities for ensuring effective implementation of the recommended quarantine measures by member countries.

VII. ACKNOWLEDGMENT

The author is indebted to Dato' Abu Bakar Mahmud, the Director-General of Agriculture, Malaysia for his encouragement and approval to produce this paper, to Dr. K. H. Chee of the Rubber Research Institute, Malaysia, and to Dr. J. F. Karpati of the U.S. Department of Agriculture, Animal and Plant Health Inspection Service for their invaluable suggestions and reviews of the manuscript. The views expressed in this paper are the author's own. They do not reflect the views of the Department of Agriculture, Malaysia.

REFERENCES

1. **Alvim, P. de T.,** A perspective appraisal of perennial crops in the Amazon basin, *Inerciencia,* 6, 139, 1981.
2. ANRPC Agreement on South American Leaf Blight (SALB), The Association of Natural Rubber Producing Countries, Jakarta, 1984.
3. **Chee, K. H. and Wastie, R. L.,** The status and future prospects of rubber diseases in Tropical America, *Rev. Plant Pathol.,* 59, 541, 1980.
4. **Chee, K. H.,** Micro-Organisms Associated with Rubber(*Hevea brasiliensis* Muell.-Arg.), Agric. Ser. Rep. 4, The Rubber Research Institute of Malays, Kuala Lumpur, 1976, 78.
5. **Chee, K. H.,** Factors affecting discharge, germination and viability of spores of *Microcyclus ulei, Trans. Br. Mycol. Soc.,* 66, 499, 1976.
6. **Chee, K. H.,** South American Leaf Blight of *Hevea brasiliensis:* spore dispersal of *Microcyclus ulei, Ann. Appl. Biol.,* 84, 147, 1976.
7. **Chee, K. H.,** Improved control of South American Leaf Blight of *Hevea* in Brazil, *J. Plant Prot. Trop.,* 1, 1, 1984.
8. **Chee, K. H.,** personal communication, 1985.
9. **Chee, K. H.,** The biology of *Microcyclus ulei,* in *Proc. IRRDB Workshop on Collection of Germplasm from South America,* Rubber Research Institute of Malaysia, Kuala Lumpur, 1979.
10. **Chee, K. H.,** The suitability of environmental conditions in Asia for the spread of South American Leaf Blight of *Hevea* rubber, *Planter,* 56, 445, 1980.
11. Food and Agricultural Organization, Plant Protection Agreement for the Asia and Pacific Region, Rome, 1956.
12. **Hilton R. N.,** South American Leaf Blight, a review of the literature relating to its depredations in South America, its threat to the Far East, and the methods available for its control, *J. Rubber Res. Inst. Malaya,* 14, 287, 1955.
13. International Rubber Study Group, Area under plantation rubber (000 hectares), *Rubber Stat. Bull.,* 39, 44, 1984.

14. **Kang, S. M. and Zam, A. K.,** Quarantine aspects of the introduction into Malaysia of an oil palm insect pollinator, in *Proc. Int. Conf. Plant Prot. Tropics,* Kuala Lumpur, Malaysia, 1982, 615.
15. **Karpati, J. F.,** Plant quarantine on a global basis and the Int. Plant Prot. Convention (IPPC), presented at the 13th session of the Asia and Pacific Plant Protection Comm., Manila, April 18 to 23, 1983.
16. **Lim, T. M., Gasparotto, Silva, H. M., Trindade, D. R., Castro, A. M. G., and Souza, A. R.,** Diseases in relation to rubber cultivation, presented at the ANRPC/IRRDB Workshop on SALB, CEPLAC, Itabuna, Brazil, August 27 to September 1, 1984.
17. **Liyanage, A. de S.,** Long distance transport and deposition of spores of *Microcyclus ulei* in Tropical America — possibility, presented at the 4th Meet. ANRPC Technical Committee on SALB, Chiangmai, Thailand, January 14 to 16, 1982.
18. **Paranjothy, K., Lim, T. M., Othman, R., Hashim I., Abdul Aziz, S.A.K., and Leong, S. K.,** Transfer of *Hevea* seed germplasm, in *Proc. IRRDB Workshop on Collection of Germplasm from South America,* Rubber Research Institute of Malaysia, Kuala Lumpur, October 9 to 11, 1979.
19. **Rao, B. S.,** South American Leaf Blight: chances of introduction and likely behaviour in Asia, *J. Rubber Res. Inst.,* 50, 218, 1973.
20. **Reddy, D. B.,** Plant Protection Committee for the South East Asia and Pacific region — 25 years of service 1956 — 1980, Plant Prot. in Asia and Pacific — an overview, presented at the 12th session Plant Protection Committee for the South East Asia and Pacific Region, Chiangmai, Thailand, October 27 to November 3, 1980.
21. Report of the 1st Meet. of the ANRPC Technical Committee on SALB, Colombo, January 6 to 8, 1975.
22. Report of the 13th Session of the Asia and Pacific Plant Prot. Comm., Manila, April 18 to 23, 1983, FAO, United Nations.
23. Rubber Res. Inst. of Malays. and Dept. of Agric. Malays., Contingency Plan for the Eradication and Control of Dangerous Pest (Plant Quarantine Act 1976), South American Leaf Blight (*Microcyclus ulei* (P. Henn.) Arx.) 1985. (Unpublished)
24. **Wycherley, P. R.,** Introduction of *Hevea* to the Orient, *Planter,* 44, 1, 1968.

Chapter 10

FIRE BLIGHT OF POME FRUITS

T. van der Zwet

TABLE OF CONTENTS

I. INTRODUCTION

In 1880, fire blight of pear and apple became the first plant disease attributed to a bacterium (*Erwinia amylovora* [Burr.] Winsl. et al.) when Burrill [1] made his important announcement at the University of Illinois. He also called the disease the "anthrax" of fruit trees, since he considered the organism to resemble the anthrax bacillus or the "vibrion butyrique of Pasteur and the *Bacillus amylobacter* of van Tieghem." It was not until 1882 that Burrill published the description of the bacterial species and named it *Micrococcus amylovorus*.[12] The generic name of the blight pathogen was successively changed to *Bacillus* and *Bacterium*, and in 1920 Winslow et al.[67] established the genus *Erwinia*, named after the famous bacteriologist Erwin F. Smith. Today, *Erwinia amylovora* serves as the type species of the genus *Erwinia*.

II. HOST RANGE

Besides apple (*Malus*) and pear (*Pyrus*), fire blight has been reported to occur on about 150 species in 37 genera of the family Rosaceae.[63] These are as follows:

Genus	No. species	Genus	No. species
Amelanchier	3	*Kerria*	1
Aronia	2	*Mespilus*	1
Aruncus	1	*Osteomeles*	1
Chaenomeles	2	*Peraphyllum*	1
Cotoneaster	62	*Photinia*	3
Cowania	1	*Physocarpus*	1
Crataegomespilus	1	*Potentilla*	1
Crataegus	12	*Prinsepia*	1
Cydonia	2	*Prunus*	16
Dichotomanthes	1	*Pyracantha*	8
Docynia	1	*Raphiolepis*	2
Dryas	1	*Rhodotypos*	1
Eriobotrya	1	*Rosa*	4
Exochorda	1	*Rubus*	2
Fragaria	2	*Sorbaria*	1
Geum	1	*Sorbus*	6
Heteromeles	1	*Spiraea*	3
Holodiscus	1	*Stranvaesia*	1
Kageneckia	1		

Of these genera, those that are most susceptible to fire blight and most important economically are *Cotoneaster*, *Crataegus* (hawthorn), *Cydonia* (quince), *Pyracantha*, and *Sorbus* (mountainash). Of these five genera, the many species and varieties of *Cotoneaster*, especially those with large, long leaves are the most important in the nursery trade. All other genera are of very minor economic importance.

In the genus *Pyrus*, 34 species have been reported as host for the fire blight pathogen.[63] Of these, only *P. communis* (European desert pear), *P. nivalis* (perry pear), and a few of the oriental species (*P. pyrifolia, P. betulaefolia*, and *P. ussuriensis*), producing the Oriental pear or apple pear, are of economic importance. The degree of fire blight resistance varies considerably between clones and selections within a species.[64] Also, in species hybrids, resistance varies between selections.

Similarly, in the genus *Malus*, nearly all species, which include the numerous crab apples, are susceptible to fire blight. However, clonal selections of *M.* × *atrosanguinia, M. fusca*, and *M. surgenti* have been reported as quite resistant to artificial shoot inoculation.[15] More significant for apple than for pear is the degree of blight susceptibility of certain Malling apple rootstocks. Of the most important commercially used rootstocks, M-7, MM-109, and

MM-111 are the most resistant, whereas M-9 and M-26 are the most susceptible to fire blight. In addition, very susceptible scion cultivars such as 'York Imperial', 'Jonathan', and 'Rome Beauty' propagated on these two susceptible rootstocks produce severe fire blight which often results in death of the trees. For pears, *P. ussuriensis* and *P. calleryana* as well as certain Old Home × Farmingdale (Oregon) selections have been used as rootstocks or interstems to enhance blight resistance in the tree structure.

III. SYMPTOMS AND DAMAGE

The name ''fire blight'', coined about 1817, describes the most characteristic symptom of this disease: a blackening of twigs, flowers, and foliage as though they had been swept by fire. Depending on the tree or plant part affected, many names such as blossom, twig, fruit, trunk, and collar blight have frequently been used.

Blossom blight is usually the first symptom observed in the spring, about 5 to 15 days after petal fall, depending upon temperatures. At first, blossoms appear water-soaked, then wilt and turn brownish to black. Progress of infection into the peduncle and new shoots on the suppporting branch depends on the degree of host resistance. Blighted shoots and leaves usually appear light to dark brown on apples and dark brown to black on pears. When progress of blight infection ceases, a canker is formed on the branch, scaffold limb, or trunk of the tree. The canker margin is usually marked by a definite crevice or crack.

Symptoms of twig and fruit blight can be observed throughout the growing season. The most characteristic of these is the so-called shepherd's crook, formed by the bending of the affected shoot tip. In the early stages of infection, bacterial ooze droplets can often be observed (Plate 1A).* Fruit blight is most common following a severe rain storm. Occasionally, and usually early in the morning in areas with low relative humidities, bacterial strands can be seen on the surface of infected fruit or strung like cobwebs between leaf petioles and other tree parts (Plate 1C). Trunk and collar blight occur either as extensions of blight in the main framework of the tree, directly through growth cracks in the trunk, or through infections in suckers from roots or the base of the tree trunk.

Damage to pears, apples, or other rosaceous plants may vary from the partial or total loss of the current year's crop to the death of an entire tree or shrub. Under severe blight conditions, entire blocks of young trees may have to be removed and replaced. In older orchards, severely infected trees can be brought back to production within 2 to 3 years by careful pruning, retraining, and spraying. Therefore, annual losses are erratic and very difficult to assess. Significant losses of young nursery trees due to blight have been reported.[52]

IV. LIFE CYCLE

The life cycle of the fire blight organism (*E. amylovora*) occurs almost entirely in association with the living host, without any alternate host plant. In an infected tree or shrub, the bacterium can overwinter inside the tissue (canker) and express itself in the spring in the form of ooze (Plate 1B). This bacterial ooze can be disseminated by means of insects (flies) or splashing rain. At bloom time, bacteria from this source can also be spread to open blossoms. It is unknown how effective bacterial strands are in the dissemination of fire blight. However, *E. amylovora* has been demonstrated to survive for at least 2 hr on small airborne particles at relative humidities between 40 and 90%.[51] Larger bacterial tendrils (Plate 1D) have occasionally been observed on the surface of pear fruit in the state of Washington. In addition to cankers, the blight pathogen reportedly can overwinter in diseased fruit[3,32] and occasionally in twigs and branches left on the ground after pruning.[30]

* Plate 1 appears after page 152.

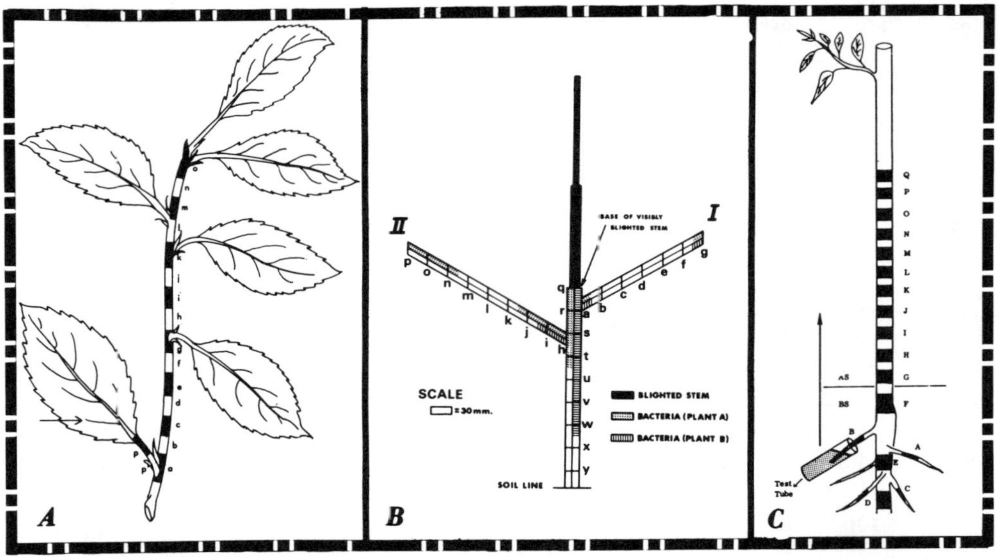

FIGURE 1. Schematic diagrams showing movement of *Erwinia amylovora* in apparently healthy pear and apple tissues. (A) From leaf surface through petiole and stem to shoot apex;[41] (B) from base of blighted shoot into side shoot developed from axillary buds after visible blight ceased to spread to main stem;[40] (C) from root tip through tem into upper part of shoot.[34]

A third form of the bacterium in the life cycle of *E. amylovora* is the epiphytic population on various plant organs. This has best been documented in California on the surface of pistils in opened pear flowers (Plate 1E). Such epiphytic bacteria have also been recovered from leaves,[61] buds,[4,7] budwood,[60] and fruit[8,61] of apparently healthy trees, as well as from certain insect species.[56,63] From Canada, however, Dueck[16] reported the absence of epiphytic *E. amylovora* from symptomless 'Wealthy' apples collected from naturally infected trees. Ever since fire blight reached England and the European continent, similar reports of epiphytic populations of *E. amylovora* were published.[9,45] However, in vitro studies in The Netherlands demonstrated survival of the bacterium between 6 hr and 2 days, when cells on glass slides were exposed to solar radiation, relative humidities over 75%, or storage at temperatures of 20 °C or higher.[42]

A fourth form of *E. amylovora* is that of endophytic bacteria residing inside visibly symptomless tissues. These bacteria reportedly can migrate both in the phloem[34] and the xylem[54] (Plate 1F). Hydathodes and glandular trichomes on leaves as well as lenticels on stems have been reported to serve as natural openings for infection.[41] Extremely little information is available regarding the presence and movement of internal bacteria within pome fruit tissues. Recent attempts to isolate the blight organism from fruit core samples of 'Red Delicious' apples in West Virginia, Washington, and Ontario were negative.[65]

V. VASCULAR MOVEMENT

Figure 1 shows various ways of the movement of *E. amylovora* in symptomless tissues following artificial inoculation. In (A), a drop of inoculum was spread over 6 mm of the serrated margin of the uninjured upper surface of the sixth leaf from the shoot apex.[41] Histological examination revealed the pathogen to be present in the entire shoot length in about 7 hr. In (B), virulent *E. amylovora* bacteria were readily isolated from internal tissues of apparently healthy shoots (developed from axillary buds) on artificially inoculated apple and pear trees in the greenhouse.[40] In (C), the blight organism spread 40 cm into the symptomless aerial part of a shoot following immersion of a decapitated root tip in a culture tube containing an aqueous suspension of *E. amylovora*.[34] Subsequent inoculation studies showed that a virulent strain moved rapidly in xylem vessels at a speed of 34 mm/hr.[53] Electron microscopy proved conclusively that systemic migration occurs in mature xylem vessels (Plate 1F) rather than in the cortex or pith.[33,54]

The most vulnerable portion in the life cycle of *E. amylovora* is the epiphytic stage between late fall and early spring, when inoculum levels are lowest and control with copper compounds is most effective. On the other hand, host tissues most sensitive to infection are open blossoms in spring and succulent shoots in early summer. Shoots and fruit are especially vulnerable when injured by windstorms or hail.[39,62] Streptomycin has proven to be very effective for control when properly and thoroughly applied. However, once fire blight becomes established through blossom, shoot, or fruit infection, control becomes nearly impossible.

VI. GEOGRAPHIC DISTRIBUTION

From its earliest days in America (late 1700s) until the mid 1900s, fire blight remained known mainly on the North American continent and had spread only to New Zealand (1920) and to a few isolated locations in Mexico and Guatemala.[63] Thus, South of the border, the disease became known as "Tizon de fuego" or "Fuego bacteriano".

Since its introduction into England in the early 1950s and The Netherlands in 1966, fire blight spread within about 10 years along the coast of the west European continent from the Baltic Sea to the Pyrenees in southern France, presumably mainly by birds. To date, the disease has moved inland to cover all of Denmark, Belgium, and The Netherlands, as well as the western half of France and West Germany. Fire blight has thus become known as "Bacterievuur" (Dutch), "Ildsot" (Danish), "Zaraza ogniowa" (Polish), "Feu bacterien" (French), and "Feuerbrand" (German). Since 1980, the disease has been reported from Egypt ("El-lafha el-nareya"), the Greek portion of Cyprus ("Vakteriako kapsimo"), Israel ("Kherakhon"), and the mainland and several islands of Greece.

In 1903, Uyeda[57] reported the presence of fire blight ("kasho-byo") in Japan. The disease was believed to have been brought over on nursery stock from America. During the next 50 years, fire blight was reportedly observed in both apple and pear orchards from several prefectures throughout Japan.[38,48,50] However, in 1974, a report from the Ministry of Agriculture in Tokyo, through the European and Mediterranean Plant Protection Organization (EPPO) in Paris, stated that all previous reports on fire blight in Japan were based on misidentification and that the country was presumed free of the disease.[25]

Fire blight is currently a serious threat to Switzerland, Austria, Italy, Spain, and all of the east European countries. To date the disease has not been reported from the two largest pome-fruit-producing countries in the southern hemisphere, Australia[2,14] and South Africa.[21] Apples and pears, as well as nearly all of the significant ornamental plants in the family Rosaceae, can be found in almost all countries with a temperate climate. Fire blight is particularly threatening to those countries with an abundance of hawthorns used for windbreaks, hedges, and dune protection and the large-leaf cotoneasters used for landscaping in gardens and parks.

VII. QUARANTINE SIGNIFICANCE

The fire blight pathogen can thus be spread in numerous ways which may or may not affect quarantine regulations. Of the natural pathways in which the bacterium can spread and which are uncontrollable by man, we should consider spread of bacterial cells, ooze, or strands carried by wind, rains, insects, and birds. By one or all three means, the fire blight organism has been known to become disseminated.[63] Naturally, spread by wind or rain is generally of short distance and dissemination by birds could be quite far. How far *E. amylovora* has been spread by means of insects or birds is not known. Bacterial strands are thought to be either wafted into wind currents and transported in the dry state over long distances or carried upward by wind currents into nearby clouds where they are dispersed in moisture and subsequently fall in raindrops at a distant location.

Of a more serious nature, and to some extent controllable by man, are the epiphytic and endophytic fire blight bacteria on the surface and inside various plant tissues. As mentioned before, *E. amylovora* has been recovered from apparently healthy tissues.[4,8,18,31,46,55,61] Aside from numerous undocumented recordings, several cases have been reported in which apparently healthy trees or shrubs showed symptoms upon arrival and possibly carried the blight organism in transit. In June 1968, Burkowicz[10] observed fire blight in the pear cultivar collection of the Research Institute of Pomology at Skierniewice, Poland. He noted the disease in the current season's growth of a single tree, which had been propagated in early spring with grafts of the cultivar 'Conference' obtained from abroad. The blighted tree was destroyed and fire blight has never been recorded since in central Poland.

In October 1981, the French Plant Protection Service detected fire blight on several plants of *Cotoneaster dammeri* var. 'Royal Beauty' in a nursery at St. Julien-en-Genevois near Lake Geneva in eastern France.[26,28] The plants were imported from The Netherlands in August 1981 and showed no symptoms at that time. In May 1983, an ornamental apple tree was found infected by *E. amylovora*. All infected plants, the noninfected imported plants, as well as all plants susceptible to fire blight in the nursery were destroyed. In Italian nurseries, during the winter months of 1980 to 1981, large numbers of buds from thousands of imported trees were examined for the presence of *E. amylovora*, as a measure to ensure keeping the bacterium out of the country.[13] Even though only one bud proved to contain *E. amylovora*, the discovery showed that contaminated buds present the potential of spreading fire blight.

These examples in Poland, France, and Italy clearly show two important matters regarding the spread and control of fire blight: (1) how *E. amylovora* can unknowingly be spread through contaminated, apparently healthy plant material from a country with fire blight to a region without the disease, and (2) how the disease can temporarily be eradicated through rapid, thorough destruction of the infected or infested and surrounding plant material. Following the occurrence in eastern France, new regulations[27] were published by the Plant Protection Service through EPPO regarding import prohibition of plant material between certain dates from countries with fire blight as well as the establishment and enforcement of certain protection zones around blight-infested areas.

Similar but even stricter regulations were enforced in The Netherlands, which delayed the spread of fire blight by several years.[44] Extremely interesting was the announcement FAO received from EPPO in May 1970 that The Netherlands considered fire blight eradicated when the disease had not been observed for 3 years.[29] However, fire blight again occurred in 1971 and subsequent years. A similiar story, but of longer duration, happened in Egypt, where fire blight was first reported in 1964.[20] Following surveys throughout the concentrated pear acreage in the Nile Delta, El-Goorani[19] reported in 1973 that the country was considered free of fire blight. However, probably due to changing weather patterns during bloom, the disease reoccurred in 1982 and 1983,[1] which apparently set the stage for large-scale epiphytotics in 1984 and 1985.[59]

Regarding the presence and control of epiphytic *E. amylovora* on fruit in Canada, Dueck[17,18] reported the complete elimination of fire blight bacteria (10^6 cells per milliliter) when mature apples were dipped for 10 min in a 1.0 *M* solution of acetic acid. Similar studies in West Virginia[58] showed complete effectiveness in vitro when 1-mℓ cell suspensions (10^8 cells per milliliter) of *E. amylovora* were added for 1 to 10 min to sodium hypochlorite (5, 10, and

20% or ethyl alcohol (50 and 100%), and 0.5 mℓ of these suspensions were plated on nutrient-yeast glucose agar. Also, fruit surfaces of 'Rome Beauty' and 'Red Delicious' apples remained unaffected after 10-min dips in the highest concentrations of these chemicals. When fruit of both cultivars were injured and immediately dipped in *E. amylovora* inoculum for 5 or 10 min, 77% of 'Red Delicious' and 94% of 'Rome Beauty' fruit became infected.[66] When noninjured fruit were dip-inoculated for 10 min, none of the 'Red Delicious' but 41% of the 'Rome Beauty' fruit became infected through the stem or calyx end.

Recent fruit-irradiation research proved that 'Rome Beauty' apples collected from severely blighted trees still contained live epiphytic *E. amylovora* cells following an irradiation dose as high as 140 krad.[37] When such fruit were treated with the sensitizing agent *N*-ethylmaleimide, however, *E. amylovora* failed to survive a dose of 150 krads. The chemical or irradiation treatments did not cause any fruit surface damage following 3 months of cold storage.

VIII. PHYTOSANITARY MEASURES

Due to very radical and well-enforced import regulations and phytosanitary measures, Australia has kept fire blight out of its territory.[2,14,47] Combined with its geographic location (an island continent with winds from the west coast across a large desert), there is no chance for fire blight to enter the country except through importation by man. A somewhat different situation exists in the Republic of South Africa, where fire blight could be introduced over land.[21] Definite examples in which fire blight was introduced by man by means of budwood or fruit over long distances are New Zealand (1920), England (1957), and Egypt (1963).

Since its introduction to the European continent in 1966, today fire blight is threatening Italy, the largest pear-producing country in the world. The immediate surrounding countries of Switzerland[22,35] and Spain[49] have published warning articles to step up their phytosanitary measures in order to prevent introduction of the disease. In general terms, these measures include:

1. Increasing the knowledge of local scientists about the disease and establishing international cooperation with fellow scientists in countries with fire blight
2. Informing the agricultural sector of the country about the potential risk of disease introduction through publications, slides, movies, and posters (Figure 2) in local languages and necessary dialects
3. Surveying the pome-fruit-producing areas and immediate surroundings, including all rosaceous host plants, for early symptoms of fire blight[5]
4. Conducting various studies related to fire blight, including monitoring for *E. amylovora*, establishing disease risk based on the Billing system,[6] and determining degree of susceptibility of domestic apple and pear varieties through cooperative efforts with countries where fire blight has become established
5. Adapting necessary legislation to new situations in the future, including emergency plans to be implemented whenever fire blight makes its appearance

Nearly all countries in Western Europe have published local leaflets or quarantine announcements through EPPO[23,24] to warn scientists and nurserymen alike about the danger of introducing susceptible host plants, both fruit trees and ornamental plants.

Whenever a country still free of fire blight wants to import and test certain rosaceous plant material, this obviously should be done with extreme caution, and such plant material should be handled exclusively in a well equipped plant quarantine or introduction station for several years.[36,43]

In conclusion, fire blight is a dangerous and complicated quarantine disease because of

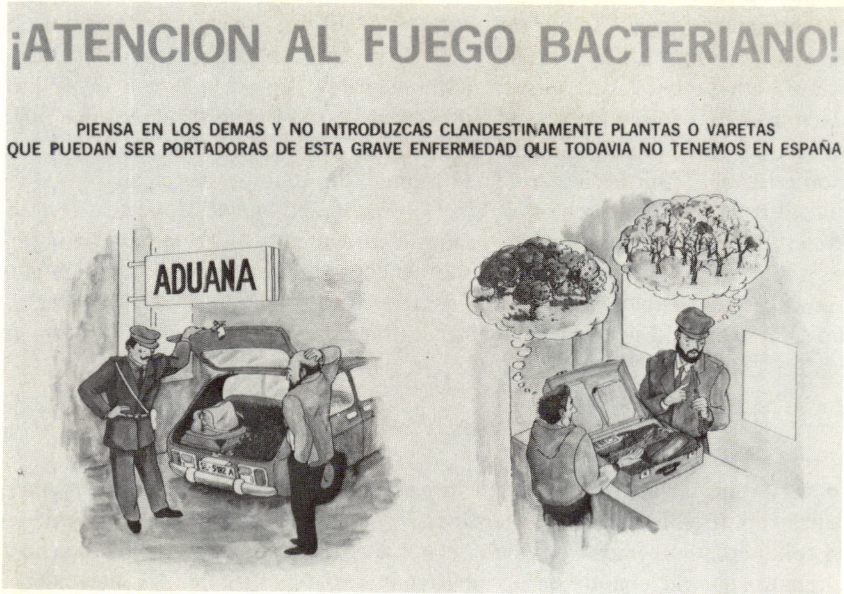

FIGURE 2. Portion of a poster, focusing attention on the danger of introducing fire blight used in Spain at custom offices, sea-, and airports by the Ministry of Agriculture, Fisheries, and Nutrition. (Photo courtesy of Ministry of Agriculture, Madrid, Spain.)

the ability of its pathogen to exist in various apparently healthy tissues. Therefore, numerous measures need to be taken to prevent its introduction to new areas where the disease does not yet occur. Once the disease becomes established on a continent, spread to ''clean'' areas is inevitable and can at best be delayed. However, spread to isolated countries can be prevented through tight adherence to phytosanitary requirements and strict enforcement of quarantine regulations.

REFERENCES

1. **Abo-El-Dahab, M. K., El-Goorani, M. A., El-Kasheir, H. M., and Shoeib, A. A.,** Severe outbreaks of fire blight in Egypt during 1982 and 1983 seasons, *Acta Hortic.*, 151, 341, 1984
2. **Anon.**, Fire blight — let's keep it out of Australia, *Tasmanian J. Agric.*, 45, 175, 1974.
3. **Anderson, H. W.,** Maintaining virulent cultures of *Erwinia amylovora* and suggestion of overwinter survival in mummified fruit, *Plant Dis. Rep.*, 36, 301, 1952.
4. **Baldwin, C. H. and Goodman, R. N.,** Prevalence of *Erwinia amylovora* in apple buds as detected by phage typing, *Phytopathology*, 53, 1299, 1963.
5. **Becker, C. G.,** Non-pear zones and blight eradication, *J. Econ. Entomol.*, 21, 485, 1928.
6. **Billing, E.,** Principles and applications of fire blight risk assessment systems, *Acta Hortic.*, 151, 15, 1984.
7. **Bonn, W. G.,** Fire blight bacteria in symptomless dormant apple and pear buds, *Can. J. Plant Pathol.*, 1, 61, 1979
8. **Bonn, W. G.,** Monitoring epiphytic *Erwinia amylovora* and the incidence of fire blight of apple and pear in southwestern Ontario, *Acta Hortic.*, 117, 31, 1981.
9. **Brulez, W. and Zeller, W.,** Seasonal changes of epiphytic *Erwinia amylovora* on ornamentals in relation to weather conditions and the course of infection, *Acta Hortic.*, 117, 37, 1981.
10. **Burkowicz, A.,** The appearance and current situation of fire blight in Poland, *Eur. Mediterr. Plant Prot. Organ. Newsl.*, 72E, 3, 1972.

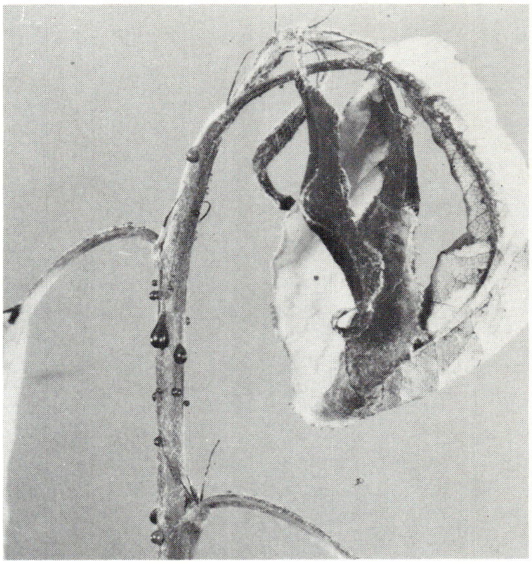

A B

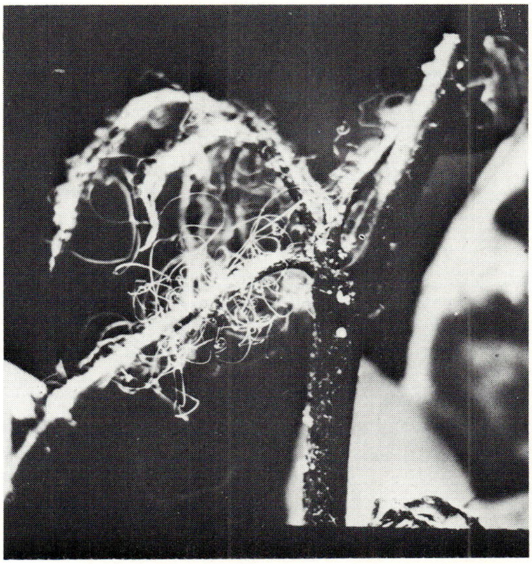

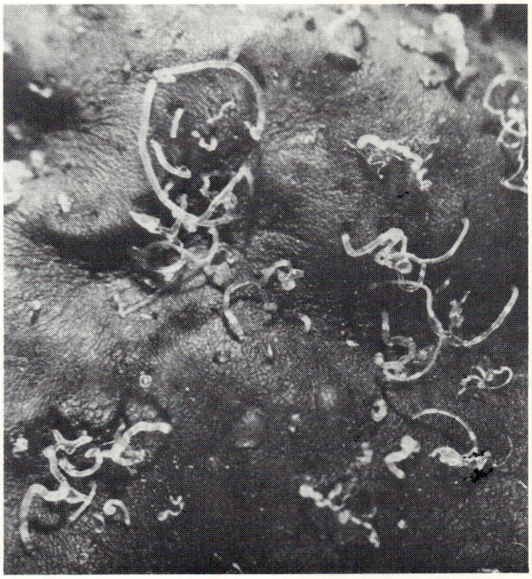

C D

Plate 1. Nature of characteristic forms of *Erwinia amylovora* on pear and apple. (A) Bacterial ooze droplets on infected succulent shoot tip; (B) profuse bacterial ooze on tree bark from nearby canker; (C) bacterial strands on pear leaf petiole; (D) bacterial tendrils on surface of infected pear fruit; (E) epiphytic bacteria on stigma of pear flower (magnification ×15,000); and (F) endophytic bacterial cells in xylem vessel of apple shoot (magnification ×26,000). (D courtesy of R. P. Covey, Tree Fruit Research Center, Wenatchee, Wash; E courtesy of S. V. Thomson, Utah State University, Logan; and F courtesy of R. N. Goodman, University of Missouri, Columbia.)

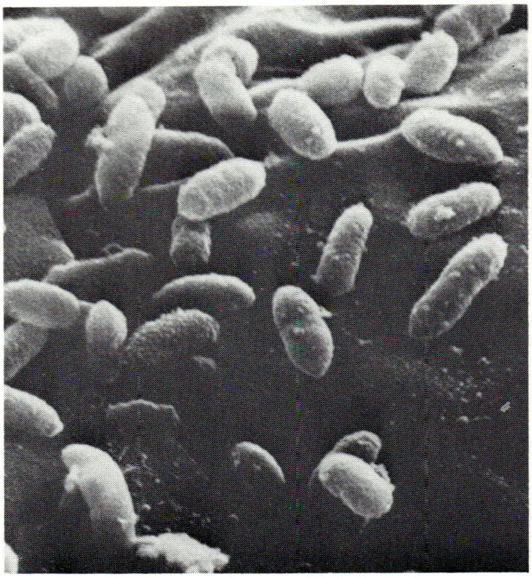

E

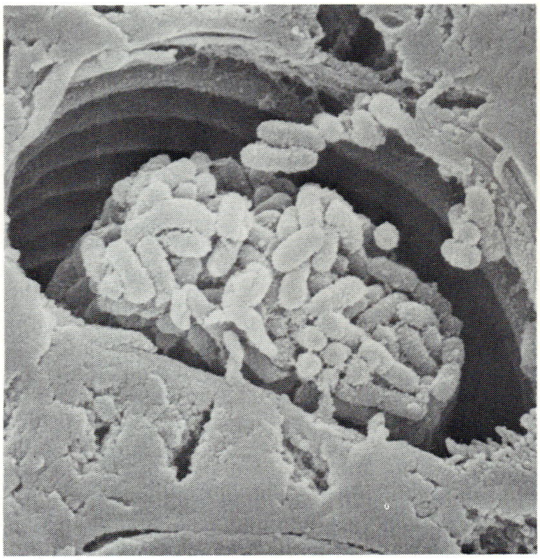

F

Plate 1 (continued)

11. **Burrill, T. J.,** Anthrax of fruit trees; or the so-called fire blight of pears and twig blight of apple trees, *Am. Assoc. Adv. Sci.*, 29, 583, 1880.
12. **Burrill, T. J.,** New species of *Micrococcus* (bacteria), *Am. Nat.*, 17, 319, 1382.
13. **Calzolari, A., Peddes, P., Mazzucchi, U., Mori, P., and Garzena, C.,** Occurrence of *Erwinia amylovora* in buds of asymptomatic apple plants in commerce, *Phytopathol. Z.*, 103, 156, 1982.
14. Commonwealth of Australia, Quarantine against pear blight in Australia, *Commonw. Austr. Gaz.*, 35, 1275, 1924.
15. **Cummins, J. N. and Aldwinckle, H. S.,** Broadening the spectrum of rootstock breeding, *Proc. 19th Int. Hortic. Congr.*, Vol. 3, 303, 1974.
16. **Dueck, J.,** Survival of *Erwinia amylovora* in association with mature apple fruit, *Can. J. Plant Sci.*, 54, 349, 1974.
17. **Dueck, J.,** Bactericidal treatment of apples for elimination of surface-borne *Erwinia amylovora, Can. J. Plant Sci.*, 54, 353, 1974.
18. **Dueck, J. and Morand, J. B.,** Seasonal changes in the epiphytic population of *Erwinia amylovora* on apple and pear, *Can. J. Plant Sci.*, 55, 1007, 1975.
19. **El-Goorani, M. A.,** Current status of fire blight in Egypt, *Plant Dis. Rep.*, 57, 646, 1973.
20. **El-Helaly, A. F., Abo-El-Dahab, M. K., and El-Goorani, M. A.,** The occurrence of the fire blight disease of pear in Egypt, *Phytopathol. Mediterr.*, 3, 156, 1964.
21. **Erskine, J. M.,** Preventing the entry of dangerous plant diseases into South Africa — the fire blight threat, *S. Afr. J. Sci.*, 71, 265, 1975.
22. European and Mediterranean Plant Protection Organization, New Swiss regulations in regard to fire blight *(Erwinia amylovora), Eur. Mediterr. Plant Prot. Organ. Rep.*, 358, 1, 1972.
23. European and Mediterranean Plant Protection Organization, Amendment of Norwegian import regulations concerning fire blight *(Erwinia amylovora), Eur. Mediterr. Plant Prot. Organ. Rep.*, 362, 1, 1972.
24. European and Mediterranean Plant Protection Organization, Fire blight legislation in Sweden, *Eur. Mediterr. Plant Prot. Organ. Rep.*, 363, 2, 1972.
25. European and Mediterranean Plant Protection Organization, No fire blight in Japan, *Eur. Mediterr. Plant Prot. Organ. Rep.*, 380, 1, 1974.
26. European and Mediterranean Plant Protection Organization, Fire blight in France, *Eur. Mediterr. Plant Prot. Organ. Rep.*, 442, 1, 1982.
27. European and Mediterranean Plant Protection Organization, New fire blight regulations in France, *Eur. Mediterr. Plant Prot. Organ. Rep.*, 444, 2, 1983.
28. European and Mediterranean Plant Protection Organization, New infection by *Erwinia amylovora* detected in the department of Loiret, *Eur. Mediterr. Plant Prot. Organ. Rep.*, 448, 1, 1983.
29. Food and Agricultural Organization, Fire blight eradicated in The Netherlands, *Plant Prot. Bull.*, 18, 123, 1970.
30. **Fulton, H. R.,** The persistence of *Bacillus amylovorus* in pruned apple twig, *Phytopathology*, 1, 68, 1911.
31. **Geenen, J., Vantomme, R., and Veldeman, R.,** The system used in Belgium to monitor *Erwinia amylovora, Acta Hortic.*, 117, 115, 1980.
32. **Goodman, R. N.,** Apple fruits a source of overwintering fire blight inoculum, *Plant Dis. Rep.* 38, 414, 1954.
33. **Goodman, R. N. and White, J. A.,** Xylem parenchyma plasmolysis and vessel wall disorientation caused by *Erwinia amylovora, Phytopathology*, 81, 844, 1981.
34. **Gowda, S. S. and Goodman, R. N.,** Movement and persistence of *Erwinia amylovora* in shoot, stem and root of apple, *Plant Dis. Rep.*, 54, 576, 1970.
35. **Grimm, R.,** Phytosanitary measures to prevent the introduction of fire blight into Switzerland, *Acta Hortic.*, 117, 89, 1981.
36. **Grimm, R.,** Der Feuerbrand heute; seine Verbreitung, seine Erkennung und die Abwehrmassnahmen in der Schweiz, *Schweiz. Z. Obst Weinbau*, 120, 284, 1984.
37. **Janisiewicz, W. J., van der Zwet, T., and Jahrling, P. B.,** Laboratory studies on the effect of gamma irradiation on *Erwinia amylovora* survival on apple fruit, *Can. J. Microbiol.*, 32, 787, 1986.
38. **Kazui, M.,** Fire blight disease of pear and apples, *Byochu-Gai Zasshi*, 9, 545, 1922.
39. **Keil, H. L., Smale, B. C., and Wilson, R. A.,** Role of injury and longevity of *Erwinia amylovora* in the epidemiology of fire blight, *Phytopathology*, 56, 464, 1966.
40. **Keil, H. L. and van der Zwet, T.,** Recovery of *Erwinia amylovora* from symptomless stems and shoots of Jonathan apple and Bartlett pear trees, *Phytopathology*, 62, 39, 1972.
41. **Lewis, S. M. and Goodman, R. N.,** Mode of penetration and movement of fire blight bacteria in apple leaf and stem tissue, *Phytopathology*, 55, 719, 1965.
42. **Maas Geesteranus, H. P. and de Vries, Ph. M.,** Survival of *Erwinia amylovora* bacteria on plant surfaces and their role in epidemiology, *Acta Hortic.*, 151, 55, 1984.
43. **Mathys, G.,** Reflections on plant health regulations and quarantine aspects of fire blight, *Acta Hortic.*, 117, 13, 1981.

44. **Meijneke, C. A. R.,** Prevention and control of fire blight, *Eur. Mediterr. Plant Prot. Organ. Bull.*, 9, 53, 1979.
45. **Miller, H. J. and van Diepen, H.,** Monitoring of epiphytic populations of *Erwinia amylovora* in The Netherlands, *Acta Hortic.*, 86, 57, 1978.
46. **Miller, T. D. and Schroth, M. N.** Monitoring the epiphytic population of *Erwinia amylovora* on pear with a selective medium, *Phytopathology*, 62, 1175, 1972.
47. **Noble, R. J.,** Plant quarantine measures. Some observations on their establishment in Australia, *Agric. Gaz. N. S. W.*, 35, 883, 1924.
48. **Okabe, N. and Goto, M.,** Bacterial plant diseases in Japan. I. A list of bacterial diseases and their pathogens, *Shizuoka Univ. Fac. Agric. Rep.*, 5, 63, 1955.
49. **Sampayo, M. and Palazon, I.,** Prevention against fire blight in an uncontaminated country, Spain, *Acta Hortic.*, 151, 335, 1984.
50. **Shiraishi, H.,** On the fire blight disease of pears, *Byochu-Gai Zasshi*, 17, 655, 1930.
51. **Southey, R. F. W. and Harper, G. J.,** The survival of *Erwinia amylovora* in airborn particles: tests in the laboratory and in the open air, *J. Appl. Bacteriol.*, 34, 547, 1971.
52. **Stewart, V. B.,** The fire blight disease in nursery stock, *N.Y. (Cornell) Agric. Exp. Stn. Bull.*, 329, 1913.
53. **Suhayda, C. G. and Goodman, R. N.,** Infection courts and systemic movement of 32P-labeled *Erwinia amylovora* in apple petioles and stems, *Phytopathology*, 71, 656, 1981.
54. **Suhayda, C. G. and Goodman, R. N.,** Early proliferation and migration and subsequent xylem occlusion by *Erwinia amylovora* and the fate of its extracellular polysaccharide (EPS) in apple shoots, *Phytopathology*, 71, 697, 1981.
55. **Sutton, T. B. and Jones, A. L.,** Monitoring *Erwinia amylovora* populations on apple in relation to disease incidence, *Phytopathology*, 65, 1009, 1975.
56. **Thomson, S. V., Schroth, M. N., Moller, W. J., and Reil, W. O.,** Occurrence of fire blight of pears in relation to weather and epiphytic populations of *Erwinia amylovora*, *Phytopathology*, 65, 353, 1975.
57. **Uyeda, E.,** On the causal bacteria of fire blight of apple, *Dainippon Nokai-ho*, 260, 1, 1903.
58. **van der Zwet, T.,** In vitro testing of various chemicals for bactericidal activity against *Erwinia amylovora*, *Phytopathology*, 74, 825, 1984.
59. **van der Zwet, T.,** Identification, symptomatology and epidemiology of fire blight on Le Conte pear in the Nile Delta of Egypt, *Plant Dis.*, 70, 230, 1987.
60. **van der Zwet, T., Bell, R. L., and Stroo, H. F.,** Long distance dissemination of *Erwinia amylovora* as resident bacteria in apparently healthy pear budwood, *Phytopathology*, 72, 711, 1982.
61. **van der Zwet, T. and van Buskirk, P. D.,** Detection of endophytic and epiphytic *Erwinia amylovora* in various pear and apple tissues., *Acta Hortic.*, 151, 69, 1984.
62. **van der Zwet, T. and Keil, H. L.,** Importance of pear tissue injury to infection by *Erwinia amylovora* and control with streptomycin, *Can. J. Microbiol.*, 18, 893, 1972.
63. **van der Zwet, T. and Keil, H. L.,** Fire blight — a bacterial disease of rosaceous plants, *U. S. Dept. Agric. Handb.*, 510, 1979.
64. **van der Zwet, T., Oitto, W. A., and Westwood, M. N.,** Variability in degree of fire blight resistance within and between *Pyrus* species, interspecific hybrids, and seedling progenies, *Euphytica*, 23, 295, 1974.
65. **van der Zwet, T., Thomson, S. V., Covey, R. P., and Bonn, W. G.,** Absence of endophytic *Erwinia amylovora* in seed and core tissues of Red Delicious apple, *Phytopathology*, 76, 1140, 1986.
66. **van der Zwet, T. and Walter, J. C.,** Significance of apple tissue injury to infection by *Erwinia amylovora*, *Phytopathology*, 75, 629, 1985.
67. **Winslow, C. E. A., Broadhurst, J., Buchanan, R. E., et al.,** The families and genera of the bacteria; Erwineae, *J. Bacteriol.*, 5, 209, 1920.

Chapter 11

POTATO SPINDLE TUBER

Luis F. Salazar

TABLE OF CONTENTS

I. NAMES OF THE DISEASE

The potato "spindle tuber" disease first described by Martin[33] has also been named "unmottled curly dwarf"[51] and potato "gothic".[29]

II. CAUSAL AGENT

The causal agent of the disease is the potato spindle tuber viroid (PSTV). PSTV was considered a virus until 1967 when Diener and Raymer[5] suggested a different nature. Its unique properties initiated studies on its structure and composition, and it is at present the best studied plant pathogen. This viroid is a 359-nucleotide circular RNA molecule devoid of the protein coat characteristic of viruses, with a high degree of intramolecular complementarity.[17] Although linear molecules, with the same infectivity as circular ones, are always present in PSTV-RNA preparations, these are believed to arise in vivo by cleavage of circular molecules.[18]

The three main strains of PSTV have been identified as severe, intermediate, and mild by the severity of the symptoms produced in potato and tomato.[49] These three strains have been shown to differ by six to ten nucleotide exchanges.[59]

III. HOST RANGE

In spite of its wide experimental host range, PSTV only affects the potato (*Solanum tuberosum* L.) in nature. Diener[6] lists 156 species in 12 families in which PSTV has been shown to replicate. Of these, 139 are members of the Solanaceae, suggesting ecological adaptation to species in this family, especially those in the genus *Solanum*. Several *Solanum* species and species in closely related genera, however, were not infected by PSTV.

Aside from potato, some tomato cultivars (*Lycopersicon esculentum* Mill) are known to react with symptoms of diagnostic value. In fact, systematic studies on PSTV were really initiated since the demonstration by Raymer and O'Brien[44] that PSTV could be transmitted by mechanical (sap) inoculation to tomatoes. The tomato cultivar most largely used was cv. Rutgers. Initial symptoms on tomato are usually rugosity, epinasty, and twisting of new leaflets.[44,45] Necrosis of midribs, secondary veins, and vascular tissue develops later and gradually becomes more severe. Infected plants are severely stunted with a rosette appearance. Severity of symptoms, provided high temperature and high light intensity are given to inoculated tomatoes, will depend on the strain of the viroid. Mild strains usually do not induce vein necrosis, and sometimes symptoms are so mild that they can only be distinguished by comparing infected with healthy plants.

Another host reacting with diagnostic symptoms is *Scopolia sinensis*.[55] Necrotic lesions are evident in inoculated leaves around 14 days after inoculation. Systemic symptoms (vein necrosis) develop later and usually lead to plant death.

IV. SYMPTOMS AND DAMAGE

Symptoms caused by PSTV on the foliage of potato are rarely dramatic and their severity is known to depend on several factors. The potato cultivar,[62,42] strain of PSTV,[57,42] environmental conditions,[12,13] and even the method of inoculation[42] influence the severity of symptoms. Severely affected plants may be stunted with upright growth due to a more acute angle between stems and petioles. Leaflets show a degree of undulation and size reduction, and foliage turns slate gray with a dull surface. The tubers may become very spindly and reduced in size and may have growth cracks. A decrease in eye depth and an increase in the number of eyes per tuber have been reported. Spindle tubers usually have a rounded transversal section and may show a discoloration which is more evident in dark-colored tubers.

Several reports have indicated severe yield losses due to PSTV infection. Experiments by Le Clerg et al.[28] showed that a 4% level of infection with PSTV reduced yields of marketable tubers by 2.6% in 31 trials with five varieties. They did not consider the relative incidence of mild and severe strains, which were unknown at the time.

Singh et al.[57] estimated that the decrease in tuber yield of the variety Saco was between 17 and 24% for mild strains and 64% for a severe strain. Based on the higher prevalence of mild strains over the severe ones and an incidence of approximately 4% of PSTV, Singh et al.[57] calculated a hypothetical loss of 1%. More recent studies by Pfannestiel and Slack[42] in 14 potato cultivars showed that decrease in tuber weight per plant depended on the potato cultivar and the length of time they were infected. In their experiments, decrease of tuber weight by a severe strain of PSTV varied from nonsignificant to 46% in the first year of infection up to 87% in the second year, and up to 97% in the third year. Mild strains did not reduce tuber weight, but in some cultivars tuber symptoms were more pronounced, obviously affecting the marketable quality of the tubers.

V. TRANSMISSION AND SPREAD

Transmission of PSTV by contact between infected and healthy plant parts was suggested in early reports on the disease.[51] Later results have conclusively demonstrated that PSTV is readily transmitted through mechanical contact,[14,36] sap inoculation,[2] and by infected tools and farm equipment.[14,35,36] Therefore, there is presently no doubt that mechanical transmission plays an important role in field dissemination of the disease.

Spindle tuber has been reported as being transmitted at low levels by grasshoppers (*Melanoplus* sp.),[15] flea beetles (*Epitrix cucumeris* and *Systena elongata*), tarnished plant bugs (*Lygus pratensis*), and by larvae of the Colorado potato beetle (*Leptinotarsa decemlineata*) and leaf beetles (*Dysonicha triangularis*).[16] These results, however, have never been confirmed and because handling of plants might have been responsible for the transmissions observed, they must be regarded with some scepticism.

Early reports on transmission of spindle tuber by aphids[51] are also doubtful because of the possibility of inadvertent transmission during handling of the plants. More recently, De Bokx and Piron[4] reported a low proportion of transmission of PSTV by means of the aphid *Macrosiphum euphorbiae* but not by *Myzus persicae* or *Aulacorthum solani*. In view of the results so far reported, we can accept the concept that PSTV is not transmitted by arthropod vectors or that it is very unlikely that they are responsible for large field dissemination.

Vertical transmission of PSTV was first reported by Benson and Singh[1] in potato and tomato. Hunter et al.[20] reported 100% seed transmission in cultivars Katahdin and Russet Sebago when both parents were infected. However, Singh[56] was able to obtain only 12% seed transmission in crosses between cultivars Saco and Red Pontiac or 6% in a sibcross of Saco when both parents were infected. In tomato, Singh obtained 6 or 9% seed transmission when only one parent was infected. In none of the experiments did PSTV infect the tubers of healthy mother plants pollinated with infected pollen. Seed transmission ranging from 0 to 100% was found by Fernow et al.[9] in open-pollinated female parents. They did not find any correlation with the cultivar or age of the seed.

In experiments carried out at CIP[46] in *Solanum acaule*, it was found that PSTV was transmitted to 23 to 35% of progeny seeds when only one parent (either male or female) was infected, or between 60 and 100% when both parents were infected.

VI. GEOGRAPHIC DISTRIBUTION

PSTV has been reported to occur in potato fields only in the northern and northeastern U.S., and Canada,[6] the U.S.S.R.,[29] and China.[58,30] In Great Britain it has been detected in and eliminated from potato accessions held by the Commonwealth Potato Collection.[53,54] The introduction and prompt elimination of PSTV into South Australia via potato breeding material from Glen Innes, New South Wales[52] has also been reported.[3] In South America, the presence of PSTV has been reported on the basis of symptoms in Argentina.[8] Its presence in Peru, Venezuela, and Brazil has also been mentioned.[58] A survey done at CIP of more than 3000 samples from native cultivars and wild species of the Andean region failed to disclose the presence of PSTV.[21] This suggested that PSTV is not a pathogen that evolved with the potato in its center of origin, but more likely adapted to the potato in other regions where it was introduced.[21]

VII. QUARANTINE SIGNIFICANCE OF PSTV TO REGIONS WHERE IT IS NOT KNOWN TO OCCUR

Based on temperature requirements for multiplication and dissemination, PSTV acquires different significance for regions with hot and cold climates. Under high-temperature regimes, more severe and earlier appearance of PSTV symptoms have been reported[12,13] and, since these are associated to higher viroid concentration,[50] one can assume that PSTV would become very dangerous for potatoes in regions with hot climates. Singh[58] has already warned us of this situation. The fact that there are no reports of resistant cultivars aggravates the situation, especially in symptomless cultivars which may act as sources of inoculum. Fortunately, PSTV is not transmitted by arthropod vectors, at least efficiently; otherwise, its dissemination would be much higher than what has been reported.

Inadvertent introduction of PSTV would not necessarily mean a high risk of dissemination, especially if it should occur in materials for research which would undoubtedly be tested during experimentation. The highest risk of introduction and subsequent dissemination is expected in materials for direct multiplication under field conditions, like imported seed stocks. Even though a testing procedure might be in force, only a few samples can be examined, and lower incidences of PSTV-infected tubers are not likely to be picked up.

In regions with cold climates, if introduction of PSTV occurred, dissemination could likely be limited by the low concentration of the viroid. However, this hypothesis has apparently not been tested.

VIII. PATHWAYS ANALYSIS

In a similar manner to viruses, especially those without arthropod vectors, it is very unlikely that PSTV can spread over great distances by natural pathways. Man-made pathways are responsible for long-distance dissemination.[25] There are two main properties of this viroid which adapt it for mechanical or contact spread. These are (1) the usually high concentration it reaches in plant tissue, particularly under high-temperature growing conditions, and (2) its high resistance to inactivation in plant extracts or sap as well as in infected farm tools and machinery. In addition, PSTV can be transmitted through botanical seed of potato and other hosts.

A. Natural Pathways

None of the pathways discussed in Section V can be responsible for a significant PSTV spread over long distances. However, natural seed dispersal and plant-to-plant contact can be responsible for the spread of PSTV within a field and, aided by man, to other fields and regions. Since potato crops are not normally planted from botanical seed (= true potato seed, TPS), this type of dissemination can only be important in regions with less developed agricultural practices where, for instance, elimination of volunteer plants is not properly conducted. In this case, viroid-infected seedlings act as alternate hosts and become highly important in conjunction with plant-to-plant contact dissemination. The utilization of TPS for planting potatoes as in some regions of China[23] may open a new man-made way of PSTV dispersal. In fact, the high dissemination of PSTV in China is thought to be in part due to production of potatoes from TPS.

B. Man-Made Pathways

1. Plants and Plant Parts

Potato tubers are, by far, the most important means of PSTV dispersal over short and long distances. Even though the origin of PSTV cannot be definitively indicated, its spread from Maine to New Jersey in the U.S. can be attributed to the seed tubers produced in Maine.[6]

Since many potato cultivars do not show visible symptoms on infected plants and tubers (tolerant cultivars), they can be responsible for PSTV survival and spread. Some PSTV-tolerant cultivars in the U.S. like La Chipper, Katahdin, and Kennebec have been suggested as important for PSTV survival.[42] It would, therefore, not be surprising that most of the spread of PSTV has occurred through infected tubers, especially those intended for use in seed production. Dispersal through the botanical seed might have occurred in China, where close to 18,000 ha of potato are planted annually with seed tubers produced from TPS.[23] The initiation and improvement of TPS technology in developing countries[23] may become an effective way to disperse PSTV if precautions and rigorous testing of mother plants, such as those applied by CIP, are not implemented.

2. Plant Propagation Procedures

The rapid multiplication techniques whereby thousands of cuttings are produced from few mother plants within a short period of time are also potentially important ways for PSTV spread. Normally these techniques are applied locally to rapidly increase a particular cultivar in one region. Nevertheless, in some cases they may be transported long distances within or outside a country or used to produce tuberlets as initial stocks for a seed-production system. The value of this agronomic practice should be recognized, but we should also be aware of the possible danger of rapidly increasing and disseminating infected stocks.

A similar situation can be found with in vitro cultures used as alternatives for multiplication. Tissue culture is carried out under rather high temperature regimes which are conducive to PSTV multiplication at high levels.[31] High-temperature treatments commonly used to eliminate viruses in combination with tissue culture would not eliminate PSTV, and would rather favor its rapid multiplication and accumulation in plant tissues.

3. Shipments of Live Pests for Scientific Purposes

In the case of PSTV, its rather unique properties among all virus-like agents make it an attractive subject of study or simply an example for scientific and teaching purposes. Those requesting plant pathogens not only use them strictly in the laboratory but commonly have a need to multiply the agent under greenhouse conditions. If precautions are not taken to prevent accidental contamination of materials, there is a risk of contamination. Evidently this pathway is not very common, and appropriate regulations can prevent spread of this agent.

4. Germplasm Banks or Collections

Many countries maintain, to a certain extent, a vegetatively propagated potato germplasm collection in which the virus or viroid content of accessions is unknown. In many instances, these collections are maintained by the same institution responsible for improving the potato in the country and often in or close to fields where other advanced clonal materials are grown. These collections constitute an efficient source of dissemination not only for PSTV, but also for viruses. There are at least two examples in which PSTV has been found in and eliminated from germplasm collections. In Great Britain, PSTV was detected in and eliminated from a few accessions of The Commonwealth Potato Collection,[53,54] and fortunately it did not infect other material being propagated. In Australia, however, introduction of PSTV into South Australia was reported to have taken place via breeding materials from Glen Innes, New South Wales.[3] In this case, again the rapid steps taken towards elimination of infected materials prevented escape to other breeding materials or commercial crops growing nearby.

The above reports illustrate the importance that germplasm collections have in disseminating PSTV if precautions are not implemented to prevent escapes.

IX. DETECTION METHODOLOGY

Since spindle tuber disease was discovered in 1922,[33] efforts have been directed toward finding a method to reliably detect diseased plants and tubers. However, it was only possible to use symptomatology until 1967 when Raymer and O'Brien[44] showed that the causal agent was capable of being transmitted mechanically to and produce diagnostic symptoms in tomato. *Scopolia sinensis*[55] has been reported as a local lesion host for PSTV, but its use has been limited. The discovery of the nature of the causal agent permitted the development of two more techniques which are preferred by programs where large-scale testing is required. In 1975, Morris and Wright[37] developed an electrophoretic procedure which was subsequently simplified.[38,43] Recently, since the production of a molecular probe by Owens and Cress,[39] there has been an increased interest and use of these reagents for detection of PSTV.[40]

In spite of this development, however, all detection techniques will continue to have some value for the detection of PSTV and will undoubtedly be used in many instances. The advantages and limitations of each technique are discussed in the following sections.

A. Symptomatology

Symptoms caused by PSTV in the foliage of infected plants are obscure and frequently can only be detected by comparison with healthy plants. On the other hand, tuber symptoms are more easily determined, but the existence of mild strains, adverse climatic conditions for symptom expression, and the prevalence of cultivars which do not show clear symptoms upon infection[42] complicate its value in diagnosis. In spite of these limitations, symptomatology can give a rapid estimate of disease incidence in cultivars that react to PSTV infection with diagnostic symptoms.

B. The Tomato Test

Mechanically inoculated tomatoes develop symptoms if grown under high temperature ($<$30°C) and high light intensity ($<$1500 fc).[50] Symptoms usually develop between 2 to 4 weeks after inoculation, and their severity depends on viroid concentration in the sample, growing conditions, and especially with the viroid strain. Severe strains of PSTV can be easily detected in tomatoes because of the severe symptoms they induce. Tomato plants infected with mild strains, however, react with very slight symptoms often indistinguishable from healthy plants. The "challenge test" reported by Fernow[10] was a refinement for the detection of mild strains using the cross-protection phenomenon. In this test a group of tomatoes are inoculated with the sample to test and 10 to 15 days later some of them are reinoculated (challenged) with a severe strain. If the reinoculated tomatoes do not show symptoms of the severe strain, it is assumed that the sample contained a mild strain. On the contrary, if tomatoes show symptoms characteristic of the severe strain, the sample was free of mild strains. The presence of severe strains is detected in the tomatoes that were not challenged.

The detection of both mild and severe strains was reported by Yang and Hooker[63] in inoculated tomatoes maintained under high temperature (<30°C) and continuous fluorescent light (<1500 fc). Under these conditions, infected tomatoes react with albinism, which is easily observed. In CIP, this technique has been successfully used for a number of years. We found that in some cases healthy tomatoes also react with some degree of albinism which was apparently caused by any condition different to those indicated above. This albinism, however, can be distinguished from that caused by PSTV due to the absence of crinkling and necrosis which is almost invariably present in plants infected with mild and severe strains, respectively.

In all cases, tomatoes should be inoculated when very young, preferably at the cotyledonary stage, with the help of an abrasive. Though PSTV in sap can infect tomatoes, it is preferable to inoculate them with phenol extracts or sap extracted and diluted in buffer (0.05 to 0.1 *M* phosphate is commonly used).

The tomato test is normally sufficiently sensitive for detecting PSTV in field-grown samples. However, its application is very expensive and laborious, it requires large space in heated greenhouses, and results can only be obtained 2 to 5 weeks after inoculation. These disadvantages make the test inappropriate for large-scale testing.

C. The Use of a Local Lesion Host

Scopolia sinensis is the only host known to react with local lesions after inoculation with PSTV. Both mild and severe strains are reported to induce local lesions between 10 and 18 days followed by systemic invasion which causes veinal necrosis, leaf drop, and eventually death of the plants.[55] *S. sinensis* requires specific reaction conditions. A temperature above 18 to 23°C, light intensity of 300 to 400 fc, and relative humidity over 60% are the most adequate conditions for symptom development. This host is said to be as sensitive as tomatoes for PSTV detection, however, it requires more care for raising seedlings and does not readily produce seeds.[19] In addition, some seed lots give raise to seedlings which are less susceptible than parents.

D. Electrophoresis

Morris and Wright[37] first described a procedure to identify the presence of PSTV in fractions of low molecular weight nucleic acids extracted from infected plants and electrophoresed in polyacrylamide gels. This method is based on the characteristic of PSTV being a low molecular weight RNA that migrates discretely in gel electrophoresis as compared to other cellular RNAs. The original procedure required 3 days for extracting nucleic acids and was later reduced to 2 days.[38] Other modifications reported by Pfannestiel et al.[43] reduced extraction to only few hours. At CIP, a procedure based on that of Pfannestiel et al.[43] was developed, and permitted the simultaneous extraction of many samples with a minimum of steps. Basically, the procedure adopted at CIP requires grinding 0.5 to 1 g of tissue in a buffer system that simultaneously allows extraction of PSTV from the nucleii, prevents oxidation of extracts, and precipitates high molecular weight nucleic acids. Proteins are denatured with phenol, and low molecular weight nucleic acids are precipitated from supernatants with ethanol. Electrophoresis is carried out for 4 hr at 100 V in polyacrylamide gels under a Tris-EDTA buffer system. After electrophoresis, nucleic acid bands are detected by staining. The utilization of silver nitrate staining instead of Toluidine blue O, methylene blue, or ethidium bromide permitted higher resolution of bands and, consequently, increased sensitivity in the detection.[34]

Electrophoresis can produce results in only 2 days, whereas the tomato test requires 2 to 4 weeks. However, its sensitivity is lower than tomatoes. Inoculation of samples on tomatoes and their subsequent detection by electrophoresis has been reported as an alternative to increase detection of PSTV in diluted or composite samples. This system has been used at CIP to detect PSTV in botanical seed lots which otherwise is not reliably detected. Experiments have shown that in certain seed lots, PSTV can be detected by the tomato-electrophoresis procedure when tomatoes are inoculated with extracts from ground seeds but not in seedlings from the same seed lots. The reason for this is not known, but other experiments suggested that PSTV was present in the endosperm of those seeds but not in the embryos.[47]

E. The Nucleic Acid Spot Hybridization Test

The determination of the 359-nucleotide sequence of PSTV[17] facilitated the molecular cloning and characterization of a double-stranded (ds) PSTVcDNA. Owens and Cress[39] synthesized cDNA fragments of PSTV and constructed one full-length copy cDNA by ligating fragments from two partially overlapping clones. The details of such work falls beyond the scope of this review. The PSTVcDNA was ligated into a plasmid and transformed into *E. coli*. Large amounts of cDNA can be produced in the plasmid replicating in *E. coli*, from where it can be purified, made radioactive, and used to detect PSTV. As originally described by Owens and Diener,[40] the procedure for testing is based on the hybridization of highly radioactive PSTVcDNA, with PSTV bound to a solid support, and subsequent detection of DNA-RNA hybrids by autoradiography.

Using this method, one can detect PSTV in foliage and tuber sprouts of infected plants. Even though the presence of sap reduces PSTV binding to the nitrocellulose membrane, a few picograms of PSTV can be detected after hybridization with radioactive recombinant DNA. The nucleic acid spot hybridization test was able to detect PSTV in mixtures of seed extracts equivalent to 1 infected seed among 80 healthy ones.[48] This test is about 1000 times more sensitive than electrophoresis, and it is suitable for testing a large number of samples with minimum expenditure of labor, materials, and laboratory space and facilities.[40] At CIP, this test has replaced electrophoresis in the routine testing of PSTV at headquarters and in the regional centers.[24] For testing materials outside headquarters, nitrocellulose membranes are spotted with the samples and sent to headquarters where hybridization and autoradiography are carried out. The utilization of a radioactive label (e.g.,^{32}P) is troublesome for countries that need to import the reagent. Therefore, attempts to replace the radioactive label by an enzymatic or fluorescent detector system have apparently met some success, and it appears that this approach could be in the future widely accepted.

X. SAFEGUARDS

A. Safeguards Against the Entry of PSTV Along Natural Pathways

As PSTV is mainly disseminated through man-made pathways, safeguards are basically those that prevent entry and dissemination by those pathways. Some of the safeguards against the entry of pests along natural pathways listed in Section VI are, however, also applicable to PSTV or other exotic pests. For instance, gathering of information about the disease becomes important if the potato is an important crop and especially if seed stocks are produced or multiplied in that country.

Similarly, some research should be conducted to develop an expedited methodology for its detection. In most cases, it is assumed that PSTV is not present in the country. This situation is usually not known or not reported for the majority of countries and, therefore, one of the studies required is to determine the occurrence and incidence of the disease in the country or region.

B. Safeguards Against the Entry of PSTV Along Man-Made Pathways

1. Exclusion

a. Rules and Regulations

Since all plant parts are infected by PSTV systemically, there is no advantage in regulating the type of plant material or age. Regulating the volume of the consignments will lower the risk of introduction of PSTV since it will allow better care in handling and more thorough testing. In this regard, in vitro plantlets become an ideal safeguard.[26] The release of only that part of the shipment that is passed through quarantine would help to reduce the risk of PSTV introduction.

Since potatoes are the only known host naturally infected by PSTV, exclusion of the host[25] is the best way of control. Both the U.S.[11] and European and Mediterranean Plant Protection Organization (EPPO) consider potatoes under the category of a prohibited plant. As stated by Kahn,[27] "prohibited" is generally defined internationally as entry not allowed for the general or commerical publics and occasionally not allowed even for government services, but may be admissible for scientific purposes under specified conditions. For example in the U.S. and for EPPO, importation for scientific and educational purposes can be effected following pre- and postentry quarantine regulations. The pre- and postentry regulations for EPPO depend on the country where the importation is required and the type or use intended for the material. For instance, from the U.S., Canada, and other areas where PSTV occurs, seed potatoes must have been grown under a potato production system known to be free of PSTV and subject to an EPPO-recommended inspection procedure for certifying freedom from PSTV. Ware potatoes from the same countries must be treated to ensure supression of germination.[7]

Paradoxically, potatoes from countries or places where PSTV has not been reported may offer higher risk, especially if potatoes are not thoroughly tested. This situation may occur because routine testing procedures are usually applied or enforced when a pathogen is known to occur. It seems this is not the situation of EPPO countries where testing is required, even though PSTV is not known to occur in the area. Practically all countries prohibit the importation of PSTV or virus isolates. However, exemptions are made by permit for scientific and eduational purposes as long as the institution requesting the isolate can provide acceptable conditions to prevent escape. Certainly, it would not be advisable to authorize the importation to institutions which handle within the same installations a seed production program.

b. Permits

A permit usually indicates pre- and postentry regulations[27] and constitutes an evidence that the importer country accepts the risk of introducing PSTV. It is common among some scientists to exchange samples of pathogen isolates without a permit being issued by quarantine officials. This is a dangerous practice and should be avoided. AT CIP, requests for PSTV and viruses without an official permit are denied.

c. Phytosanitary Certification

Most countries require a phytosanitary certificate accompanying all shipments of potato germplasm. Some countries usually provide a phytosanitary statement to acompany the exported potato material. Although a phytosanitary statement is not a legal document, it provides the importer with information about how the material was produced and, as in the case of CIP, the methodology used for pathogen detection. Because of their obvious limitations, developing countries importing potato germplasm rely on procedures applied by exporting countries to guarantee freedom of PSTV. The case of CIP, which is located in a developing country, constitutes a good example of safe international transfer of germplasm.[61] Exports from headquarters are accompanied not only by a Peruvian Phytosanitary Certificate, but also by a phytosanitary statement which provides information about production of materials as well as procedures used for testing. Imported materials, on the other hand, are not released until tests have shown their freedom from important pathogens.

d. Inspection

Before potato germplasm introductions are released from plant quarantine facilities, they should be tested for PSTV. In the U.S., they are tested by gel electrophoresis of RNA purifications or by the nucleic acid spot hybridization test (NASH).[11] At EPPO, partially purified sap extracts of seedlings, plants, or cuttings are inoculated to tomato cv. Sheyenne. Tomato test plants are kept at 24 to 30°C under high light intensity for at least 5 weeks before extracting RNA and electrophoresing on polyacrylamide gels. Alternatively, potato leaf sap can be tested by NASH.[7]

The routine application of electrophoresis and NASH techniques in developing countries is usually not possible due to the lack of equipment and reagents. Diagnosis in these countries is either done by visual observation of symptoms or inoculation of some samples to tomato. Since symptomatology is highly insensitive and the tomato test and electrophoresis laborious and expensive, efforts should be directed to contract PSTV testing by NASH in specialized laboratories.

e. Treatment

There are no simple treatments to eliminate PSTV from infected plants. Tissue culture techniques alone or in combination with heat therapy have not shown any PSTV eradication.[60] This is understandable, because PSTV replicates at higher rates under high-temperature conditions[50] such as those used for tissue culture and heat therapy. The only alternative is to free plants by a combination of cold treatment and meristem culture.[31] PSTV-free meristems can be obtained from infected plants grown for 3 to 6 months at temperatures between 6 to 10°C. A higher frequency of PSTV-free plantlets are obtained if apical domes rather than meristems are excised.[32] However, because apical domes have a need to regenerate through callus tissue, they are susceptible to somatic mutations. Meristems and shoot tips can reduce this problem and although less efficient than apical domes for eliminating PSTV, they can be successfully used. This procedure has also been successfully used to eradicate chrysanthemum chlorotic mottle and chrysanthemum[41] stunt viroids from chrysanthemum.[41]

When it is intended to use this method, there is a need for having a reliable procedure for PSTV detection. Hence the NASH test[40] would be extremely useful. As stated by Lizarraga et al.,[31] elimination of PSTV by cold treatment should only be used to recover universally infected valuable germplasm.

XI. RESPONSES SHOULD THE PEST BECOME ESTABLISHED

A. Surveys

Surveys to determine the occurrence of PSTV in a country should be conducted with some periodicity depending on the frequency of introduction and dissemination of germplasm or other potato materials into the country. These surveys should use the most sensitive methods of detection available. Some countries, where PSTV is considered to be absent, may find that PSTV does occur in some materials. Under these circumstances, the first reaction is close to panic. The immediate conclusion is usually to believe in a recent introduction of PSTV without realizing that it might have remained undetected in those materials until new sensitive methods of detection were applied or until it had time to spread. This was evident at CIP when the NASH test was used on a large scale. Some cultivars native from the Andean region were shown to carry PSTV at a concentration below the level of detection by electrophoresis or tomato. Therefore, one is inclined to believe that the introduction of the NASH test as a preferred method of diagnosis will dramatically change the restricted geographic distribution of PSTV in the near future.

B. Eradication, Suppression, Containment

Once PSTV is found in some materials, it seems obvious that the first measure that should be taken is to continue testing all potato material that has had a chance to become infected. The elimination of infected materials and the application of hygiene procedures should help reduce the risk of recontamination. At CIP, for instance, hygiene procedures are based on disinfecting tools and operators hands with a 10% solution of calcium hypochlorite. Other forms of hypochlorite or other chemicals were not effective in destroying PSTV infectivity.[22] In practice, two outbreaks of PSTV occurred at CIP in stem cuttings under rapid multiplication when calcium hypochlorite was not used. Soil used in sand beds for rapid multiplication appears to also be a source of PSTV because root pieces from infected stem cuttings remained infective after steam or chemical disinfection.

ACKNOWLEDGMENT

The author wishes to thank Drs. P. Gregory and E. R. French, and Mr. J. E. Bryan of the International Potato Center, Lima, Peru, for reviewing the manuscript and for their valuable suggestions.

REFERENCES

1. **Benson, A. P. and Singh, R. P.,** Seed transmission of potato spindle tuber virus in tomato, *Am. Potato J.*, 41, 294, 1964.
2. **Bonde, R. and Marriam, D.,** Studies on the dissemination of the potato spindle tuber virus by mechanical inoculation, *Am. Potato J.*, 28, 558, 1951.
3. **Cartwright, D. N.,** Potato spindle tuber viroid survey-South Australia, *Australas. Plant Pathol.*, 13, 4, 1984.
4. **De Bokx, J. A. and Piron, P. G. M.,** Transmission of potato spindle tuber viroid by aphids, *Neth. J. Plant Pathol.*, 87, 31, 1981.
5. **Diener, T. O. and Raymer, W. B.,** Potato spindle tuber virus: plant virus with properties of a free nucleic acid, *Science*, 158, 378, 1967.
6. **Diener, T. O.,** *Viroids and Viroid Diseases*, John Wiley & Sons, New York, 1979.
7. European and Mediterranean Plant Protection Organization, Specific quarantine requirements, *EPPO (Eur. Mediterr. Plant Prot. Organ.) Publ. Ser. B.*, 86, 53 pp., 1982.
8. **Fernandez Valiela, M. V. and Calderoni, A. V.** The search for potato growing areas in the Argentine Republic, *Actas Inst. Micol.*, 2, 60, 1965.
9. **Fernow, K. H., Peterson, L. C., and Plaisted, R. L.,** Spindle tuber virus in seeds and pollen of infected potato plants, *Am. Potato J.*, 47, 75, 1970.
10. **Fernow, K. H.,** Tomato as a test plant for detecting mild strains of potato spindle tuber virus, *Phytopathology*, 57, 1347, 1967.
11. **Foster, J. A.,** Regulation of plant germplasm imported into the United States, *EPPO Bull.*, 11, 156, 1981.
12. **Goss, R. W. and Peltier, G. L.,** Further studies on the effect of environment on potato degeneration diseases, *Nebr. Agric. Exp. Stn. Res. Bull.*, 29, 32 pp., 1925.

13. **Goss, R. W.** The symptoms of spindle tuber and unmottled curly dwarf on the potato, *Nebr. Agric. Exp. Stn. Res. Bull.*, 47, 39 pp., 1930.
14. **Goss, R. W.**, Transmission of potato spindle tuber by cutting knives and seed piece contact, *Phytopathology*, 16, 299, 1926.
15. **Goss, R. W.**, Transmission of potato spindle tuber by grasshoppers (Locustidae), *Phytopathology*, 18, 445, 1928.
16. **Goss, R. W.**, Insect transmission of potato virus diseases, *Phytopathology*, 20, 136, 1930.
17. **Gross, H. J., Domdey, H., Lossow, C., Jank, P., Raba, M., Alberty, H., and Sanger, H. L.**, Nucleotide sequence and secondary structure of potato spindle tuber viroid, *Nature*, 273, 203, 1978.
18. **Hadidi, A. and Diener, T. O.**, *In vitro* synthesis of potato spindle tuber viroid: kinetic relationship between the circular and linear forms, *Virology*, 86, 57, 1978.
19. **Hanneman, R. E. and Singh, R. P.**, Seed production in the virus indicator plant *Scopolia sinensis*, *Can. Plant Dis. Surv.*, 52, 60, 1972.
20. **Hunter, D. E., Darling, D. H., and Beale, W. L.**, Seed transmission of potato spindle tuber virus, *Am. Potato J.*, 46, 247, 1969.
21. International Potato Center (CIP), Annual Report, 1980, CIP, Lima, Peru, 1981, 198 pp.
22. International Potato Center (CIP), Annual Report, 1982, CIP, Lima, Peru, 1983, 148 pp.
23. International Potato Center (CIP), Potatoes for the Developing World, CIP, Lima, Peru, 1984, 150 pp.
24. International Potato Center (CIP), Annual Report, 1984, CIP, Lima, Peru, 1984, 167 pp.
25. **Kahn, R. P.**, The host as a vector: exclusion as a control, in *Pathogens, Vectors and Plant Diseases*, Harris, K. and Maramorosch, K., Eds., Academic Press, New York, 1982, 123.
26. **Kahn, R. P.**, Tissue culture applications for plant quarantine, in *Practical Tissue Culture Application*, Maramorosch, K. and Hirumi, H., Eds., Academic Press, New York, 1979, 186.
27. **Kahn, R. P.**, Plant quarantine: principles, methodology and suggested approaches, in *Plant Health and Quarantine in International Transfer of Plant Genetic Resources*, Hewitt, W. B. and Chiarappa, L., Eds., CRC Press, Boca Raton, Fl., 1977, 289.
28. **Le Clerg, E. L., Lombard, P. M., Eddins, A. H., Cook, H. T., and Campbell, J. C.**, Effect of different amounts of spindle tuber and leaf roll on yields of Irish potatoes, *Am. Potato J.*, 21, 60, 1944.
29. **Leont'eva, Yu-A.**, Identification of potato gothic, *Izv. Kuikyshev. Khoz. Inst.*, 14, 279, 1964; *Rev. Appl. Mycol.* 45, 273, 1966.
30. **Liu, A.**, Experimental results on separation of potato spindle tuber viroid (PSTV-RNA), *Acta Phytopathol. Sin.*, 10, 103, 1980.
31. **Lizarraga, R. E., Salazar, L. F., Roca, W. M., and Schilde-Rentschler, L.**, Elimination of potato spindle tuber viroid by low temperature and meristem culture, *Phytopathology*, 70, 754, 1980.
32. **Lizarraga, R. E., Salazar, L. F., and Schilde-Rentschler, L.**, Effect of meristem size on eradication of potato spindle tuber viroid, in *Research for the Potato in the Year 2000*, Hooker, W. J., Ed., International Potato Center, Lima, Peru, 1983, 118.
33. **Martin, W. H.**, "Spindle tuber", a new potato trouble. Hints to potato growers, *N.J. State Potato Assoc.*, 3, 4 pp., 1922.
34. **Martinez, E.**, La tinción de geles en electroforesis con nitrato de plata y su applicación para identificar viroides, Tesis Biol. Univ. Nac. Pedro Ruiz Gallo, Lambayeque, Peru, 1984, 39 pp.
35. **Manzer, F. E. and Merriam, D.**, Field transmission of the potato spindle tuber virus and virus X by cultivating and hilling equipment, *Am. Potato J.*, 38, 346, 1961.
36. **Merriam, D. and Bonde, R.**, Dissemination of spindle tuber by contaminated tractor wheels and by foliage contact with diseased plants, *Phytopathology*, (Abstr.), 44, 111, 1954.
37. **Morris, T. J. and Wright, N. S.**, Detection on polyacrylamide gel of a diagnostic nucleic acid from tissue infected with potato spindle tuber viroid, *Am. Potato J.*, 52, 57, 1975.
38. **Morris, T. J. and Smith, E. M.**, Potato spindle tuber disease: procedures for the detection of viroid RNA and certification of disease-free potato tubers, *Phytopathology*, 67, 145, 1977.
39. **Owens, R. A. and Cress, D. E.**, Molecular cloning and characterization of potato spindle tuber viroid cDNA sequences, *Proc. Natl. Acad. Sci. U.S.A.*, 77, 5302, 1980
40. **Owens, R. A. and Diener, T. O.**, Sensitive and rapid diagnosis of potato spindle tuber viroid disease by nucleic acid hybridization, *Science*, 213, 670, 1981.
41. **Paludan, N.**, Elimination of chrysanthemum chlorotic mottle and chrysanthemum stunt viroids, *6th. Int. Symp. Virus Diseases of Ornamental Plants*, Cornell University, Ithaca, N. Y., June 1984.
42. **Pfannenstiel, M. A. and Slack, S. A.**, Response of potato cultivars to infection by the potato spindle tuber viroid, *Phytopathology*, 70, 922, 1980.
43. **Pfannenstiel, M. A., Slack, S. A., and Lane, L. C.**, Detection of potato spindle tuber viroid in field-grown potatoes by improved electrophoretic assay, *Phytopathology*, 70, 1015, 1980.
44. **Raymer, W. B. and O'Brien, M. J.**, Transmission of potato spindle tuber virus to tomato, *Am. Potato J.*, 39, 401, 1962.

45. **Raymer, R. W., O'Brien, M., and Merriam, D.,** Tomato as a source of an indicator plant for the potato spindle tuber virus, *Am. Potato J.*, 41, 311, 1964.
46. **Salazar, L. F.,** unpublished data, 1981.
47. **Salazar, L. F., Balbo, I., and Owens, R. A.,** unpublished results.
48. **Salazar, L. F., Owens, R. A., Smith, D. R., and Diener, T. O.,** Detection of potato spindle tuber viroid by nucleic acid spot hybridization: evaluation with tuber sprouts and true potato seed, *Am. Potato J.*, 60, 587, 1983.
49. **Sanger, H. L.,** Biology, structure, functions and possible origin of viroids, in *Nucleic Acids and Proteins in Plants,* Vol. 2, Parthier, B. and Boulter, D., Eds., Springer-Verlag, Berlin, 1982, 368.
50. **Sanger, H. L. and Ramm, K.,** Radioactive labelling of viroid RNA, in modification of the information content of plant cells, in *Proc. 2nd John Innes Symp. July 1974,* Norwich, England, Markham, R. et al., Eds., 1975, 250.
51. **Schultz, E. S. and Folsom, D.,** Transmission, variation, and control of certain degeneration diseases of Irish potatoes, *J. Agric. Res.*, 25, 43, 1923.
52. **Schwinghamer, M. W., Scott, G. R., Mallinson, F. K., Tesoriero, L. A., and Morrison, W. L.,** Potato spindle tuber viroid: an extensive infection in the New South Wales potato breeding programme, Abstr. *4th Int. Plant Pathol.*, Conf., Melbourne, Aust., 1983, 122.
53. Scottish Plant Breeding Station, 51st Annual Report, Edinburgh, 1972, 62.
54. Scottish Plant Breeding Station, 55th Annu. Rep., Edinburgh, 1976.
55. **Singh, R. P.,** A local lesion host for potato spindle tuber virus, *Phytopathology,* 671, 1034, 1971.
56. **Singh, R. P.,** Seed transmission of potato spindle tuber virus in tomato and potato, *Am. Potato J.*, 47, 225, 1970.
57. **Singh, R. P., Finnie, R. F., and Bagnall, R. E.,** Losses due to the potato spindle tuber virus, *Am. Potato J.*, 48, 262, 1971.
58. **Singh, R. P.,** Viroids and their potential danger to potatoes in hot climates, *Can. Plant Dis. Surv.*, 63, 13, 1983.
59. **Semancik, J. S.,** Small pathogenic RNA in plants — the viroids, *Annu. Rev. Phytopathol.*, 17, 461, 1979.
60. **Stace-Smith, R. and Mellor, F. C.,** Eradication of potato spindle tuber virus by thermotherapy and oxillary bud culture, *Phytopathology,* 60, 1857, 1970.
61. **Thresh, J. M.,** The orgins and epidemiology of some important plant virus diseases, *Appl. Biol.*, 5, 1, 1980.
62. **Werner, H. O.,** The spindle tuber disease as a factor in seed potato production, *Nebr. Agric. Exp. Stn. Res. Bull.*, 32, 128 pp., 1926.
63. **Yang, T. C. and Hooker, W. J.,** Albinism of potato spindle tuber viroid-infected Rutgers tomato in continuous light, *Am. Potato J.*, 54, 519, 1977.

Chapter 12

PESTS AND PATHOGENS OF CASSAVA

J. C. Lozano and Barry L. Nolt

TABLE OF CONTENTS

I. INTRODUCTION

Cassava (*Manihot esculenta* Crantz) is grown throughout the tropical regions of the world. Its starchy roots are a major energy source for over 400 million people. A member of the Euphorbiaceae, cassava was domesticated in the Americas[69] more than 5000 years ago.[39]

The cassava plant is a perennial shrub whose leaves are formed at active apices and consist of an elongated petiole and a palmatic leaf blade. The plant exhibits apical dominance, producing a single stem with petioles borne on raised structures, giving the stem a characteristic knobby appearance. When the main apex enters the reproductive stage, apical dominance is broken, and two or four axillary buds immediately below the reproductive structure develop and branching occurs. Carbohydrates accumulate in the root parenchyma to form swollen storage organs. The growing season ranges from 8 to 24 months depending on ecological conditions. The cassava plant can be grown from seed, but is usually propagated vegetatively from stem cuttings. Cassava is grown commercially at altitudes between sea level and 2000 m.[23]

During the last decade, the importance of increasing cassava production and research has been realized because of the shortage of carbohydrates in tropical regions. Many tropical countries lack foreign exchange to import sufficient carbohydrate sources, and too often population growth in cassava-producing countries is above the world average. Cassava is regarded as one of the best crops for carbohydrate production, even on the very poor soils typical of many tropical countries. Consequently, the exchange of cassava-propagating material as botanical seed, stem cuttings, or meristem-cultured plantlets has increased considerably with the aim of improving cassava production. Such exchange may be between continents (i.e., the Americas with Asia, Australia, and Africa), between countries, or between regions in the same country (i.e., the transitional areas between Campo Cerrado and the Amazonas regions of Brazil). The risk of introducing pests and pathogens into new areas is very high and unfortunately has already taken place, resulting in severe crop losses and famine in some regions.

II. DISEASE AND PEST COMPLEXES

Approximately 60 fungi, 5 bacteria, and 4 viruses have been reported as cassava pathogens.[45,53] In addition, several mycoplasma- and virus-like diseases occur in the Americas.[25,33] Most cassava pathogens appear to have evolved with the host in the Americas,[48,51] with the apparent exceptions of African cassava mosaic virus (ACMV), which occurs in Africa and the Indian subcontinent,[59,61] and cassava brown streak virus (CBSV), reported from East Africa.[7] Fortunately, most disease agents are still confined to their purported centers of origin as a result of isolated cropping systems in each region and the limited movement of germplasm from the Americas.[48] Nevertheless, there have been several recent introductions of pathogens into new areas within the Americas[4] and from the Americas to other continents.[3,44]

Over 200 species of insects and mites have been reported as cassava pests, but the majority are of minor economic importance.[2] Spider mites, thrips, stemborers, whiteflies, mealybugs, scales, and hornworms represent the most important cassava pests.[1,2] All of the 17 most important pests are found in the Americas; 12 of these are also present in Africa and 5 can be found in Asia. The green spider mite (*Mononychellus tanajoa*) and the mealybug (*Phenacoccus manihoti*) both originated in the Americas but now cause devestating yield losses in Africa.[3,54] The white scale (*Aonidomytilus albus*), possibly of American origin, is now found worldwide.[3] *Bemisia tabaci*, the whitefly vector of ACMV, is found throughout Africa and Asia but is not a serious pest of cassava in the Americas.[1,2]

III. QUARANTINE PROBLEMS

The exchange of cassava stem cuttings is probably the major means of disseminating cassava pathogens and pests. The introduction of green spider mites, mealybugs, and the bacterial blight pathogen into Africa are important examples of this.[3-44] The bacterial blight pathogen (*Xanthomonas campestris* pv. *manihotis*) can survive in the xylem vessels of infested stems for months.[44] Cassava viruses and mycoplasmas are efficiently harbored in stem cuttings from infected plants and readily transferred to plants propagated from infested cuttings.[46] *Mononychellus tanajoa*, *P. manihotis*, and *A. albus* can survive for months by feeding on the lateral buds of stem cuttings.

The green spider mite was introduced from the Americas (probably Brazil) into Africa before 1960. It was first reported in Uganda in 1971 and then spread rapidly to over 60% of the cassava-growing areas of Africa. This pest is now found in a wide belt from Mozambique northward through Zaire and the Central African Republic to the coastal regions of West Africa in Senegal and Guinea Bissau.[32]

Green spider mites reproduce rapidly in a hot, dry climate. Development is completed in 8 to 12 days, and each female mite produces an average of about 50 eggs in a 2- to 3-week period. The larvae, both nymphal stages and the adults, feed on young leaves, causing severe chlorosis and up to 95% leaf-area reduction during the dry season when population levels are very high. In severe infestations the entire apical shoot dries up and dies. As the food supply on infested plants is depleted, the females drop silken threads 15 to 30 cm long and can drift in the wind for long distances. This mechanism accounts for the rapid spread of the green spider mite in Africa of up to 375 km/year.[1,2]

The cassava mealybug was probably introduced into Africa around 1970,[1,54,55] but may have been present earlier.[56] Spread has been rapid because of favorable environmental conditions[41] and the lack of natural predators. Severe attacks of cassava mealybugs can cause almost complete leaf loss and up to 80% reductions in root yield. The reduction in leaf canopy allows weeds to develop, causing further crop loss by competition. Although the plant may recover with the onset of the rainy season, the resulting weakened planting material will produce a more susceptible plant the following season[1,2]

The cassava mealybug is a parthenogenetic species, developing from egg to adult in 27 days under tropical conditions. The adults live about 20 days, during which time they produce as many as 400 eggs, most of them in the first 10-day period. The mealybug attacks the growing points of the plant first, producing a stunted, bunched effect in the terminal shoots. A toxin present in the salivary juice contributes to this leaf distortion. Very young plants may be killed, and any attacked plant is significantly weakened.[1]

It has been estimated that these two pests cause economic losses of nearly $2 billion/year in Africa. In some areas, losses are so severe that farmers are beginning to abandon the crop altogether despite its long history as a famine food.[31]

Cassava bacterial blight (CBB), caused by *X. campestris* pv. *manihotis*, was introduced into Africa via infected cassava cuttings, possibly in the early 1960s. In 1973 it was first recognized as a significant disease on the African continent in Nigeria and later in Zaire.[65] In 1975 and 1976, the disease was found in other African and Asian countries.[44]

Bacterial blight has been considered one of the most important diseases of cassava in Africa and in many regions of America. Losses of 100% have been reported on susceptible clones growing in environmental conditions that favor the disease. Although it has been reported in almost all cassava-growing countries, there are still many areas within these countries that remain free of the disease; nevertheless, CBB remains a threat unless its introduction is prevented.[44-46]

It is suspected that cassava viruses have also spread into new cassava-growing regions via the interchange of infected vegetative material. It has not been possible to substantiate claims of introduction because, with the exception of ACMV and cassava common mosaic virus (CCMV), cassava viruses and virus-like agents have not been sufficiently characterized.

A. Pathogens and Pests Disseminated via Botanical Seed

Several cassava pathogens can be disseminated by the interchange of botanical seeds. They can be divided into two broad groups: those that infest the seed and those that infect it.

Information regarding the dissemination of pathogens via infested seed does not currently exist in the literature, but the possibility cannot be overlooked. Cassava seeds are produced in a tricapsular fruit, and infestation may occur after fruit dehiscense when the seeds are released. If the seeds fall to the ground, the probability of seed infestation is higher than when the seeds are collected and stored under controlled conditions. Pathogens of cassava that can probably infest the seeds and survive on them are those producing abundant muscilagenous propagules, such as *Colletotrichum* spp., *Phoma* spp., *Diplodia* spp., and *X. campestris* pv. *manihotis*. In many cases, the infestation of the storage containers could be much riskier than of the seed per se.

Pathogens that infect cassava seeds include *X. campestris* pv. *manihotis*,[16,28] *Diplodia manihotis*, *Fusarium* spp., and *Cladosporium* spp.[50] However, considering the limited information available at present, this does not preclude the possibility of other fungal and bacterial pathogens.

A high percentage of botanical seeds collected from CBB-affected plantations are symptomless carriers of *X. campestris* pv. *manihotis*. Similarly, seeds collected from many apparently healthy plantations are also affected by fungal pathogens and saprophytes. When infections by CBB or fungal pathogens are severe, the seeds show deformations, necrotic areas on the cotyledons and endosperm, and corrugation of the testa. These seeds have a very low germination rate. In CBB-infected seeds without visible symptoms, the bacterial pathogen is generally located in the embryo in a dormant state, which is overcome soon after germination. The seeds germinate normally, but disease symptoms are visible along the stem or leaves.[16,28,46]

Information on seed transmission of cassava viruses and virus-like agents is incomplete. Neither ACMV or CCMV is transmitted via cassava seeds;[25,27] however, determination of the seed transmissibility of all cassava viruses is essential for the safe interchange of botanical seeds. This task will be facilitated by the eventual characterization and development of sensitive indexing methods for each of the cassava viruses.

In relation to insects, there are few insects that attack cassava seeds, so the risk of disseminating these pests is relatively low.[3] Adult female fruitflies *(Anastrepha manihoti* and *A. pickeli)* oviposit on fruits,[1-3] and larvae feed on immature seeds, consuming them completely. The risk of dissemination is minimal as intact fruits are generally not distributed. This may be one of the reasons why cassava fruitflies are still confined to the Americas.[3]

B. Pathogens and Pests Disseminated through Vegetative Planting Material

The dissemination of cassava pathogens and pests through vegetative material is much more probable than through botanical seeds. During the long vegetative cycle of the cassava plant, the stem, which is normally used for propagation, is exposed to infestation and infection by fungal, bacterial, mycoplasmal, and viral pathogens, as well as by insects and mites. Lignified stems can also be invaded by most cassava pathogens without showing visible or easily noticeable symptoms.[53]

1. Pathogens

Considering the type of invasion and damage caused, cassava pathogens can be classified as localized or systemic.

a. Localized Pathogens

These pathogens are unable to invade the plant systemically; thus damage is localized to limited areas or portions of the plant. Characteristic symptoms are usually the formation of cankers, galls, necrosis of the epidermal or cortical areas (yellow to blackish-brown coloration), and deterioration of the pith. Pith, cortical, or epidermal degradations can be specific to a given causal agent or to a complex induced by several pathogens.[52,53]

The most common symptoms caused by localized pathogens are galls (*Agrobacterium tumefasciens*), cankers (*Elsinoë brasiliensis*), epidermal and cortical ulcers (*Colletotrichum* and *Phoma* spp.), pith and cortical degradation (*E. carotovora* pv. *carotovora*), and bud necrosis (*X. campestris* pvs. *manihotis* and *cassavae*).[53]

b. Systemic Pathogens

Some pathogens are able to invade susceptible plants systemically. Generally, a systemic invasion does not produce visible symptoms on the lignified mature stem portions, making it almost impossible to identify diseased material. Symptoms are usually noticeable on leaves, very young unlignified branches, or on the roots. Systemic pathogens, such as viruses, do not produce visible symptoms on all clones but still cause a significant reduction in vigor and yield.[52] From the standpoint of quarantine, systemic pathogens are very important and merit special attention. The following systemic pathogens can be disseminated by the use of infected vegetative material.

i. Fungal pathogens

Diplodia manihotis and *Fusarium* spp. are the most important fungal pathogens affecting cassava production. These fungi induce vascular streaking along the stem, which results in wilting. In the case of *D. manihotis*, epidermal symptoms are only noticeable on the very young portion of the stem, which makes it very difficult to detect diseased vegetative material.[17-19] A similar syndrome is also noticeable in the case of infections induced by species of *Verticillium* and *Fusarium*.[18,19] Although *E. brasiliensis*, the causal agent of superelongation disease, is not a systemic pathogen, it primarily infects the stem epidermal tissues, causing minute cankers that are difficult to identify visually.[77]

ii. Bacterial pathogens

Xanthomonas campestris pv. *manihotis*, the causal agent of CBB, is one of the most important pathogens of cassava. The distribution of infected vegetative planting material from diseased plantations has been the main means of disseminating this pathogen over long distances into Africa, Asia, and other regions where it did not exist before. This is due mainly to the lack of visible symptoms on lignified stems and the ability of the pathogen to survive in the invaded tissues for a very long time.[12,13,47] However, the report that *X. campestris* pv. *cassavae* can be disseminated through the use of infected cuttings because of its systemic invasion needs further confirmation.[57]

iii. Mycoplasmal pathogens

Mycoplasma-like organisms (MLOs) have been associated with several cassava diseases in Brazil and Colombia.[25-33] The MLOs described from Brazil induce severe witches' brooming, yellowing, vein clearing, and stunting.[53] The MLO from Colombia affects the floral parts, causing an antholysis syndrome (metamorphosis of floral parts: normal flower to a completely vegetative leafy branch), a progression of virescence, hypertrophy, and phyllody.[33] The insect vectors involved in the transmission of the cassava MLOs are not known, but all are disseminated efficiently via cuttings from infected plants. The incidence and distribution of the cassava mycoplasmal diseases is low, but the root yield of infected plants can be reduced substantially even when only the floral parts are affected.[20] MLOs have been successfully eradicated from cassava cuttings by thermotherapy in a hot-water bath[25] and from infected plants by chemotherapy with tetracycline and streptomycin.[33] It is possible to detect MLOs in the sieve tubes of infected plants with the aid of an electron microscope.

iv. Viral pathogens

All viruses and virus-like agents of cassava can be disseminated via vegetative material from infected mother plants. The identification of virus-free plants based solely on freedom from typical symptoms is unreliable under field conditions because the expression of disease symptoms can be suppressed by environmental conditions, particularly high temperature, and confounded by pest and pathogen damage or symptoms due to nutritional disorders. If possible, cassava plants should be assessed for freedom from virus symptoms only under optimal conditions for symptom expression such can be obtained in the greenhouse. Nevertheless, several virus-like agents found in Latin America remain symptomless in most cassava clones, but can be detected by grafting to a susceptible, symptom-producing clone in the greenhouse. Although apparently free, the yield and vigor of infected plants are significantly reduced.[16-20]

Only two cassava viruses, ACMV in Africa and CCMV from Latin America, have been adequately characterized. Two viruses, CBSV in East Africa and cassava veinal mosaic virus (CVMV) in Brazil, have been isolated from infected plants, but their characterization is incomplete.[7,25] Of the virus-like diseases reported from Latin America, cassava frog skin disease (FSD) in the Amazon basin, Caribbean mosaic disease (CMD) on the north coast of Colombia, and latent mosaic diseases are considered quarantine threats.[49]

African cassava mosaic disease[35,74] caused by ACMV[8] is the most serious cassava disease on the African continent in terms of geographic distribution and economic losses.[75] ACMV has also been found in southern India,[58] but to date there have been no reports of its occurrence in Latin America. ACMV is a geminivirus composed of paired virus particles, 20 × 30 nm in size,[5] containing a circular single-stranded DNA (ssDNA) genome.[29] The virus is transmitted by the whitefly *Bemisia tabaci*, with a rapid rate of dissemination in some regions.[11,40] Recently the ACMV genome has been completely sequenced[73] and found to consist of two circular ssDNA molecules, which are both required for infectivity.[72] ACMV has been purified and an antiserum produced that is suitable for use in the enzyme-linked immunosorbent assay (ELISA).[9,71]

CCMV, first reported from Brazil[24] and later from Peru[25] and Colombia,[19] is probably present in all cassava-growing regions of Latin America. Although the incidence of CCMV-infected plants is low, yield losses per plant can be as high as 60%.[25] As there are no known vectors reported, the main means of dissemination is via infected planting material.[25] CCMV is a member of the potexvirus group with elongated rod-shaped particles 15 nm wide by 495 nm long.[38] The virus reaches high concentrations in most host species and can be readily purified for antiserum production. Effective control measures include the use of virus-free planting material, clean propagating tools, and roguing of infected plants.[25]

CBSV was first reported from Tanzania in 1936[62] and is also found in Kenya.[6] Although severely infected roots are not marketable,[35] the incidence and economic importance of CBSV is low in comparison to ACMV. The elongated rod-shaped particles associated with the cassava brown streak disease[36,43] have an average length of 650 nm and are similar to viruses in the carlavirus group;[6] however, their role in the etiology of the disease is still unknown.

CVMV was first reported from Brazil in 1940,[24] but its incidence in Brazil is thought to be very low.[25] Spherical virus particles, 40 to 50 nm in diameter, have been associated with CVMV-infected plants.[37] CVMV has been purified and an antiserum produced. The virus particles contain a double-stranded DNA genome,[42] making CVMV a tentative member of the caulimovirus group.[30] There are no known vectors of CVMV, and the main source of dissemination is via infected planting material.[25] As the incidence of CVMV and field transmission are very low, roguing infected plants is the recommended control practice.

Frog skin disease was first reported from southern Colombia in 1971.[12] Roots from infected plants are reduced in size and are usually not marketable because of the presence of deep cortical cracks or scale-like formations on the epidermis. There are no distinctive symptoms on the aboveground parts of most clones. The etiology of FSD is unknown; but elongated rod-shaped as well as spherical virus-like particles have been observed in leaf sap from infected plants. The causal agent has not been mechanically transmitted, but it is graft transmissible and can be disseminated via infected planting material. There is inconclusive evidence that whiteflies may be involved in field transmission.

A severe mosaic disease of cassava grown in the Caribbean coast region of Colombia was described in 1981.[16] Susceptible plants are stunted with deformed curled leaves, which exhibit a brilliant yellow mosaic. Thus far, no virus-like particles have been associated with the disease; however, the etiological agent is readily graft transmissible and is disseminated via infected planting material. Field transmission into plots of healthy plants is rapid, suggesting the existence of a biological vector.[21]

In 1982 a mosaic-producing agent was detected in several apparently symptomless cassava clones by grafting to scions of a susceptible indicator clone.[17] An elongated rod-shaped virus can be mechanically transmitted from symptomless plants. Its role in the etiology of the disease is under investigation.[20]

2. Pests

The most important pests are disseminated via stem cuttings. The green spider mite (*M. tanajoa*), the mealybug (*P. manihotis*), and the white scale (*A. albus*) were probably disseminated from the Americas to other continents by this means. Of the arthropods that attack cassava, those able to survive by feeding on lateral buds on the stem are the ones that can be disseminated on stem cuttings. The most important pests are mites, scales, mealybugs, and stemborers.

a. Mites

Forty-six species of mites have been reported attacking cassava; but in most cassava-growing regions, mites are of little economic importance. However, under certain environmental conditions, they can cause severe losses, reducing yields by up to 80%.[3,10,13,63] The most destructive species is the green mite *M. tanajoa*. All the important mite species are found naturally in the Americas, but *Tetranychus telarius* appears to be confined to Cuba.

Mites are important in regions with extended dry periods of 3 or more months. Their life cycle is very short; but because of their high multiplication rate, populations increase rapidly during the dry season.[1-3] Mites feed on the leaves, leaving yellowish-white dots. Leaf tissues are eventually destroyed; and during severe attacks, leaf fall and plant death can result. *Mononychellus* spp. feed on very young leaflets, even on buds; the nymphs are minute and difficult to detect. Mites can easily be disseminated by interchanging stem cuttings.

b. Thrips

Several species of thrips attack cassava in the Americas. The most important is *Frankliniella williamsi*, which can reduce yields by up to 28%.[3,70] It could be a potentially serious pest if it were introduced into Africa and Asia. *Retithrips syriacus* affects cassava plantations in India and Australia; *Euthrips manihoti* appears to be restricted to Brazil.

When *F. williamsi* attacks the stems and petioles of cassava, the infested parts become suberized and the stem internodes are shortened. General symptoms are very similar to those of cassava viral diseases. Larvae and adults of *F. williamsi* and *F. manihoti* can survive on stem buds; thus the interchange of cassava cuttings may disseminate these pests.

c. Scales

Several scale species attack cassava stems in many parts of the world. The most important species are *Aonidomytilus albus*, which is distributed worldwide,[1-3] and *Saissetia* sp.[3] A high scale population can cause serious damage to the plant. *A. albus* induces leaf fall and, when the whole stem is covered, it can reduce yield by 19%.[3,14] The most important damage, however, is the reduction in the quality of planting material resulting from stem bud attack.[3,76]

d. Mealybugs

Severe attacks by these insects have been reported in Brazil and Colombia, as well as in Africa. Losses of up to 80% have been calculated.[3] *Phenacoccus manihoti* and *P. herreni* attack the very young shoots first, then the petioles, and finally the expanded leaves, resulting in shortened internodes, rolled leaves, and delayed development of new leaves.[3,41]

Adult females move to lateral buds on the stem; therefore, cuttings from affected plants can disseminate these insects. As *P. manihoti* has only been reported in Africa, Brazil, Paraguay, and Bolivia, and *P. herreni* only in Brazil and Colombia, it is advisable to take sanitary precautions to avoid their dissemination into unaffected regions.[1-3]

IV. SANITARY MEASURES TO MINIMIZE RISK OF DISSEMINATING PATHOGENS AND PESTS THROUGH PROPAGATIVE MATERIAL

The following sanitary measures, independent of others legally established by quarantine regulations, could reduce the risk of disseminating pathogens and pests through propagative material of cassava. Their effectiveness depends on the strict application of such measures by both the donor and recipient.

A. Vegetative Propagative Material

1. The recipient country or institution must carefully analyze and evaluate the responsibilities and competence of donors as regards compliance with established quarantine regulations. Any introduction should be approved only after a careful analysis of the need and genetic advantages for such an introduction.
2. The recipient country should be cautious in the introduction of cassava propagating material from countries or areas where exotic diseases exist. For example, because of the undetermined origin of ACMV, cassava vegetative material should only be imported from Africa, India, or any other country where the disease is known to be present after adequate virus indexing has been done.
3. The vegetative propagating material should be collected from plantations apparently free of systemic disease. More than one inspection should be made at the appropriate time before collecting such material in order to determine the apparent sanitary condition of the plantation. Such inspections should be made during the optimum climatic conditions for disease development; for example, during the middle to the end of the wet seasons for diseases such as CBB, superelongation, and *Diplodia* stem rot, or during the optimum conditions established for the expression of viral symptoms. Before collection, vigorous, healthy plants should be identified.
4. Cassava cuttings should not be introduced to or exported from any country or any geographic region. Vegetative material should be introduced only as meristem cultures and only after sufficient tests have been carried out to determine the sanitary conditions of the introduced material. Meristems should be cut from shoots sprouted on cuttings taken from apparently healthy plants from plantations free from known viral, mycoplasmal, and virus-like diseases. The cuttings should be sprouted and maintained at a high temperature for several weeks before obtaining meristems for culture to minimize

the risk of virus contamination.[18] Any suspect material should be eliminated by autoclaving (120 °C and 20 lb pressure for 45 min) or by burning. All introduced materials should receive a postentry quarantine treatment for at least a period of 1 year (growing cycle). During this period, the sanitary conditions of the introduced materials should be observed carefully and periodically; at least three periodic bioassays should be carried out in addition to any other available test that can help in the detection of systemic diseases of cassava. It should be kept in mind that there are mosaic-carrier varieties of cassava that do not show any visible symptoms even though they are virus infected and can therefore become the focus of dissemination of diseases when introduced into clean areas. These diseases could be a threat to the cassava varieties in the region and even to other crop species. As information on systemic diseases of cassava in relation to their identification, detection, and epidemiology is very limited, the introduction of vegetative material should be kept at a minimum, and clones should be imported only when absolutely necessary.

B. Botanical Seed Material

There are several seed-borne fungal and bacterial pathogens in most of the cassava-growing areas which represent potential quarantine risk to cassava in areas where the infested/infected material is introduced. The eradication of the CBB pathogen from seed has been attempted by hot water[66] and dry-heat treatments.[17] However, the former treatment appears to be of low efficiency,[28] and the latter did not control seedborne fungal pathogens.[18,19] It appears that the best way to avoid disseminating cassava seed-borne pathogens can be by following these recommendations carefully:

1. Collecting apparently healthy mature fruits from healthy plants and drying them before seed removal
2. Selecting seeds with normal size and shape
3. Discarding seeds that float in water
4. Treating high-density seeds (approximately 100 seeds) in water (300mℓ in a 600-mℓ Pyrex® beaker), placed centrally in the cavity (volume 39975 cm^3) of a microwave oven at full power (1400 W heating power, 2.450 MHz) for 120 sec
5. Dusting microwave-treated seeds with Arasan® (tetramethyl-thiuran disulfide) before packing in sterile paper bags. As the optimum temperature obtained with the microwave treatment can be modified by several factors, the system should be set according to the equipment available[50]

C. Virus-Detection Methods Available for Indexing Cassava

Various virus-detection techniques are currently available for indexing cassava for the presence of viruses and virus-like agents. These techniques differ considerably with respect to sensitivity, ease of manipulation, cost, and labor inputs. The choice of a particular technique or combination of techniques depends on the facilities and personnel available and the level of confidence required for certifying a plant virus free. An institute involved in the shipment of cassava germplasm between continents or between countries would be expected to make the greatest effort to increase the probability of ensuring freedom from all known viruses.

Cassava virus-detection methods can be based on the observation of symptoms or on the detection of virus particles and viral products. Detection based on symptomatology includes the observation of characteristic symptoms on cassava, on inoculated indicator plants, or on grafted susceptible scions. Cassava virus particles or viral products can be detected using virus-specific antiserum in the ELISA or immunosorbent electron microscopy (ISEM) tests. The isolation of double-stranded RNA (dsRNA) and nucleic acid hybridization tests permits the detection of cassava viruses without the use of virus-specific antiserum.

The reliability of detection methods based on the observation of symptoms can be increased by observing plants maintained under optimal conditions for symptom expression. For example, the symptoms of CMD are not expressed at temperatures above 28°C.[16,17] In this case, plants are grown in a controlled environment for symptom development. However, in the case of FSD, root symptoms are readily apparent under field conditions. At the Centro Internacional de Agricultura Tropical (CIAT),[21] for example, the roots of plants selected for virus indexing are routinely checked for symptoms of FSD in the field, and cuttings taken from these plants are sprouted at 25 to 30°C in the glasshouse to enhance the symptom expression of other viruses. It is also possible to select apparently ACMV-free clones in Kenya based on the absence of visible mosaic symptoms.[6]

The bioassay of mechanically transmissible cassava viruses to indicator hosts is a sensitive indexing method if a very susceptible bioassay host is available, virus concentration in the test plant is high, and environmental conditions are optimal for symptom expression. The Nigerian isolate of ACMV produces a severe, systemic infection in inoculated *Nicotiana benthamiana* plants.[32] The Kenyan isolate of CBSV can be bioassayed on *N. debneyi*.[6] Graft indexing is a very sensitive method if a highly susceptible indicator clone is used in the graft. The native Colombian clone Secundina has been selected at CIAT for its high susceptibility to the frog skin, Caribbean mosaic, and latent diseases in Latin America. When Secundina is grafted onto an infected rootstock, a moderate to severe mosaic symptom is expressed by Secundina.[15] Grafting provides a method for indexing viruses and virus-like agents that are not mechanically transmissible, e.g., CMD. Although a graft-indexing program requires minimal facilities and training, the procedure is labor intensive and indexing results are not available for several weeks. Another major constraint is the difficulty in maintaining virus-free stocks of the indicator clone.

Sensitive serological tests are available for indexing cassava for viruses that have been isolated, purified, and an antiserum produced. ELISA[22] is a highly sensitive, efficient, and rapid method for detecting ACMV[71] and CCMV[20] in cassava. The ISEM[26,67] test can also be used for detecting ACMV[71] and CCMV.[20] ELISA is suited to a large-scale virus-indexing program, so hundreds of plants can be tested in a day with results available within 36 hr. ISEM can only be used by personnel with access to an electron microscope facility. The preparation of test material and examination of grids is simple and rapid, but the use of the electron microscope is expensive, so it can only be used for a limited number of samples. Although ISEM is not as sensitive as ELISA, it has the advantage of providing a virus identification within several hours.

Nucleic acid or spot hybridization and isolation of viral-specific dsRNAs have recently been applied for detecting cassava viruses. Neither test is based on the use of virus-specific antiserum. Spot hybridization[64] has been adapted for detecting ACMV in cassava.[68] The procedure is based on the use of a radioactively labeled DNA molecule that is complementary to the viral genome to "probe" spots of leaf sap for the presence of viruses. The test is highly sensitive and suited for processing large numbers of samples. Isolation of dsRNAs as a plant virus-detection method[60] is being tested for detecting cassava viral infections.[21] Although dsRNA molecules are not normally a component of normal plant cells, they are readily produced during the replication of many plant viruses. Therefore, the detection of dsRNA in a plant can be used as a general indicator for viral infections. This technique is particularly suitable for detecting uncharacterized viruses for which an antiserum or nucleic acid probe is not available. The extraction and analysis of dsRNA are somewhat laborious, making the test more suited for indexing a limited number of mother plants than as a general screening method.

V. GENERAL CONCLUSIONS

The continued development of improved cassava hybrids through conventional plant breeding methods, the enhancement of the yield potential of native clones through thermotherapy and meristem culture, and the application of plant biotechnology to genetic improvement are increasing the amount of cassava vegetative material and botanical seeds interchanged between continents or countries. Inherent in the potential benefits to be gained from foreign material are the real risks posed by introducing exotic pathogens and pests. These risks can be minimized through a better understanding of the etiological agents, the development and deployment of improved eradication and indexing methods, and closer collaboration among quarantine officials and research scientists in the countries of origin and destination.

ACKNOWLEDGMENTS

The authors are very grateful to Robert Ziegler, pathologist from CIAT's Rice Program, for his valuable suggestions on the organization of this paper, and to Trudy Brekelbaum, Special Projects, for her editorial assistance.

REFERENCES

1. **Bellotti, A. and van Schoonhoven, A.,** Cassava Pests and their Control, Cassava Information Center, Centro Internacional de Agricultura Tropical, Cali, Colombia, 1976, 71.
2. **Bellotti, A. and van Schoonhoven, A.,** Mite and insect pests of cassava, *Annu. Rev. Entomol.*, 23, 39, 1978.
3. **Bellotti, A. and Vargas, O.,** Diseminación de las plagas de la yuca por semilla sexual, asexual y por la raíz seca almacenada, in *Proc. Primer Taller Latinoamericano sobre Germoplasma de Yuca y Papa,* Roca, W. M., Hershey, C. D., and Malamud, O. S., Eds., Centro Internacional de Agricultura Tropical, Cali, Colombia, 1982, 117.
4. **Bellotti, A., Reyes, J. A., and Varela, A. M.,** Observations on cassava mealybugs in the Americas; their biology, ecology and natural enemies, in *Proc. 6th Symp. Int. Soc. Tropical Root Crops,* CIP, Lima, Peru, 1984, 339.
5. **Bock, K. R.,** Maize streak, cassava latent and similar viruses, *3rd Int. Congr. Virol.*, Abstracts, 1975, 93.
6. **Bock, K. R.,** Crop virology research project (R3177), Kenya Agricultural Research Institute, ODA, London, 1984, 136.
7. **Bock, K. R. and Guthrie, E. J.,** Recent advances in research on cassava viruses in East Africa, in *Proc. African Cassava Mosaic Workshop,* IDRC, Ottawa, Canada, 1976, 11.
8. **Bock, K. R. and Woods, R. D.,** Etiology of African cassava mosaic disease, *Plant Dis.*, 67, 994, 1983.
9. **Bock, K. R., Guthrie, E. J., and Meredith, G.,** Distribution, host range, properties and purification of cassava latent virus, a geminovirus, *Ann. Appl. Biol.*, 90, 361, 1978.
10. **Byrne, D.,** Studies of Resistance to the Mites *Mononychellus tanajoa* (Bondar) and *Mononychellus caribbeanae* (McGregor) in Cassava (*Manihot esculenta* Crantz), Ph.D. dissertation, Cornell University, Ithaca, N.Y., 1980, 174.
11. **Chant, S. R.,** Studies on the transmission of cassava mosaic virus by *Bemisia* spp. (Aleyrodidae), *Ann. Appl. Biol.*, 46, 210, 1958.
12. CIAT, Cassava, in Annual Report 1974, Centro Internacional de Agricultura Tropical, Cali, Colombia, 1975, 73.
13. CIAT, Cassava, in Annual Report 1975, Centro Internacional de Agricultura Tropical, Cali, Colombia, 1976, 63.
14. CIAT, Cassava, in Annual Report 1976, Centro Internacional de Agricultura Tropical, Cali, Colombia, 1977, 76.
15. CIAT, Cassava, in Annual Report 1979, Centro Internacional de Agricultura Tropical, Cali, Colombia, 1980, 107.
16. CIAT, Cassava, in Annual Report 1980, Centro Internacional de Agricultura Tropical, Cali, Colombia, 1981, 99.
17. CIAT, Cassava, in Annual Report 1981, Centro Internacional de Agricultura Tropical, Cali, Colombia, 1982, 268.

18. CIAT, Cassava, in Annual Report 1982, Centro Internacional de Agricultura Tropical, Cali, Colombia, 1983, 229.
19. CIAT, Cassava, in Annual Report 1983, Centro Internacional de Agricultura Tropical, Cali, Colombia, 1984, 288.
20. CIAT, Cassava, in Annual Report 1984, Centro Internacional de Agricultura Tropical, Cali, Colombia, 1985, 270.
21. CIAT, Cassava, in Annual Report 1985, Centro Internacional de Agricultura Tropical, Cali, Colombia, unpublished data.
22. **Clark, M. F. and Adams, A. N.,** Characteristics of the microplate method of enzyme-linked immunosorbent assay for the detection of plant virus, *J. Gen. Virol.*, 34, 475, 1977.
23. **Cock, J. H.,** *Cassava: New Potential for a Neglected Crop*, Westview Press, Boulder, Colo., 1985, 191.
24. **Costa, A. S.,** Observacoes sobre o mosaico comun e o mosaico das nervaduras da mandioca (*Manihot utilissima* Pohl), *J. Agron.*, 3, 239, 1940.
25. **Costa, A. S. and Kitajima, E. W.,** Studies on virus and mycoplasma diseases of the cassava plant in Brasil, in *Proc. Cassava Mosaic Workshop*, IITA, Ibadan, Nigeria, 1972, 18.
26. **Derrick, K. S.,** Quantitative assay for plant viruses using serologically specific electron microscopy, *Virology*, 56, 652, 1973.
27. **Dubern, J.,** Quelques propriétés de la mosaique africaine du manioc. I. La transmission, *Phytopathol. Z.*, 96, 25, 1979.
28. **Elango, F. and Lozano, J. C.,** Transmission of *Xanthomonas manihotis* in seed of cassava *(Manihot esculenta), Plant Dis.*, 64, 784, 1980.
29. **Harrison, B. D., Barker, H., Bock, K. R., Guthrie, E. J., Meredith, G., and Athinson, M.,** Plant viruses with circular single-stranded DNA, *Nature*, 270, 760, 1977.
30. **Hull, R.,** Caulimovirus group, in *CMI/AAB Descriptions of Plant Viruses*, No. 295, Commonwealth Agricultural Bureaux, 1984.
31. IITA, Biological Control of Cassava Mealybug and Green Spider Mites, Information Series No. 16, International Institute of Tropical Agriculture, Ibadan, Nigeria, 1974, 25.
32. IITA, Annual Report, Virology unit, International Institute of Tropical Agriculture, Ibadan, Nigeria, 1980, 134.
33. **Jayasinghe, U., Pineda, B., and Lozano, J. C.,** Antholysis in cassava (*Manihot esculenta* Crantz) possibly caused by mycoplasma-like organisms, *Phytopathol. Z.*, 109, 295, 1984.
34. **Jennings, D. L.,** Observations of virus diseases of cassava in resistant and susceptible varieties. I. Mosaic disease, *J. Exp. Agric.*, 28, 23, 1960.
35. **Jennings, D. L.,** Breeding for resistance to cassava viruses in East Africa, in *Proc. Cassava Mosaic Workshop*, International Institute of Tropical Agriculture, Ibadan, Nigeria, 1972, 40.
36. **Kitajima, E. W. and Costa, A. S.,** Elongated particles found associated with cassava brown streak, *East Afr. Agric. For. J.*, 30, 28, 1964.
37. **Kitajima, E. W. and Costa, A. S.,** Particulas esferoidais associadas do virus do mosaico das nervuras da mandioca, *Bragantia*, 25, 211, 1966.
38. **Kitajima, E. W., Wetter, C., Oliveira, A. R., Silva, D. M., and Costa, A. S.,** Morfologia do virus do mosaico comun da mandioca, *Bragantia*, 24, 247, 1965.
39. **Leon, J.,** Origin, evolution, and early dispersal of root and tuber crops, in *Proc. 4th Symp. Int. Soc. Tropical Root Crops*, Centro Internacional de Agricultura Tropical, Cali, Colombia, 1977, 277.
40. **Leuchner, K.,** Whiteflies: biology and transmission of african mosaic disease, in *Proc. Cassava Protection Workshop*, Brekelbaum, T., Bellotti, A., and Lozano, J. C., Eds., Centro Internacional de Agricultura Tropical, Cali, Colombia, 1977, 51.
41. **Leuschner, K. and Nwanze, K.,** Preliminary observations of the mealybug (Hemiptera: Pseudococcidae) in Zaire, in *Proc. Cassava Protection Workshop*, Brekelbaum, T., Bellotti, A., and Lozano, J. C., Eds., Centro Internacional de Agricultura Tropical, Cali, Colombia, 1977, 195.
42. **In, M. T. and Kitajima, E. W.,** Purificao e serologia do vírus do mosaico das nervaduras da mandioca, *Fitopatol. Brasil.*, 5, 419, 1980.
43. **Lister, R. M.,** Mechanical transmission of cassava brown streak virus, *Nature*, 183, 1588, 1959.
44. **Lozano, J. C.,** Bacterial blight of cassava, *PANS (Pest Artic. News Summ.)*, 21, 38, 1975.
45. **Lozano, J. C. and Booth, R. H.,** Diseases of cassava *(Manihot esculenta Crantz), PANS (Pest Artic. News Summ.)*, 20, 30, 1974.
46. **Lozano, J. C. and Jayasinghe, U.,** Problemas fitopatológicos de la yuca diseminados por semilla sexual y asexual, in *Proc. Primer Taller Latinoamericano Sobre Germoplasma de papa y yuca*, Roca, W. M., Hershey, C. D., and Malamud, O. S., Eds., Centro Internacional de Agricultura Tropical, Cali, Colombia, 1982, 295.
47. **Lozano, J. C. and Sequeira, L.,** Bacterial blight of cassava in Colombia. II. Epidemiology and control, *Phytopathology*, 64, 83, 1974.

48. **Lozano, J. C., Byrne, D., and Bellotti, A.,** Cassava/ecosystem relationships and their influence on breeding strategy, *Trop. Pest Manage.*, 26, 180, 1980.
49. **Lozano, J. C., Jayasinghe, U., and Pineda, B.,** Viral diseases affecting cassava in the Americas, *Cassava Newsl.*, CIAT, Cali, Colombia, 7, 2, 1983.
50. **Lozano, J. C., Laberry, R., and Bermudez, A.,** Use of a microwave treatment to eradicate seed-borne pathogens found in cassava botanical seed, in *Proc. 7th Symp. Int. Soc. Tropical Root Crops,* Guadeloupe, 1985, 66.
51. **Lozano, J. C., Hershey, C. H., Bellotti, A., and Zeigler, R.,** A comprehensive breeding approach to pest and disease problems of cassava, in *Proc. 6th Symp. Int. Soc. Tropical Root Crops,* CIP, Lima, Peru, 1984, 315.
52. **Lozano, J. C., Toro, J. C., Castro, A., and Bellotti, A.,** Production of Cassava Planting Material, Series GE-17, Cassava Information Center, Centro Internacional de Agricultura Tropical, Cali, Colombia, 1977.
53. **Lozano, J. C., Bellotti, A., Reyes, J. A., Howeler, R., Leihner, D., and Doll, J.,** Field Problems in Cassava, Series 07EC-1 (2nd ed.), Centro Internacional de Agricultura Tropical, Cali, Colombia, 1981.
54. **Lyon, W. F.,** A plant-feeding mite *Mononychellus tanajoa* (Bondar) (Acarina: Tetranychidae) new to the African continent threatens cassava (*Manihot esculenta* Crantz) in Uganda, East Africa, *PANS (Pest Artic. News Summ.*, 19, 36, 1973.
55. **Lyon, W. F.,** A green cassava mite recently found in Africa, *Plant Prot. Bull.*, 22, 11, 1974.
56. **Mallamaire, A.,** Les insects nuisibles au manioc en Afrique noire, in *Congrés du Manioc et de Plantes Féculentes Tropicales,* Inst. Colonial, Marseille, 1949, 72.
57. **Maraite, H. and J. Weyns, J.,** Distinctive physiological, biochemical and pathogenic characteristics of *Xanthomonas manihotis* and *X. cassavae,* in *Proc. Int. Symp. Diseases Tropical Crops,* Maraite, H. and Meyer, J. A., Eds., U.C.L., Louvain-la-Neuve, Belgium, 1978, 316.
58. **Menon, M. R. and Raychaudhuri, S. P.,** Cucumber: an herbaceous host of cassava mosaic virus, *Plant Dis.*, 54, 34, 1970.
59. **Mohanraj, V. and Narayanasamy, P.,** Hastening symptom expression for early detection of cassava mosaic virus infection, *Indian Phytopathol.*, 34, 372, 1981.
60. **Morris, T. J., Dodds, J. A., Hillman, B., Jordan, R. L., Lommel, S. A., and Tamaki, S. J.,** Viral specific dsRNA: diagnostic value for plant virus disease identification, *Plant Mol. Biol. Rep.*, 1, 27, 1983.
61. **Nichols, R. F. W.,** Virus diseases of cassava, in *Annu. Rep. East African Agricultural and Forestry Research Organization,* 1950, 18.
62. **Nichols, R. F. W.,** The brown streak disease of cassava, *East Afr. Agric. J.*, 15, 154, 1950.
63. **Nyiira, Z. M.,** Advances in research on the economic significance of the green cassava mite *(Mononychellus tanajoa)* in Uganda, in *Proc. Int. Exchange and Testing of Cassava Germplasm in Africa,* Terry, E. R. and MacIntyre, R., Eds., International Institute of Tropical Agriculture, Ibadan, Nigeria, 1975, 27.
64. **Owens, R. A. and Diener, T. O.,** Sensitive and rapid diagnosis of potato spindle tuber viroid disease by nucleic acid hybridization, *Science,* 213, 670, 1981.
65. **Persley, G. J.,** Distribution and importance of cassava bacterial blight in Africa, in *Cassava Bacterial Blight,* Persley, G., Terry, E. R., and MacIntyre, R., Eds., IDRC/IITA, Ibadan, Nigeria, IDRC-096e, 1976.
66. **Persley, G. J.,** Studies on the survival and transmission of *Xanthomonas manihotis* on cassava seed, *Ann. Appl. Biol.*, 93, 159, 1979.
67. **Roberts, I. M.,** Factors affecting immunosorbent electron microscopy (ISEM), *Report of the Scottish Horticultural Research Institute for 1980,* 1981, 106.
68. **Robinson, D. J., Harrison, B. D., Sequeira, J. C., and Duncan, G. H.,** Detection of strains of African cassava mosaic virus by nucleic acid hybridization and some effects of temperature on their multiplication, *Ann. Appl. Biol.*, 105, 483, 1984.
69. **Rogers, D. J.,** Studies of *Manihot esculenta* Crantz and related species, *Torrey Bot. Club Bull.*, 90, 1, 1963.
70. **van Schoonhoven, A. and Peña, J. E.,** Estimation of yield losses in cassava following attack from thrips, *J. Econ. Entomol.*, 69, 514, 1976.
71. **Sequeira, J. C. and Harrison, B. D.,** Serological studies on cassava latent virus, *Ann. Appl. Biol.*, 101, 33, 1982.
72. **Stanley, J.,** Infectivity of the cloned geminivirus genome requires sequences from both DNAs, *Nature,* 305, 643, 1983.
73. **Stanley, J. and Gay, R. M.,** Nucleotide sequence of cassava latent virus DNA, *Nature,* 301, 260, 1983.
74. **Storey, H. H. and Nichols, R. F. W.,** Studies of the mosaic diseases of cassava, *Ann. Appl. Biol.*, 25, 790, 1938.
75. **Terry, E. R. and Hahn, S. K.,** The effect of cassava mosaic disease on growth and yield of a local and an improved variety of cassava, *Trop. Pest Manage.*, 26, 34, 1980.

76. **Vargas, O.,** The white scale (*Aonidomytilus albus* Cockerell) on cassava, in *Proc. Cassava Workshop,* Brekelbaum, T., Bellotti, A., and Lozano, J. C., Eds., Centro Internacional de Agricultura Tropical, Cali, Colombia, 1978, 199.
77. **Zeigler, R. S., Lozano, J. C., and Alvarez, E.,** Summary of recent research on the superelongation disease of cassava, in *Proc. 6th Symp. Int. Soc. Tropical Root Crops,* CIP, Lima, Peru, 1984, 363.

Chapter 13

PLANT QUARANTINE DIAGNOSTIC PROBLEMS: VIRUSES

R. Stace-Smith and R. R. Martin

TABLE OF CONTENTS

I. INTRODUCTION

The main agricultural regions of the world are dependent to a large extent on introduced crops, and those crops that arise from indigenous species often account for a minority of the hectarage. To achieve this state, agriculturists have been encouraged to engage in extensive intercontinental distribution of plant materials. The scale of traffic has increased with improved modes of transportation and with more funds being made available for crop plant improvement. All of this activity has taken place in a vacuum insofar as the distribution of plant viruses is concerned, with the result that the viruses that infected a considerable portion of the vegetative plant material have been inadvertently distributed over wide geographical areas. Once distributed and established in their new homeland, the virus inoculum, a key factor necessary for the development of an epidemic, was in place. Field spread of the virus would not necessarily take place, however, unless an efficient vector was also introduced with the planting stock or was indigenous.

In either event, one cannot turn back the clock. The drive for increasing agricultural production has accelerated the dissemination of crop improvement or plant propagation material on a global scale, particularly in developing countries that have embarked on a plan of crop improvement or agricultural diversification. Such transfer has been encouraged by technical assistance programs and international agricultural research centers. How serious are the problems which are associated with the distribution of plant viruses along with plant materials which took place before the necessary quarantine infrastructure existed at either the country of origin or the receiving country? This question cannot be answered with any degree of certainty because there are few scientific data available on the intercontinental spread of plant viruses, particularly in the early years of germplasm collecting. Some evidence of quarantine hazards associated with plant movement in recent years would indicate that much of the material moved before the 1960s was infected, often with devastating consequences.[38]

Many plant viruses cause epidemics and pandemics, particularly if they spread from their endemic area of origin into countries where they have not existed before. In a review of "catastrophic" plant diseases, Klinkowski[40] identified five diseases caused by viruses and three by mycoplasma-like organisms (MLOs) which were sufficiently devastating in their consequences to warrant inclusion. These diseases generally followed a typical course — they began, rose to cause an epidemic, expanded into a pandemic, and subsided into a stage that was tolerable. The three MLOs that were included — peach yellows, phony peach, and pear decline — are considered to be of minor importance today. The five viruses selected — sugarcane mosaic, cacao swollen shoot, plum pox, sugar beet yellows, and tobacco veinal necrosis — are still considered serious virus problems. Perhaps one could generalize from this limited information that MLOs are more likely to progress through a typical disease cycle than are viruses.

The objective of plant quarantine is to prevent or delay the establishment of plant pests in areas where they are not known to occur. The basic principles[7] that should be observed when promulgating plant quarantine measures are (1) the quarantine measure must be based on sound biological grounds, i.e., the pest must pose a substantial threat to an industry; (2) quarantine measures must be designed for crop protection, not for trade protection; (3) the subject area must be carefully and thoroughly researched before quarantine measures are recommended; (4) the quarantine measures must be derived from adequate legislation; and (5) the quarantine measures should be modified as conditions change or further facts become available.

While it is essential to develop quarantine measures based on the principles outlined, it is difficult in practice to adhere to these principles in controlling the introduction and spread of plant viruses, primarily because of the many problems associated with accurate detection and diagnosis of virus infections. In this chapter, we shall attempt to identify the major diagnostic problems and relate them to the quarantine principles.

In considering the content of this chapter, we had to decide whether to confine our examples to those diseases caused by the true viruses or broaden the scope to include virus-like agents (VLAs) such as viroids and MLOs. Prior to 1967, these two classes of plant pathogens were assumed to be of viral etiology. Since diseases caused by the VLAs resemble those caused by viruses and virological techniques are used for detection and identification of the VLAs, we opted to include these classes of organisms in our treatment.

II. MAJOR DIAGNOSTIC PROBLEMS

A. Indicator Hosts

In the early years of plant virology, the detection of plant viruses was based solely on visual inspection of the crop plants. Symptomatology still plays a major role in plant virus detection, but over the years some dramatic improvements have been made in the reliability of symptomatology as a means of virus detection. These developments occurred because of the realization that symptomless infections in the host plants were common; hence many infections remained undetected. For those viruses that were sap transmissible, indicator species were found which showed diagnostic symptoms for those viruses that were symptomless in their original host. Those viruses that were not sap transmitted readily were either graft- or insect transmitted to a more reliable indicator host. Thus over the years, considerable improvements have been made using host reactions as a means of virus detection. For quarantine purposes, host reactions have the advantage over all other techniques in that a program can be administered by relatively unskilled personnel with unsophisticated facilities. For these reasons, most quarantine programs around the world rely heavily on symptomatology as their method of choice for detecting virus infections.

Despite the advantages of using host reactions for virus detection, there are some obvious limitations, at least with some host plant-virus combinations. Some of these limitations may be overcome with improved indicator hosts, but others may require the use of other detection techniques to supplement symptomatology. Problems associated with virus indexing based upon symptomatology are many and varied, depending to a large extent on the crop and the specific virus or viruses that the indexing procedure is expected to detect. Some of the problems that may be encountered are as follows:

1. Virus may be symptomless in the crop host as well as in the indicator hosts.
2. Transmission of virus from the woody host to the herbaceous indicator host may be difficult.
3. Time between inoculation and symptom development may be months or even years.
4. Symptoms are sometimes subtle, and training and experience are required before readings can be made with confidence.
5. Symptoms induced by a specific virus may be mimicked by another virus.
6. Some of the readings may include false positives or false negatives because of lack of familiarity with the subtleties of symptomatology.
7. Symptoms vary with plant species, crop variety, and with the age and physiological condition of the plant at the time infection occurs.
8. Environmental factors are important for symptom development, e.g., high light intensity may be critical.
9. Mixed infections, common in many crops, may result in incomplete or incorrect diagnosis.
10. Insect damage may alter or mimic virus symptoms.
11. Pesticides (herbicides, fungicides, insecticides) may cause symptoms that can be confused with virus symptoms.

12. Technique is time consuming and expensive with respect to manpower and plant growth facilities when testing a large number of samples.
13. The same virus is capable of inducing a confusing variety of host responses, rendering identification doubtful.
14. A highly infectious virus may be introduced into the test system as a contaminant during experimental transmission.

B. Serology

Symptoms on plants are almost always inadequate on their own to provide a positive diagnosis of an unknown virus disease. Considering the limitations of virus diagnosis based upon host plant reactions, it is obvious that host reactions must be supplemented with information obtained from other procedures. Of the various procedures that might be used as an aid to virus diagnosis, the one most widely employed involves the application of a serological technique. From the earliest serological studies on plant viruses, it became apparent that the antigenic specificity of individual viruses would be extremely useful in virus diagnosis. In contrast to symptomatology, serology is more reproducible and has the capability of providing a more reliable and definitive answer. The antigenic properties of a virus provide the single most useful criterion for plant virus identification. In recent years, the use of increasingly sensitive serological methods has made it possible to detect even those viruses that are present in very low concentrations in the infected tissues. The enzyme-linked immunosorbent assay (ELISA) technique in particular has become widely used for the large-scale testing of field samples. The sensitivity of detection by ELISA is equal to or possibly superior to that of detection based on infectivity tests, and has the added advantage of specificity. Further, sensitive serodiagnosis has made it possible to detect virus directly in woody host materials.

Van Regenmortel[82] has compiled a list of those plant viruses that have been identified by serological means. By about 1980, when his literature search was completed, antisera had been prepared against some 262 plant viruses. This list is being expanded each year, and will undoubtedly continue to expand as new viruses are found or improved techniques are devised to adapt serodiagnosis to viruses that were previously considered to have too few virions in the infected tissues for serological detection.

The advantages of serodiagnosis have been widely acclaimed and, indeed, a plant virologist faced with the challenge of identifying an unknown virus has no choice but to use serology as a part of the assay. However, the identification of unknown viruses may be exceedingly complex, involving a wide range of physical, chemical, and biological procedures.[29] The equipment and expertise associated with virus identification would normally reside in a research rather than a quarantine facility. While quarantine facilities are equipped to detect most viruses in a given crop, they are not usually able to handle the identification of all viruses that might be expected to infect a particular crop. More often, quarantine officials are more concerned with detection and may take quarantine action without identification. If in the course of their assays they isolate a virus that is unknown to them, they would normally pass this unknown entity over to an associated research agency to study further. With this distinction in mind, a quarantine facility does not require a well-equipped virus laboratory in order to do routine detection and diagnosis. All that is required is a supply of antisera against those specific viruses for which indexing is being done, together with the modest equipment and supplies required for the test procedure.

For reliable identification of an unknown virus sample, it is also necessary to react the test antiserum with an extract from healthy plants and with the known homologous virus antigen. Inclusion of the "known" positive control may present a problem in view of the risks associated with maintaining a quarantinable virus in a laboratory that is not a secure containment facility.

Obtaining a consistent supply of antiserum is a problem to any laboratory that has to rely on an outside source. In the past, these laboratories depended on the goodwill and cooperation of other scientists. When a virus disease agent is being investigated, its serological properties are almost invariably studied and, as a by-product of this study, a supply of antiserum is produced. Small quantities of this antiserum are usually available from the researcher upon request. The requirement for small quantities of antisera has also been filled in part by the American Type Culture Collection in Rockville, Md., who charge a nominal fee for available sera. However, this service is limited in both quantity and range of antisera offered. It does not supply large volumes for diagnostic work. Immunological reagents for the diagnosis of a few specific viruses have recently been prepared and marketed by commercial companies. This commercial service, particularly in the form of ELISA kits, will probably expand to include many viruses of interest to quarantine and certification agencies. Alternatively, those agencies charged with the responsibility of quarantine and certification may be required to establish a separate facility for production of their antiserum requirements.

Assuming that specific antisera are available, and recognizing the many advantages associated with serodiagnosis for plant virus detection, one should still be aware of the limitations of the procedure. One or more of the following problems may arise during serodiagnosis:

1. Some commonly used serological procedures, such as the tube precipitin test, may not be as sensitive as assays involving host reactions.
2. Nonspecific precipitation reactions occur with the sap of some plants.
3. Major differences in the pathogenicity of strains may not be reflected in the antigenic properties of the corresponding virions.
4. End-to-end aggregation of some rod-shaped plant viruses can markedly affect assay results.
5. False positives may result when virus antisera contain antibodies to host plant antigens.
6. False negatives may result when there is an imbalance of reactants or when an insufficiently sensitive test is used.
7. False negatives may result when the virus is incompletely distributed within the infected plant.
8. Sensitivity of serological tests can be influenced by many variables in serological procedures.
9. A particular antiserum may react well against the strain to which it was prepared (homologous strain), but fail to react or react only weakly against other strains of the virus.
10. Tests should include "known" viruses as positive controls. If knowns are of quarantine significance and exotic, risks are involved in maintaining positive controls under secure conditions.

III. DETECTION OF VIRUSES AND MYCOPLASMA

A. Viruses

The use of plant quarantine to limit the movement of plant viruses consists of two important and distinct responsibilities. The first is to prevent dangerous virus diseases from moving into areas free of the virus (e.g., to keep plum pox virus out of North America). Rather than importing certified material year after year from the affected countries for commercial planting, the host should be prohibited commercially, with only a small number of plants being imported through quarantine. This will limit the entry of viruses which do not exist in the importing country but are present in the exporting country. Each of these plants should be indexed thoroughly and then propagated for distribution in the receiving country.

The second responsibility of quarantine involves the movement of commercial volumes of plant material such as seeds or cuttings, to be planted directly in the field in the receiving country. This material may be infected with economically important viruses that already exist in the importing country (e.g., strawberry mild yellow-edge virus is found in any strawberry-growing area of the world). Commercial importing of plant material is really a problem of certification as well as acceptance of the certification programs of the exporting country. If a country does not have a certification scheme for producing its own plants, then it must rely on that of another country. For example, all strawberry plants in British Columbia, Canada imported from the U.S. come from American certification programs. When grafted plants are imported as certified material, then both the rootstock and scions must meet the conditions of quarantine regulations.

In this section we will outline briefly the procedures available to plant quarantine officials for the testing of viruses, viroids, and mycoplasma. We can define quarantine testing as intensive indexing of a few plants for propagation in the importing country, and certification audit testing as a check on a small sample of the certified material that is imported for commercial planting. The two types of importation testing must be distinguished because the requirements of quarantine are much more important than those of audit samples.

Detection methods for viruses for the purpose of plant quarantine need not be expected to handle large numbers of samples. Therefore, labor-intensive methods are feasible. Ideally, rapid tests are desirable, but the reliability of the test should be of primary importance for plant quarantine. In addition, a single test that detects a broad range of viruses would be most desirable for economic reasons and may possibly be able to detect previously unidentified viruses or VLAs.

1. Transmission Tests

Many virus infections result in recognizable symptoms such as mosaic, stunting, mottling, color breaking in flowers, or fruit symptoms. In such cases growing the plants in quarantine will be sufficient to determine if these viruses are present. However, many other viruses are latent and require an alternate method of detection. Viruses that cause fruit symptoms may require as many as 4 years before symptoms develop.[1] In such cases, methods to shorten the time necessary for virus detection are needed.

In many cases graft transmission to sensitive indicators is still the method of choice when woody hosts are assayed for virus infections. This approach will not be replaced until faster reliable methods are available for all quarantinable viruses of a woody crop. Faster methods of detection for one or several viruses of a crop may reduce the number of grafts required but will not affect the total length of time a plant is quarantined before it can be certified free of quarantinable viruses if some viruses must still be detected by graft-transmission studies. In some cases the time for symptom development on woody indicators can be reduced considerably by conducting tests in greenhouses,[17] but for many virus indicator combinations the conditions required to accelerate symptom development are not known.

Mechanical transmission of viruses to herbaceous indicator plants has been a very useful means of assaying for latent viruses in herbaceous and woody hosts. Immature leaves are usually the best source of inoculum when transmitting viruses from woody hosts since viruses often cannot be transmitted after terminal growth has ceased.[19]

2. Serological Tests

Antisera are available for most viruses that are transmissible mechanically from woody crop plants. However, with woody plant material, virus concentrations are generally too low to detect with gel diffusion assays and therefore mechanical transmission to herbaceous hosts is still required, with gel diffusion being used as a confirmatory test. The ability to detect viruses in most plant material became available with the development of the ELISA technique.[9] The ELISA procedure is relatively simple, rapid, very sensitive, requires small amounts of antiserum, is adapted readily to large- or small-scale testing and does not require any expensive equipment.[8] These characteristics have led to rapid adoption of the ELISA test for many diagnostic applications in plant pathology.[3,8,44] ELISA has many desirable characteristics to recommend it for virus detection. However, it does require an antiserum specific for each virus to be tested which can be a problem for quarantine personnel since antisera are available for only about one third of all virus and VLAs that affect fruits, vegetables, and ornamentals. Strain specificity has also been reported to be a problem with ELISA, but this has been partially overcome with the use of antibody specific conjugates, e.g., protein A conjugates and anti-antibody conjugates. ELISA is an ideal test for audit samples and can be used to assay seed for seedborne viruses as well as leaf tissue for the presence of specific viruses.

The monoclonal antibody (MCA) technology provides the means of producing antibodies of a predefined specificity.[41] The procedure for making MCAs is well established.[22,63] The MCA technology provides the means of producing an unlimited supply of specific antibody to meet the increasing requirements for high-quality antisera. MCAs can be developed which are highly strain specific[21] or which can detect all isolates of a virus,[48] depending on the selection criteria used when screening potential antibody-producing cell lines (hybridomas). For purposes of plant quarantine or certification programs, a broad spectrum antibody is desirable. In creating this antibody, use can be made of the ability of the MCA technology for production of antibodies against synthetic peptides. With this methodology, conserved protein sequences can be synthesized, linked to a carrier, and then used to immunize mice. After fusion of spleen β-lymphocytes from the immunized mice with cultured myeloma cells, the resulting hybridomas are screened for antibody activity against the synthetic peptide. If the synthetic peptide consists of regions of homology known to exist between viruses within a group or between viruses from wide-ranging groups, there is potential for making broad spectrum antibodies.[16,32,71]

Another approach to making virus-specific antibodies involves the expression of viral genes in *Escherichia coli* and using these expressed proteins for production of monoclonal antibodies. MCAs with broad specificity are selected. Both the synthetic peptide approach and the expression of viral genes in *E. coli* have the advantage that any viral gene (non-structural as well as structural) can be used to make antibodies. Regions of homology in nonstructural proteins between wide-ranging virus groups such as bromoviruses, cucumoviruses, tobamoviruses, alfalfa mosaic virus, and Sindbis virus have been reported,[32,71] as well as regions of homology between nonstructural proteins of cowpea mosaic virus and picornaviruses.[16] Broad-spectrum antibodies will be very useful in plant quarantine and certification programs.

3. Immunosorbent Electron Microscopy

An electron microscope (EM), a useful instrument for plant virus studies, is generally not available to plant quarantine workers. However, where such equipment is available, the usefulness of the EM in conjunction with serology is now widely recognized. The most widely used technique, immunosorbent electron microscopy (ISEM), involves coating an EM grid with a support film and then coating this film with a layer of virus-specific antibodies which is subsequently used to trap specific virus particles from a suspension placed in contact with the coated grid. Details on the technique have been reviewed recently.[55] A 2-to 10,000-fold increase in particle numbers per grid may be obtained as compared to comparable leaf dip procedures in which antisera is not used. The antigen is recognized in the EM by its morphology and staining properties or it can be further labeled by a second incubation with a more concentrated antibody preparation, giving the individual particles an easily recognized halo of γ-globulin. Some advantages of the method are (1) characteristic particles are seen in the microscope; (2) false positives are minimal; (3) sensitivity is often high; (4) background is low, rendering particles more visible; (5) low titer antiserum may be used; (6) antisera may be used without fractionation or conjugation; and (7) only small volumes of antisera and virus extracts are required.

4. Complementary DNA Probes

The use of complementary DNA (cDNA) probes has been adapted recently for detection of plant viruses[50] and viroids.[64] The principles and techniques for making and using cDNA probes have been reviewed.[24,47,52,84] The sensitivity of cDNA probes is similar to that of ELISA, and the use of cDNA probes is the only rapid detection method available for viroids and virus infections devoid of particles.[31,64] However, for plant quarantine, speed is not of primary concern, and a general method based on visualization of the viroid nucleic acid may be more suitable.[58,60] cDNA probes require the use of a radioactive label(^{32}P), which has drawbacks for routine use in plant quarantine or diagnostic laboratories. ^{32}P has a short half-life, and many laboratories are not equipped for work with radioactive isotopes. However, testing at a centralized facility as is done for PSTV in North America (Agdia Inc.) alleviates the problem of working with radioisotopes in many laboratories and, in addition, less radioisotope is wasted because many samples are done at one location in a short period of time. Fewer cDNA probes are available than antisera, so this technology does not at present have as many applications as serology in plant virus detection. One potential advantage of cDNA probes is the development of group-specific probes. If affinities of the virus groups correlate with evolutionary events, then some genetic information may be conserved within a virus group or family. DNA probes specific for conserved genes or regions of genes may be useful for detecting any virus within a virus group or family and thus be valuable for plant quarantine or certification purposes. The main advantage of cDNA probes at present is the ability to assay for pathogens devoid of protein shells, i.e., viroids and particle-deficient viruses. Hybridization assays using cDNA probes offer many of the same advantages as ELISA and can be used very effectively for detecting specific pathogens.

5. Double-Stranded RNA

Another method of virus detection is based on the presence of virus-specific double-stranded RNA (dsRNA) which has been isolated from plants, animals, insects, bacteria, and fungi[12,57,70] infected with RNA viruses. These dsRNAs are the replicative form of single-stranded RNA viruses or genomes of double-stranded RNA viruses. The use of dsRNA for virus detection is based on the assumption that plants not infected with RNA viruses do not contain high molecular weight dsRNA ($> 0.1 \times 10^6$ daltons). The presence of high molecular weight dsRNA has been used to identify plant diseases of suspected viral etiology[36,56] and can be used to detect virus infections in woody as well as herbaceous hosts. The method is relatively labor intensive compared to serodiagnostic or cDNA hybridization methods, but provides an answer far sooner than graft transmissions.

dsRNA analysis is a procedure that can be used to detect many different viruses and has potential for plant quarantine and certification programs where small numbers of samples are tested. This method has proved very useful for detecting raspberry leaf spot virus, an uncharacterized aphid-borne virus of raspberry, and has also readily detected multiple virus infections in raspberry plants.[43] The dsRNA analysis is used currently in the British Columbia raspberry certification program for recurrent screening of mother plants. Infection by three different viruses has been proposed to explain the three distinct dsRNA patterns obtained from avocado plants.[36] This was the first report of virus-like agents other than tobacco mosaic virus in avocado. Lettuce big-vein-disease-specific dsRNA has been isolated recently from roots of infected plants,[56] indicating that this widespread disease of lettuce is caused by an RNA virus. These examples illustrate the potential of dsRNA analysis as a detection method for viruses that have not been characterized.

The procedure for isolating and detecting dsRNA is described here because it is a relatively new method and has potential for plant quarantine and certification programs. The method used most commonly for isolation of dsRNA from plant tissue is that of Morris and Dodds.[58] In this procedure, total nucleic acid is extracted from fresh or frozen tissue by homogenizing the sample in glycine phosphate sodium chloride buffer containing sodium dodecyl sulfate, mercaptoethanol, phenol, and chloroform pentanol. The aqueous phase is collected by low speed centrifugation and adjusted to 15% ethanol. Cellulose powder is added, and the slurry is shaken on ice to bind the dsRNA to the cellulose. The cellulose is washed with sodium chloride, Tris, EDTA (STE) buffer containing 15% ethanol to remove DNA and ssRNA from the cellulose, and then the dsRNA is eluted using STE. Traces of DNA are removed from the dsRNA by the addition of $MgCl_2$ and digestion with DNase. The dsRNA is then precipitated with ethanol and analyzed by electrophoresis on 5% polyacrylamide or 1% agarose gels.[12]

Unlike dsRNA analysis, other methods of virus detection, including serodiagnosis, cDNA, and biological assays, require previous knowledge of the viruses that can infect the host before the viruses can be adequately assayed. An unrecognized virus may be symptomless in the repertoire of indicators used in biological assays for virus testing, and serological and cDNA assays require specific probes for each virus that is included in a virus testing scheme. For the many viruses which have not been purified, specific antisera and cDNA probes are not available. The dsRNA analysis of plant material can be used to detect the presence of most characterized and uncharacterized viruses.

This discussion of dsRNA began with the premise that high molecular weight dsRNA does not occur in virus-free plants. However, recent examples of dsRNA found in healthy plants emphasize the necessity of using proper controls with the dsRNA method, as is true for any method of virus detection. There are reports of high molecular weight dsRNA in apparently healthy plants of *Rubus* spp.,[43] *Phaseolus vulgaris*,[83] and *Vicia faba*.[26] The dsRNA was found in all *Rubus* spp. and cultivars that were tested. In *P. vulgaris* it was found in the Black Turtle Soup cultivar as well as other black-seeded beans, but not in some other cultivars. These results point to the need for adequate controls, to the level of variety or clone, for interpretation of dsRNA analysis.

In addition to its application in virus detection and diagnosis, dsRNA has the potential to serve as a starting point for the production of cDNA or serological probes. The cloning of cDNA from dsRNA templates can be accomplished with the procedures used to clone reo- and rotaviruses.[6,23] If serological tests are preferable to hybridization assays for virus detection, viral proteins can be produced by expression of a hybrid DNA containing the cDNA sequence as well as an inducible operon. Expression of this hybrid DNA is then controlled by the regulatory system of the operon. The most studied inducible operon is the lac operon.[54] In using this operon to make hybrid DNA, the foreign DNA is inserted in the β-galactosidase structural gene and the resulting fusion protein has amino acid sequences of both the β-galactosidase structural gene and the inserted gene product.[87] This protein can then be used to produce antibodies specific for the viral part of the fusion protein. Proper expression of the inserted viral DNA depends on the orientation and reading frame of the insert. There are several families of three-expression vectors available whose polylinker sequences are shifted by one base pair each, so that any piece of an open reading frame can be inserted in phase into one of the three vectors in a family.[86] If the sequence of the entire open reading frame of the viral gene is known, the desired expression vector can determined.

The ideal method of virus detection for plant quarantine purposes would identify any virus in a plant reliably, rapidly, and easily. Of the methods currently available, dsRNA analysis for detection of RNA viruses comes closest to this ideal. Research on the application of the technique for specific quarantinable viruses or VLAs is required to ensure that they can be detected. Once this research is completed, a simple standardized technique will be available for detection of many quarantinable viruses.

B. Mycoplasma-Like Organisms

1. Electron Microscopy

Since the first report of MLOs associated with plant disease,[13] many yellows and witches' broom diseases have been attributed to them. Plants infected with MLOs often contain pleomorphic bodies in their phloem elements, and the presence of these bodies has frequently been used to confirm the presence of MLOs. Visualization requires embedding, sectioning, staining, and then electron microscopy, which is both labor intensive and expensive. In addition, false negative results are possible with electron microscopy. Since this technique visualizes only a small portion of the plant, MLOs may escape detection.

2. Fluorescence Microscopy

Several fluorochromes have been used to detect MLOs in plant tissue by fluorescence microscopy. Aniline blue staining was found to give a positive correlation between the time of symptom development and the appearance of abnormal fluorescence in phloem tissue of *Vinca rosea*.[34] With some MLOs causing symptoms, no abnormal fluorescence was detected in sections stained with aniline blue. This dye does not stain MLOs, but instead has been used as an indirect test to stain excess callose that forms in sieve tubes of MLO-infected plants.[34] Since the abnormal staining with aniline blue usually coincides with the onset of symptoms, the test is unreliable for detection of latent MLO infections.

Several DNA binding fluorochromes are useful for MLO detection in tissue cultured cells. Upon adding 4′-6-diamidino-2-phenylindole (DAPI) or the bisbenzimidazole derivative Hoechst 33258 to uninfected tissue cultured cells, highly fluorescing nuclei were obtained with no cytoplasmic fluorescence.[74] When mycoplasma infected cells were stained with DAPI, discrete fluorescing foci were seen in the cytoplasm.

Application of these DNA binding stains for the detection of plant MLOs was first reported by Seemüller.[79] With DAPI staining, 5 of 5 pear decline-infected trees and 14 of 16 apple proliferation-infected trees were diagnosed correctly, while 23 healthy trees were diagnosed as healthy. When the same trees were tested using aniline blue, 12 of the 16 apple proliferation trees were correctly diagnosed. Seemüller[79] found that Hoechst 33258 was also useful in detection of plant MLOs, but gave a higher background fluorescence that complicated the identification of diseased samples. DAPI staining has been used to follow the occurrence of MLOs in apple proliferation- and pear decline-infected trees throughout the winter months, showing that MLOs are very rare in aerial plant parts but abundant in roots during the winter.[77] With blueberry stunt disease, 24 of 25 and 19 of 25 infected plants were correctly identified with the DAPI and aniline blue stains, respectively.[76] In the same study, 25 of 25 and 10 of 25 healthy plants were correctly identified with DAPI and aniline blue, respectively.

The DAPI staining method has potential as a broad-spectrum tool for MLO detection, although localized distribution of MLOs in the host plant can be a problem.[76,77] Accurate identification requires a worker who has had considerable experience with the technique to be able to distinguish between autofluorescence and DAPI staining. There is still a need for an objective broad-spectrum test for MLO detection in plants.

3. Complementary DNA Probes

Such a test may be on the horizon with the use of cDNA probes specific for conserved genes within the genus *Mycoplasma*. Ribosomal RNAs (rRNAs) are ideal candidates for producing universal probes because they are distributed universally, exhibit constant function, and appear to change sequence very slowly. Phylogenetic relationships of prokaryotes including *Mycoplasma* have been based on the sequence of rRNAs.[15] Göbel and Stanbridge[21] reported a broad-spectrum mycoplasma detection assay based on DNA:DNA hybridization, with cloned fragments of the rRNA operon of *Mycoplasma hyorhinus* as a probe. These workers made a cDNA clone of rRNA which, depending upon the conditions of hybridization, could be used to detect prokaryotic organisms or, more specifically, mycoplasmas. This work was done using animal mycoplasmas. The applicability of the test for detecting plant MLOs has not been determined, but if all mycoplasmas and MLOs have a common origin, this probe should be very useful for MLO detection. It is this type of broad-spectrum, sensitive, and objective test for pathogen detection that will be of most use for plant quarantine testing.

4. Serological Tests

Monoclonal antibodies specific for aster yellows MLO are available and can be used to detect aster yellows in plant sap.[45] In quarantine situations, the presence of an MLO can be verified with a general probe and then specific probes, such as these aster yellows monoclonal antibodies, can be used to determine whether or not the MLO detected is a quarantinable pathogen. A two-test system using a general probe followed by a specific probe would be ideal for plant quarantine testing. However, with MLOs specific probes are currently available for only aster yellows and a reliable general probe is the best test at present.

IV. CASE STUDIES OF VIRUSES OF QUARANTINE SIGNIFICANCE

Several hundred virus diseases of plants have been described and apparently new ones are reported each year. With so many virus diseases recognized and described, and undoubtedly many other agents yet to be isolated and characterized, it is virtually impossible for a virologist involved in quarantine work to keep abreast of those virus diseases of potential quarantine significance and to know how best to detect such viruses. Possibly the most difficult task of all is to decide which of the many viruses that are of concern should be cited in quarantine regulations. The general consensus in plant pathology and entomology textbooks in which quarantine is discussed as a separate topic is that quarantine action should (1) be implemented when man rather than nature is the prime mover, (2) be based on biology or pest risk analysis, and (3) constitute the least drastic action which provides the necessary safeguards.[37]

We thought it would be appropriate in this review to examine in some detail a few of the viruses, MLOs, and VLAs that are considered to be of quarantine significance. Rather than arbitrarily selecting those viruses that might constitute good examples, we opted to make use of the results of a survey of quarantine regulations compiled by Kahn,[37] in which he identified the ten most frequently cited viruses and VLAs in quarantine regulations of 125 countries. We examined the literature on these viruses and VLAs, particularly with respect to geographic distribution, mode of transmission, economic significance, and detection problems.

A. Viruses

1. Plum Pox

Plum pox (also called Sharka) causes conspicuous symptoms on the fruit of sensitive cultivars of plum, peach, and apricot.

The virus was first noticed in Bulgaria in 1915 and it slowly spread from there to most countries in Europe. Until recently, records of occurrence were based upon the appearance of characteristic fruit symptoms, so it is probable that the virus was introduced and became established in a particular country before being detected. Losses have been particularly heavy in Yugoslavia and less severe in the countries of northern Europe. However, even in those countries where the direct damage has not been extensive, considerable time and money has been spent on virus detection surveys and attempts at virus eradication or containment.

Plum pox virus has several features that contribute to its success as a pathogen and to the failure of plant quarantine officials in curbing its spread: (1) it is efficiently vectored by not one but several aphid species, which means that most aphid species that feed on an infected source would be capable of transmitting the virus; (2) it has a narrow natural host range, including cultivated species and those that are used in hedgerows or grow wild; (3) the virus cannot be detected by visual examination of plum trees except in the fruit of susceptible cultivars and by the time the fruit appears, infected nursery stock could be distributed widely; (4) the virus is symptomless in many *Prunus* hosts, some of which are used as plum rootstocks; (5) the virus is unevenly distributed in its host, and detection is thus unreliable by use of serological techniques or woody indicators;[11] and (6) seed transmission is known to occur in plum, apricot, and peach.[62]

The introduction and spread of plum pox in England has been well documented,[11,66] despite a determined effort being made by plant protection officials to exclude this damaging virus from Britain. The virus was first isolated from a young Italian prune tree at East Malling Research Station in 1965. A search was made for the source of infection and the only one found was a plum tree grown on rootstock originating from a German nursery in 1962. This tree and all plum rootstocks imported from this source were destroyed. Cropley[11] reported that field observation and transmission tests indicated that this introduction of plum pox had been eradicated. Infection was again found at East Malling in 1970, and this time suspected plum propagation material had been widely distributed to nurseries using imported rootstock. Wherever the disease was found, infected trees were destroyed. Infections were found in 13 orchards and 4 nurseries by the end of 1974, and evidence suggested that infected material had been distributed from these nurseries.[66] It was considered impossible to inspect all nursery stock effectively because of the lack of sufficiently trained manpower. Inspection of nurseries producing certified stock was intensified and, wherever infection was found, steps were taken to eradicate the disease.

A major difficulty was identification of infected plants in nurseries. Symptom expression is variable and symptomless infections are common. Traditional methods of graft inoculation to peach and sap inoculation to *Chenopodium foetidum* were expensive with regard to glasshouse space, time consuming, and often unreliable. Many of these problems were alleviated by the development of the ELISA technique.[9] A large increase in nursery infections was confirmed in 1977 and 1978, but this increase was attributed largely to improved detection techniques and an increase in the number of nurseries inspected.

The plum pox containment program in England illustrates the extreme difficulties that are faced by establishment of a newly detected virus disease. It also illustrates the hopelessness of the situation when the virus is difficult to detect and has an efficient means of dissemination.

2. *Citrus Tristeza*

Citrus tristeza virus (CTV) is thought to have originated in southern Asia or Africa and was introduced into South America in the 1920s with disastrous consequences. The virus has virtually eliminated the citrus industry in some areas of Argentina, Brazil, and Uruguay. The virus now occurs worldwide and is an important economic problem in citrus crops. It is transmitted most efficiently by the tropical citrus aphid *Toxoptera citricidus* and less efficiently by a few other aphid species. Tristeza is principally a disease of sweet orange or other citrus scions budded on sour orange rootstock. Most sweet oranges, mandarins, and tangerines budded on sweet orange, Rangpur lime, and mandarin are unaffected. This finding has led to the use of tristeza-tolerant rootstocks for basic control.[10]

The virus did not cause epidemics in citrus-growing areas until (1) infected propagative material was introduced into areas where the efficient vector was present, (2) the vector *T. citricidus* was introduced into areas where the inoculum was present, or (3) the development (in Israel) of a highly transmissible viral mutant that was vectored by a locally important aphid, *Aphis gossypii*.[2]

Experience with CTV illustrates the fact that epidemiology is more readily applied to discerning the patterns of diseases that have occurred in the past than in predicting future trends. Analysis of the change in CTV behavior from a nonspreading phase to an epidemic phase in Israel implies that mutants may arise that are efficiently transmitted by insects prevalent in areas where the virus has not previously constituted a serious threat.

The spread of CTV in Israel emphasizes the critical importance of virus strain on virus epidemiology and provides an example as to why plant quarantine officials are concerned about the introduction or the natural occurrence of new strains. The combination of virus-infected source plants and the efficient vector *T. citricidus* has resulted in the near complete spread of CTV into citrus-growing areas where the aphid occurs. In contrast, the melon aphid *Aphis gossypii* ranges from being a nonvector to a fairly efficient vector. In Israel, the tropical citrus aphid does not occur and although the melon aphid is prevalent in the citrus groves, indexing tests made during 1954 to 1959 indicated that there was no natural spread of CTV into trees established from virus-free budwood over a period of 2 decades. This situation changed beginning in 1970, when indexing tests revealed limited natural spread attributed to the melon aphid vector. The evidence indicates that most CTV isolates introduced into Israel and other citrus-growing regions in the Mediterranean area and North America from the Far East and South Africa were transmissible by *T. citricidus* but not by *A. gossypii*, and the presence of these isolates suppressed by cross protection the multiplication and spread of *A. gossypii*-transmissible mutants. Under certain conditions, cross protection fails or is incomplete, and transmissible mutants are acquired by *A. gossypii* and transferred to healthy trees.[2]

3. Strawberry Mild Yellow-Edge

Strawberry mild yellow-edge virus (SMYEV) is probably worldwide in its distribution, being reported from virtually every country where strawberries are grown commercially. The virus is usually symptomless, or at the most causes a slight marginal chlorosis of the young leaves of commercial strawberry cultivars. While losses attributed to infection with SMYEV alone are not too serious, the virus normally occurs as a complex with strawberry mottle virus or strawberry crinkle virus or both, and the complex infection is of serious economic importance, reducing the useful life of the planting and yield by up to 75%.[53] The complex with strawberry mottle or strawberry crinkle causes a disease known as xanthosis, yellows, or yellow-edge. Symptoms are characterized by marginal chlorosis and cupping of the young leaves and premature reddening of the older leaves. Plants are dwarfed and yield is much reduced. The severity of the symptoms varies with the susceptibility of the cultivar and the number and severity of the interacting component virus.

The virus occurs only in *Fragaria* spp. It is efficiently transmitted by strawberry aphids belonging to the genus *Chaetosiphon*, primarily *C. fragaefolii*. Since commercial strawberry cultivars do not show distinctive symptoms, the best indicator hosts are selected clones of the wild *F. vesca*. The virus is not sap transmissible — experimental transmissions are done using grafting or by means of the aphid vector. The virus is of the persistent or circulative type. Experimental aphid transmission is usually erratic, although some strains of the virus are more readily transmitted than others. The causal virus has properties similar to the luteoviruses and although not formally included as a member of the group,[49] it is considered to be a possible member.

Control of the virus is linked to replacing virus-infected fields with healthy planting material supplied through certification schemes. Countries having a significant strawberry industry have devised mechanisms for producing and propagating virus-free stocks, often using in vitro culture in at least the early phases of the scheme. Control of the aphid vector is difficult and since virus-infected strawberry plants are usually grown in the vicinity of fields planted with virus-free stock, it is only a matter of time until a newly planted field becomes contaminated with SMYEV and the other strawberry viruses vectored by *C. fragaefolii*. If the certification scheme is less than perfect, resulting in virus-infected stock being planted at random within the newly planted field, the time required for extensive virus spread is appreciably reduced.

The question of why a virus such as SMYEV is included in the quarantine regulations of any country might be asked. This is not an easy question to answer in view of our understanding of the biology and epidemiology of this virus. It is known to occur in all areas where strawberry is grown as a commercial crop, and efforts to eradicate the virus from any area have not been effective. It therefore appears inappropriate to attempt to control the movement of this virus into a strawberry-growing area by applying quarantine measures.

B. Mycoplasma

1. Apple Proliferation

The most reliable symptom of apple proliferation is the premature development of axillary buds which grow instead of remaining dormant until the next season, giving rise to "witches' broom". Abnormal shoots are usually more numerous near the apex of the main shoot, whereas normal laterals arise near the base of the secondary shoots. Leaves on affected trees are finely and irregularly serrated. During the "shock stage", trees bear only small, worthless fruits without much taste and color. Trees may bear small fruit for several consecutive years and then symptoms may disappear, sometimes to reappear later. Other infected trees show mild symptoms or are symptomless.[5,78]

The causal MLO has been reported from most fruit-growing countries in continental Europe, where it has been transmitted experimentally only by grafting. The disease is often disseminated in scion wood, and nearly symptomless trees may yield a high proportion of infected buds. Assaying in woody indicators is the most frequently used technique for the selection of healthy clonal material. Symptoms are slow to develop, but with sensitive apple indicators, they show in June following September inoculation.[20] The vector has not been established, although evidence suggests that the pathogen is transmitted by leafhoppers and that it may exist in herbaceous hosts.[42] It appears that apple proliferation and apricot chlorotic leaf roll are caused by the same pathogen.[46] It is a potentially serious pathogen, and attempts to control its spread by quarantine measures are fully justified.

2. Peach Yellows

Buds in affected peach trees unfold prematurely in the spring, producing leaves that are dwarfed and yellow. The tree develops slender willowy shoots that grow upright from the main limbs, giving it a bushy appearance. Severely affected trees die 2 to 3 years after the onset of symptoms. Fruit produced on affected trees matures earlier than healthy fruit, may be larger than normal, and is of poor quality, usually having bitter taste. Symptoms on apricot, almond, and plum are similar to those in peach. The MLO is transmitted by the plum leafhopper *Macropsis trimaculata*. The incubation period of trees in the field is usually 1 to 3 years. Since plum is the preferred host of the plum leafhopper, and because some plum cultivars are symptomless carriers, it is assumed that natural spread of the disease is largely due to transmission from plum to peach by the leafhopper vector.

Peach yellows disease is not economically important today, although several outbreaks have occurred in eastern North America during the 19th and early 20th centuries.[68] As a quarantine problem, it should be considered of minor significance.

3. Strawberry Witches' Broom

Diseased plants have long, usually erect, stiff spindly petioles bearing leaflets that are smaller and lighter green than normal. There is a tendency for the stems of the individual leaflets to be longer and for the midveins to be broader and lighter in color than in healthy plants. The great number of leaves with long petioles and the multiple-branched crowns gives a bushy appearance to diseased plants. Flower stocks are scarce and are spindly and unfruitful. In some cultivars, the brooming or bushy character of the disease is much more pronounced than in other cultivars.

Since strawberry witches' broom was described as a new disease in Oregon in 1927,[88] the disease has practically disappeared from the Pacific Northwest. Similar although not identical diseases have been reported from other parts of the world.[4] Because the diseased plants are conspicuous and readily detected, they may be rogued out of commercial plantings. Since there is little or no in-field spread, witches' broom and similar disorders are of little economic importance. The disease is probably caused by an MLO that occurs in native vegetation and is occasionally transmitted to strawberry plants by viruliferous leafhoppers. As a quarantine problem, it is of minor significance.

4. Strawberry Green Petal

The most characteristic symptom of the disease in strawberry occurs in the flowers. The sepals are enlarged and degrees of sterility occur while fruit that is set seldom ripens. Plants are only slightly stunted, show vein chlorosis, slight bronzing of older leaves, and some crown proliferation.

Green petal is probably the most important leafhopper-transmitted MLO in Britain in that it causes sterility in clover seed crops.[69] The leafhoppers *Euscelis* spp. and *Macropsis* spp. are primarily responsible for spreading the disease in clover and although prevalent in strawberry fields, they do not appear to infect strawberry plants with MLOs. The less numerous *Aphrodes bicinctus* appears to be primarily responsible for vectoring green-petal disease in strawberry. In England, the disease has occasionally caused severe losses in runner beds but, in general, ocurrence is sporadic and of minor economic importance in strawberry.[51] There appears to be little justification for including this pathogen in quarantine regulations.

5. Potato Witches' Broom

Potato witches' broom has been reported from North America, Europe, and Asia. Different MLO pathogens may be involved in different geographic areas.

Symptoms are similar in different regions and in different cultivars. Plants produce multiple basal and axillary branches bearing simple leaves which may be chlorotic. Current-year symptoms appear late in the growing season, if at all, and consist of an upright habit of growth, rolling of leaflets, and some chlorotic or anthocyanin discoloration of apical leaves. Affected hills produce tubers that appear normal during the year of infection but give rise to plants with advanced symptoms the following year. Such tubers usually produce hair sprouts, and the plants that result have many small tubers and an abnormally high number of sprouts.

The MLO is transmitted by a few species of leafhoppers. The leafhoppers are unable to acquire the pathogen from potato, and spread to potato occurs when the vector feeds on other infected hosts such as perennial legumes. Generally the disease is of minor importance and does not justify elaborate control measures.[85]

C. Viroid

1. Potato Spindle Tuber

Potato spindle tuber viroid (PSTV) has been endemic for many years in potato plantings in the U.S. and Canada, but except in a few germplasm collections, is not present in Europe. It is therefore understandable that PSTV is high on the list of pathogens included in quarantine regulations in those countries that are free from the disease.

Affected potato plants produce stems and blossom pedicels that are slender and remain upright. Leaflets are somewhat smaller than normal, have fluted margins, and tend to curve inward and overlap the terminal leaflet. Leaves near the ground are shorter and upright. Tubers are elongated, round in cross section, and have pointed ends. Tuber symptoms may also include russet skins that become smooth, red skins becoming pink, and surface cracking parallel to the long axis. Diagnosis is a problem because not all tubers from diseased plants show all or even any of these symptoms, and some tubers from healthy plants resemble spindle tubers. Attempts to reduce PSTV by sorting out the suspected diseased tubers is therefore not effective. Further, mild strains with indistinct symptoms outnumber severe strains so many infections are missed in surveys based upon visual inspections.

PSTV is readily transmitted by pollen and true seed. It is transmitted to a lesser extent by leaf contact between healthy and diseased plants and it may be transmitted occasionally by chewing insects. The major method of perpetuating and distributing PSTV is in seed potato stocks, where mild infections are not detected during field inspections. The more severe strains of PSTV can be detected by sap inoculations from potato to sensitive tomato seedlings where distinctive leaf curling, stunting, and necrosis develop. This technique is not a reliable method for detection of the mild strains that occur in field-grown potatoes. Detecting mild strains of PSTV presented a problem until Fernow[14] developed the double-inoculation technique. It was based on the principle that a tomato plant inoculated with a mild strain of PSTV would be protected from subsequent challenge inoculation with a severe strain.

While the double-inoculation technique was effective, it had disadvantages in that it was time-consuming and did not produce obvious and typical symptoms in all tomato plants following inoculation with the challenge strain.[30] A major improvement in the detection technique, which involved extraction of cellular nucleic acids and their separation by polyacrylamide gel electrophoresis, was developed by Morris and Wright.[60] This technique, with minor modifications,[67] resulted in a widely used diagnostic aid for indexing elite and basic seed stocks in certification programs and for indexing *Solanum* germplasm. Further improvement in the procedure[75] and development of the nucleic acid spot hybridization technique[65] increased speed, reliability, and interpretation of sample processing and facilitated field indexing for PSTV.

Because of the economic importance of PSTV in North America, measures to prevent its spread to other potato-growing areas have been established, and restrictions in the movement of potatoes between countries have been imposed.[30] For example, foreign stocks of potatoes are imported into the U.K. under license where they are grown in quarantine, and all nuclear stock of foreign origin is routinely tested.

Transmissibility of PSTV in true seed presents a special problem to potato breeders, since the viroid occurs in lines of *Solanum* spp. originating in South America and used in breeding programs throughout the world.[30,65] Recognition of this problem coupled with a rigorous indexing and seed renewal program could result in elimination of PSTV from breeding lines. Conversely, in those breeding programs where no attention is given to PSTV indexing, the viroid could become widely distributed. It is desirable to screen accession lines being used at present in breeding programs, replace those lines that are infected, and screen new accessions before incorporating them into a collection. For this reason, inclusion of the pathogen in quarantine regulations is desirable.

D. Virus-Like Agent

1. Rose Wilt

Rose wilt is a little-studied mycoplasma-like disease of roses. It was first found in Australia,[25] and the same or similar diseases were later recorded from New Zealand,[18] the U.K.,[35] and Bulgaria.[39] Established plants show general debility, usually associated with dieback and lack of apical dominance on some shoots. Many buds break lower down the stem producing rosetted growth with small leaves. Among red oil-producing plants in Bulgaria, sterility and phyllody of flowers often occur. The presence of these symptoms, together with the difficulties of sap transmission, has led to the suggestion that the causal agent is an MLO.[39] However, MLOs have not been detected by electron microscopy, nor is there any evidence that the causal agent in Bulgaria is the same as that causing the disease in other areas of the world. Lacking positive identification of the causal agent, it appears appropriate to consider it a virus-like agent. One reason so little is known about the agent is that the disease is of minor importance, which raises the question as to whether it warrants inclusion in quarantine regulations.

V. CONCLUDING REMARKS

As was previously stated, the general consensus in plant pathology and entomology textbooks in which quarantine is discussed as a separate topic is that quarantine action should (1) be implemented when man rather than nature is the prime mover, (2) be based on biology or pest risk analysis, and (3) constitute the least drastic action which provides necessary safeguards.[37] In addition to the three principles outlined above, a fourth might be added, namely, that it is imperative for plant quarantine policy to be kept under review not only with the object of intensifying action but also of relaxing restrictions where this can be done safely.[81] While authors of textbooks are not necessarily correct in their interpretation of the decision-making process, we have no hesitation in agreeing in principle with these three basic criteria. Each country of course has the prerogative of setting its own quarantine regulations, but most would subscribe to the same principles. For example Chock,[7] a spokesman for the International Plant Protection Convention, states, "Plant quarantine measures aim to prevent the introduction and spread of pests to new areas and are justified only if the pest has little or no chance of spreading naturally." The European and Mediterranean Plant Protection Organization (EPPO), whose initial aim was to harmonize and improve phytosanitary regulations in the EPPO region, works on the principle that regulations should have a firm scientific basis and should be as up to date as possible.[81] The U.S. is philosophically committed to all plant pest measures being based on biology rather than on economic or political pressure.[73] Similarly in Australia, Morschel[61] notes that the first essential activity for the quarantine service of any country is a proper survey of the plant disease and pest situation, both within the country and throughout the world, so that potential pathogens and pests inimical to it can be determined by deduction. However, all authors recognize the difficulty in predicting how an exotic plant disease or pest will perform in a new environment since a significant pest or disease in one country may be harmless in a new environment, or vice versa.

Keeping in mind the criteria that should be used to include specific pests in quarantine, it was our objective to determine how closely they were adhered to with the viruses and VLAs reviewed above. We shall restrict our observations and comments to those ten agents that were examined in our case studies on the assumption that since these were the ones most frequently cited in quarantine regulations, they represent the viruses and VLAs that are of most concern to quarantine officials.

The first criterion for imposing quarantine regulations, i.e., man rather than nature acting as the prime mover, is amply satisfied, not only for these few agents but for virtually all of the quarantinable viruses. Viruses in general do not survive for prolonged periods in the absence of living plants or living vectors. Those that lack an efficient vector are spread exceedingly slowly in nature. Those that possess an efficient vector are capable of extending their geographic distribution by natural means if their vector can be transmitted over long distances by air currents or by its own means of locomotion. Long-distance spread can also be accomplished naturally by the cumulative effect of a series of short-distance moves over a prolonged period of time. However, this natural extension of geographical range is miniscule compared to the rapid intercontinental movement that can be accomplished by man.

The second criterion (that quarantine regulations should be based on biology or pest risk analysis) is not as readily satisfied. When a virus has been widely distributed to the point where it is considered to have worldwide distribution, such as is the case with strawberry mild yellow-edge, one can hardly make the claim that inclusion in quarantine regulations of some 20 countries is based on sound biology. Some of the MLOs (peach yellows, strawberry witches' broom, strawberry green petal, and potato witches' broom) have not been widely distributed but are considered to be of relatively minor importance in those countries where they do occur; hence there is little justification to impose quarantine regulations to prevent their introduction into other countries.

The viroid that is included in the list (potato spindle tuber) does not naturally spread readily in field-grown potatoes and can be kept under control by applying seed potato certification and phytosanitary regulations. It is worth noting, for instance, that PSTV has never been found in commercial stocks or advance selections in the U.K., a country that has an efficient scheme for testing all clonal mother plants that are used for potato propagation. The general consensus is that PSTV should be treated as a minor disease which can be kept under control by proper seed certification and seed production programs.[73] Despite the fact that spindle tuber is a minor disease provided adequate precautions are taken, it is essential that all stages of a potato certification program be monitored to ensure that PSTV is excluded. Since indexing of initial stocks used in certification schemes is often the responsibility of quarantine officials, there is ample justification for including PSTV in quarantine regulations.

We have difficulty in accepting the criterion that quarantine action should constitute the least drastic action which provides the necessary safeguards. In surveying the regulations of 125 countries Kahn[37] found that 292 different names of viruses were tabulated with a total of 645 citations. We made no attempt to assess the threat potential associated with the 292 different viruses that are included in regulations but, based on our assessment of the 10 most frequently cited ones, there is little doubt that exclusion as a quarantine strategy is grossly overused. It appears that quarantine officials are able to incorporate specific viruses in their list of published quarantine regulations with relative ease and rather than risk the possible consequences of being too lenient, they have tended to err on the safe side and cite many pathogens that cannot be justified on purely biological grounds.

In conclusion, we feel that the fourth criterion (review of plant quarantine policy with a view to relaxing restrictions where it can be done safely) is overdue with respect to viruses. A list of virus and virus-like agents found in quarantine regulations has been published.[33]

REFERENCES

1. **Anon.**, Detection of virus and virus like diseases of fruit trees, *Acta Hortic.*, 130, 319, 1983.
2. **Bar-Joseph, M.**, Cross protection incompleteness: a possible cause for natural spread of citrus tristeza virus after a prolonged lag period in Israel, *Phytopathology*, 68, 1110, 1978.
3. **Bar-Joseph, M. and Garnsey, S. M.**, Enzyme-linked immunosorbent assay (ELISA): principles and applications for diagnosis of plant viruses, in *Plant Diseases and Vectors: Ecology and Epidemiology*, Maramorosch, K. and Harris, K. F., Eds., Academic Press, New York, 1981, 35.
4. **Boone, D. M.**, Witches'-broom and multiplier diseases of strawberry, in *Virus Diseases of Small Fruits and Grapevines*, Frazier, N. W., Ed., University of California Press, Berkeley, 1970, 25.
5. **Bovey, R.**, Apple proliferation, in *Virus Diseases of Apples and Pears*, Posnette, A. F., Ed., Commonwealth Agricultural Bureaux, Farnham Royal, England, 1963, 63.
6. **Cashdollar, L. W., Chmelo, R., Esparza, J., Hudson, G. R., and Joklik, W. K.**, Molecular cloning of the complete genome of reovirus serotype 3, *Virology*, 133, 191, 1984.

7. **Chock, A. K.,** The international plant protection convention, in *Plant Health,* Ebbels, D. L. and King, J. E., Eds., Blackwell Scientific, Oxford, 1979, 1.
8. **Clark, M. F.,** Immunosorbent assays in plant pathology, *Annu. Rev. Phytopathol.,* 19, 83, 1981.
9. **Clark, M. F. and Adams, A. N.,** Characteristics of the microplate method of enzyme-linked immunosorbent assay for the detection of plant viruses, *J. Gen. Virol.,* 34, 475, 1977.
10. **Costa, A. S. and Müller, G. W.,** Tristeza control by cross protection: a U.S.-Brazil cooperative success, *Plant Dis.,* 64, 538, 1980.
11. **Cropley, R.,** The identification of plum pox (Sharka) virus in England, *Plant Pathol.,* 17, 66, 1968.
12. **Dodds, J. A., Morris, T. J., and Jordan, R. L.,** Plant viral double-stranded RNA, *Annu. Rev. Phytopathol.,* 22, 151, 1984.
13. **Doi, Y., Teranaka, M., Yora, K., and Asuyama, H.,** Mycoplasma- or PLT group-like micro-organisms found in the phloem elements of plants infected with Mulberry dwarf, Potato witches' broom, Aster yellows, or Paulownia witches' broom, *Ann. Phytopathol. Soc. Jpn.,* 33, 259, 1967.
14. **Fernow, K. H.,** Tomato as a test plant for detecting mild strains of potato spindle tuber virus, *Phytopathology,* 57, 1347, 1967.
15. **Fox, G. E., Stackebrandt, E., Hespell, R. B., Gibson, J., Maniloff, J., Dyer, T. A., Wolfe, R. S., Balch, W. E., Tanner, R. S., Magrum, L. J., Zablen, L. B., Blakemore, R., Gupta, R., Bonen, L., Lewis, B. J., Stahl, D. A., Luehrsen, K. R., Chen, K. N., and Woese, C. R.,** The phylogeny of prokaryotes, *Science,* 209, 457, 1980.
16. **Franssen, H., Leunissen, J., Goldbach, R., Lomonossoff, G., and Zimmern, D.,** Homologous sequences in non-structural proteins from cowpea mosaic virus and picornaviruses, *EMBO J.,* 3, 855, 1984.
17. **Fridlund, P. R.,** Certification and supporting phytopathology, in *Methods in Fruit Breeding,* Moore, J. N. and Janick, J., Eds., Purdue University Press, West Lafayette, Ind., 1983, 398.
18. **Fry, P. R. and Hammett, K. R. W.,** Rose wilt virus in New Zealand, *N. Z. J. Agric. Res.,* 14, 735, 1971.
19. **Fulton, R. W.,** Techniques of investigating stone fruit viruses, in Virus Diseases and Noninfectious Disorders of Stone Fruits in North America, Agricultural Handb. 437, U.S. Department of Agriculture, Washington, D.C., 1976, 1.
20. **Gheorghiu, E.,** A potential indicator for detection of virus and mycoplasma that infect fruit trees, *Acta Phytopathol. Acad. Sci. Hung.,* 15, 175, 1980.
21. **Göbel, U. B. and Stanbridge, E. J.,** Cloned mycoplasma ribosomal RNA genes for the detection of mycoplasma contamination in tissue cultures, *Science,* 226, 1211, 1984.
22. **Goding, J. W.,** Antibody production by hybridomas, *J. Immunol. Meth.,* 39, 285, 1980.
23. **Gorziglia, M., Cashdollar, L. W., Hudson, G. R., and Esparza, J.,** Molecular cloning of a human rotavirus genome, *J. Gen. Virol.,* 64, 2585, 1983.
24. **Gould, A. R. and Symons, R. H.,** A molecular biological approach to relationships among viruses, *Annu. Rev. Phytopathol.,* 21, 179, 1983.
25. **Grieve, B. J.,** 'Rose wilt' and 'dieback'. A virus disease of roses occurring in Australia, *Aust. J. Exp. Biol. Med. Sci.,* 8, 107, 1931.
26. **Grill, L. K. and Garger, S. J.,** Identification and characterization of double-stranded RNA associated with cytoplasmic male sterility in *Vicia faba, Proc. Natl. Acad. Sci. U.S.A.,* 78, 7043, 1981.
27. **Halk, E. L., Hsu, H. T., Aebig, J., and Franke, J.,** Production of monoclonal antibodies against three ilarviruses and alfalfa mosaic virus and their use in serotyping, *Phytopathology,* 74, 367, 1984.
28. **Hamdorf, G.,** The detection of plum pox virus (PPV) by indicator plants and enzyme-linked immunosorbent assay (ELISA), *Acta Hortic.,* 130, 151, 1983.
29. **Hamilton, R. I., Edwardson, J. R., Francki, R. I. B., Hsu, H. T., Hull, R., Koenig, R., and Milne, R. G.,** Guidelines for the identification and characterization of plant viruses, *J. Gen. Virol.,* 54, 223, 1981.
30. **Harris, P. S., Miller-Jones, D. N., and Howell, P. J.,** Control of potato spindle tuber viroid: the special problems of a disease in plant breeders' material, in *Plant Health,* Ebbels, D. L. and King, J. E., Eds., Blackwell Scientific, Oxford, 1979, 231.
31. **Harrison, B. D., Robinson, D. J., Mowat, W. P., and Duncan, G. H.,** Comparison of nucleic acid hybridisation and other tests for detecting tobacco rattle virus in narcissus plants and potato tubers, *Ann. Appl. Biol.,* 102, 331, 1983.
32. **Haseloff, J., Goelet, P., Zimmern, D., Ahlquist, P., Dasgupta, R., and Kaesberg, P.,** Striking similarities in amino acid sequence among nonstructural proteins encoded by RNA viruses that have dissimilar genomic organization, *Proc. Natl. Acad. Sci. U.S.A.,* 81, 4358, 1984.
33. **Havens, O.,** Plant Pests of Phytosanitary Significance to Importing Countries and States, 4th ed. (revised by Q. L. Holdeman), California Department of Food and Agriculture, Sacramento, 1984, 93.
34. **Hiruki, C., Giannotti, J., and Dijkstra, J.,** A comparative study of fluorescence in stems of *Vinca rosea* infected with mycoplasmas of different plant origins, *Neth. J. Plant Pathol.,* 80, 145, 1974.
35. **Ikin, R. and Frost, R. R.,** Virus diseases of roses. I. Their occurrence in the United Kingdom, *Phytopathol. Z.,* 79, 160, 1974.

36. **Jordan, R. L., Dodds, J. A., and Ohr, H. D.,** Evidence for viruslike agents in avocado, *Phytopathology,* 73, 1130, 1983.
37. **Kahn, R. P.,** The host as a vector: exclusion as a control, in *Pathogens, Vectors, and Plant Diseases: Approaches to Control* Harris, K. F. and Maramorosch, K., Eds., Academic Press, New York, 1982, chap. 7.
38. **Kahn, R. P., Monroe, R. L., Hewitt, W. B., Goheen, A. C., Wallace, J. M., Roistacher, C. N., Nauer, E. M., Ackerman, W. L., Winters, H. F., Seaton, G. A., and Pifer, W. A.,** Incidence of virus detection in vegetatively propagated plant introductions under quarantine in the United States, 1957—1967, *Plant Dis. Rep.,* 51, 715, 1967.
39. **Khristova, D.,** [Rose wilt virus], in *Virusni bolesti po rasteniyata,* Sofia, Bulgaria, 1974, 77; as cited in *Rev. Plant Pathol.,* 55 (Abstr.), 4749, 1976.
40. **Klinkowski, M.,** Catastrophic plant diseases, *Annu. Rev. Phytopathol.,* 8, 37, 1970.
41. **Köhler, G. and Milstein, C.,** Continuous cultures of fused cells secreting antibody of predefined specificity, *Nature,* 256, 495, 1975.
42. **Kunze, L.,** Spread of apple proliferation in a newly established apple plantation, *Acta Hortic.,* 67, 121, 1976.
43. **Kurppa, A. and Martin, R. R.,** Use of double-stranded RNA for detection and identification of virus diseases of *Rubus* species, *Acta Hortic.,* 186, 51, 1986.
44. **Lawson, R. H.,** Controlling virus diseases in major international flower and bulb crops, *Plant Dis.,* 65, 780, 1981.
45. **Lin, C. P. and Chen, T. A.,** Comparison of monoclonal antibodies and polyclonal antibodies in detection of the aster yellows mycoplasmalike organism, *Phytopathology,* 76, 45, 1986.
46. **Llacer, G., Sanchez-Capuchino, J. A., Casanova, R., and Forner, J. B.,** Graft transmission of apricot chlorotic leaf roll from an apple tree affected by proliferation, *Acta Phytopathol. Acad. Sci. Hung.,* 15, 239, 1980.
47. **Maniatis, T., Fritsch, E. F., and Sambrook, J.,** *Molecular Cloning,* Cold Spring Harbor Laboratory, Cold Spring Harbor, N.Y., 1982.
48. **Martin, R. R. and Stace-Smith, R.,** Production and characterization of monoclonal antibodies specific to potato leaf roll virus, *Can. J. Plant Pathol.,* 6, 206, 1984.
49. **Matthews, R. E. F.,** Classification and nomenclature of viruses, *Intervirology,* 17, 140, 1982.
50. **Maule, A. J., Hull, R., and Donson, J.,** The application of spot hybridization to the detection of DNA and RNA viruses in plant tissues, *J. Virol. Meth.,* 6, 215, 1983.
51. **McGrew, J. R. and Posnette, A. F.,** Leafhopper-borne virus diseases of strawberry, in *Virus Diseases of Small Fruits and Grapevines,* Frazier, N. W., Ed., University of California Press, Berkeley, 1970, 26.
52. **Meinkoth, J. and Wahl, G.,** Hybridization of nucleic acids immobilized on solid supports, *Anal. Biochem.,* 138, 267, 1984.
53. **Mellor, F. C. and Frazier, N. W.,** Strawberry mild yellow-edge, in *Virus Diseases of Small Fruits and Grapevines,* Frazier, N. W., Ed., University of California Press, Berkeley, 1970, 14.
54. **Miller, J. H. and Reznikoff, W. S.,** *The Operon,* Cold Spring Harbor Laboratory, Cold Spring Harbor, N.Y., 1978.
55. **Milne, R. G. and Lesemann, D.-E.,** Immunosorbent electron microscopy in plant virus studies, in *Methods in Virology,* Vol. 8, Maramorosch, K. and Koprowski, H.. Eds., Academic Press, New York, 1984, 85.
56. **Mirkov, T. E. and Dodds, J. A.,** Association of double-stranded ribonucleic acids with lettuce big vein disease, *Phytopathology,* 75, 631, 1985.
57. **Morris, T. J.,** Isolation of viroid and viroid-like RNA from plant tissue, *Phytopathology,* 73, 792, 1983.
58. **Morris, T. J. and Dodds, J. A.,** Isolation and analysis of double-stranded RNA from virus-infected plant and fungal tissue, *Phytopathology,* 69, 854, 1979.
59. **Morris, T. J. and Smith, E. M.,** Potato spindle tuber disease: procedures for the detection of viroid RNA and certification of disease-free potato tubers, *Phytopathology,* 67, 145, 1977.
60. **Morris, T. J. and Wright, N. S.,** Detection on polyacrylamide gel of a diagnostic nucleic acid from tissue infected with potato spindle tuber viroid, *Am. Potato J.,* 52, 57, 1975.
61. **Morschel, J. R.,** Controlling the movement of plant diseases and pests into, out of, and within Australia, in *Plant Health,* Ebbels, D. L. and King, J. E., Eds., Blackwell Scientific, Oxford, 1979, 35.
62. **Németh, M. and Kölber, M.,** Additional evidence on seed transmission of plum pox virus in apricot, peach and plum proved by ELISA. *Acta Hortic.,* 130, 293, 1983.
63. **Oi, V. T. and Herzenberg, L. A.,** Immunoglobulin-producing hybrid cell lines, in *Selected Methods in Cellular Immunology,* Mishell, B. B. and Shiigi, S. M., Eds., W. H. Freeman, San Francisco, 1980, 351.
64. **Owens, R. A. and Diener, T. O.,** Sensitive and rapid diagnosis of potato spindle tuber viroid disease by nucleic acid hybridization, *Science,* 213, 670, 1981.
65. **Owens, R. A., Smith, D. R., and Diener, T. O.,** Measurement of viroid sequence homology by hybridization with complementary DNA prepared *in vitro, Virology,* 89, 388, 1978.

66. **Pemberton, A. W.,** Campaigns against plant virus diseases in England and Wales, in *Plant Health,* Ebbels, D. L. and King, J. E., Eds., Blackwell Scientific, Oxford, 1979, 239.
67. **Pfannenstiel, M. A., Slack, S. A., and Lane, L. C.,** Detection of potato spindle tuber viroid in field-grown potatoes by an improved electrophoretic assay, *Phytopathology,* 70, 1015, 1980.
68. **Pine, T. S. and Gilmer, R. M.,** Peach yellows, in Virus Diseases and Noninfectious Disorders of Stone Fruits in North America, Agricultural Handb. 437, U.S. Department of Agriculture, Washington, D.C., 1976, 91.
69. **Posnette, A. F. and Ellenberger, C. E.,** Further studies of green petal and other leafhopper-transmitted viruses infecting strawberry and clover, *Ann. Appl. Biol.,* 51, 69, 1963.
70. **Ralph, R. K.,** Double-stranded viral RNA, *Adv. Virus. Res.,* 15, 61, 1969.
71. **Rezaian, M. A., Williams, R. H. V., Gordon, K. H. J., Gould, A. R., and Symons, R. H.,** Nucleotide sequence of cucumber mosaic virus RNA 2 reveals a translation product significantly homologous to corresponding proteins of other viruses, *Eur. J. Biochem.,* 143, 277, 1984.
72. **Rochow, W. F. and Duffus, J. E.,** Luteoviruses and yellows diseases, in *Handbook of Plant Virus Infections and Comparative Diagnosis,* Kurstak, E., Ed., Elsevier, Amsterdam, 1981, chap. 7.
73. **Rohwer, G. G.,** Plant quarantine philosophy of the United States, in *Plant Health,* Ebbels, D. L. and King, J. E., Eds., Blackwell Scientific, Oxford, 1979, 23.
74. **Russell, W. C., Newman, C., and Williamson, D. H.,** A simple cytochemical technique for demonstration of DNA in cells infected with mycoplasmas and viruses, *Nature* 253, 461, 1975.
75. **Salazar, L. F., Owens, R. A., Smith, D. R., and Diener, T. O.,** Detection of potato spindle tuber viroid by nucleic acid spot hybridization: evaluation with tuber sprouts and true potato seed, *Am. Potato J.,* 60, 587, 1983.
76. **Schaper, U. and Converse, R. H.,** Detection of mycoplasmalike organisms in infected blueberry cultivars by the DAPI technique, *Plant Dis.,* 69, 193, 1985.
77. **Schaper, U. and Seemüller, E.,** Condition of the phloem and the persistence of mycoplasmalike organisms associated with apple proliferation and pear decline, *Phytopathology,* 72, 736, 1982.
78. **Schmid, G.,** Effects of tetracycline injections against apple proliferation, *Acta Hortic.,* 130, 237, 1983.
79. **Seemüller, E.,** Investigations to demonstrate mycoplasmalike organisms in diseased plants by fluorescence microscopy, *Acta Hortic.,* 67, 109, 1976.
80. **Singh, R. P. and Crowley, C. F.,** Successful management of potato spindle tuber viroid in seed potato crop, *Can. Plant Dis. Surv.,* 65, 9, 1985.
81. **Smith, I. M.,** EPPO: the work of a regional plant protection organization, with particular reference to phytosanitary regulations, in *Plant Health,* Ebbels, D. L. and King, J. E., Eds., Blackwell Scientific, Oxford, 1979, 13.
82. **Van Regenmortel, M. H. V.,** *Serology and Immunochemistry of Plant Viruses,* Academic Press, New York, 1982, chap. 8.
83. **Wakarchuk, D. A. and Hamilton, R. I.,** Cellular double-stranded RNA in *Phaseolus vulgaris, Plant Mol. Biol.,* 5, 55, 1985.
84. **Watson, J. D., Tooze, J., and Kurtz, D. T.,** *Recombinant DNA: A Short Course,* Scientific American Books, New York, 1983.
85. **Wright, N. S., Raine, J., and Valenta, V.,** Witches' broom, in *Compendium of Potato Diseases,* Hooker, W. J., Ed., American Phytopathological Society, St. Paul, Minn., 1981, 92.
86. **Wu, R., Guo, L.-H., Stepien, P. P., Shen, T., Tso, J. Y., Brousseau, R., Narang, S., and Thomas, D. Y.,** Expression vectors for fused proinsulin production in *E. coli* and yeast, *Proc. Int. Congr. Genetics,* New Delhi, 1983.
87. **Young, R. A. and Davis, R. W.,** Efficient isolation of genes by using antibody probes, *Proc. Natl. Acad. Sci. U.S.A.,* 80, 1194, 1983.
88. **Zeller, S. M.,** Preliminary studies on witches' broom of strawberry, *Phytopathology,* 17, 329, 1927.

Chapter 14

PLANT QUARANTINE DIAGNOSTIC PROBLEMS: BACTERIA

E. L. Civerolo

TABLE OF CONTENTS

I. INTRODUCTION

Plant quarantine laws are generally intended to preclude or restrict the movement of plant pathogens to and into new areas by regulating the movement and introduction of plant material. Most plant pathogens of agricultural crops are disseminated long distances and into new areas primarily by means of infected or contaminated plant material. The effectiveness of plant quarantine laws, as well as other plant disease management strategies, are dependent upon rapid and accurate disease diagnosis and/or detection and identification of plant pathogens. Furthermore, as a result of increasing movement of plants, plant parts, and plant propagative materials, including seeds, in international commerce, there is a need for effective, reliable, accurate, sensitive, and rapid methods for detection and identification of plant pathogenic bacteria.[58] Plant disease diagnosis and pathogen identification are closely related.

Phytopathogenic prokaryotes include readily cultivable bacterial species of *Agrobacterium, Arthobacter, Clavibacter, Corynebacterium, Curtobacterium, Erwinia, Nocardia, Pseudomonas, Rhodococcus, Streptomyces,* and *Xanthomonas,* as well as fastidious bacteria (e.g., citrus greening and greening-like disease agents), *Mollicutes* (mycoplasma-like organisms, *Spirolasmas* spp.), and rickettsia-like bacteria. They cause or are associated with a wide range of plant diseases and abnormal conditions. These include leaf and fruit spots, blights, stripes, streaks, cankers, fasciations, galls, soft rots and/or vascular wilts, yellows-type diseases (general or marginal chlorosis, growth reduction, small and/or distorted leaves, shortened internodes, proliferations of axillary buds, stunting), phyllody, leaf scorching, reduced vigor, internal tissue discoloration and plant death.[2,35,63] These plant abnormalities cannot always be reliably diagnosed by symptoms only. In addition, phytopathogenic bacteria are capable of living and multiplying on surfaces of tissues of healthy plants in the absence of disease.[36,40] These epiphytic bacteria (*epipabs* sensu[36]) can be sources of inocula for infection and subsequent disease development.

Ideally, reliable disease diagnosis and/or pathogen detection or identification techniques should be simple, accurate, rapid, specific, and sensitive. In addition, it is desirable for plant health to detect populations of phytopathogenic bacteria and bacteria present before symptom development, and to determine bacteria-free plant material including propagating material.[33] Methodologies for diagnosing disease caused by phytopathogenic bacteria, as well as detecting and identifying these pathogens, include a variety of common and sophisticated techniques. In addition, improvements in existing techniques and development of new techniques are continually being made.

II. MAJOR DIAGNOSTIC PROBLEMS

A. Symptoms

Symptomatology is a primary means of presumptive disease diagnosis; however, the reliability of symptomatology for disease diagnosis is limited. Plant diseases caused by phytopathogenic prokaryotes cannot always be diagnosed based on symptoms alone.[35] Similar or identical symptoms may be caused by different agents. Distinctive symptoms may not always be apparent. Atypical symptoms may also occur. Host responses (i.e., symptoms) resulting from infection by phytopathogenic prokaryotes vary greatly.[18] Accurate disease diagnosis, therefore, often requires experience and knowledge of the host, pathogen, and environment. Symptom variation is related to such factors as the relative or differential resistance of different hosts, distribution of the pathogen within the host, specific pathotype of the pathogen involved, and environmental conditions. Many important phytopathogenic bacteria are seedborne or remain viable in plant residues.

B. Pathogen Isolation

A major problem in isolation of phytopathogenic bacteria is the presence of saprophytic bacteria and other microorganisms. These usually grow faster on primary isolation plates on nonselective common nutrient agar media than bacterial plant pathogens. Furthermore, some fastidious bacteria plant pathogens such as mycoplasma-like organisms associated with yellows-type diseases and fastidious Gram-positive phloem-inhibiting bacteria, have not been successully cultivated in vitro. Nevertheless, isolation of these organisms can be expected because of recent developments in isolation and cultivation of other fastidious phytopathogenic bacteria.

C. Pathogen Identification

Identification of phytopathogenic prokaryotes is generally based on Koch's postulates. Thus, a specific organism must be consistently associated with an infectious disease, the organism must be obtained in pure culture, the organism from pure culture must cause disease in artificially inoculated hosts (pathogenicity test), and the organism must be reisolated from the artificially inoculated host. This process of pathogen isolation, purification, testing for pathogenicity and reisolation is time consuming and laborious. In addition, some fastidious phytopathogenic prokaryotes cannot be isolated on artificial nutrient media. Other techniques for detection and/or identification of phytopathogenic prokaryotes directly in clinical samples may be insensitive or relatively nonspecific.

D. Standarization

A variety of common and sophisticated techniques is available for the detection and identification of phytopathogenic bacteria. Nevertheless, there is generally a lack of universally accepted standard techniques or protocols for these purposes. Classification of phytopathogenic bacteria is ideally based on genetic, physiological, and pathogenic characteristics; however, not all descriptions of these pathogens include all of these features.[33] A few key characteristics can be used to provide support for presumptive identificaiton of a phytopathogenic bacterial phytopathogen.

III. DIAGNOSIS OF BACTERIAL PLANT DISEASES

A. General

Time, resources, and immediate needs or objectives are factors that affect the amount of diagnostic work that can and should be done with any particular sample. The needs for rapid and presumptive diagnoses are distinguishable from the needs for more time-consuming confirmatory diagnoses.[35]

Appropriate disease control methods can be recommended on the basis of a correct presumptive diagnosis. Confirmatory diagnosis may be required if a new or different host or pathogen is involved. Alternatively, accuracy of the diagnosis may be essential for certain quarantine and other regulatory aspects of bacterial plant disease management.

B. Sample Collection

Samples of diseased material should be representative of all of the disease symptoms, involving all plant parts, and should include the earliest stage(s) of the disease if possible. Phytopathogenic bacteria are usually isolated more readily from samples containing the earliest stage of infection.

Diseased material samples are usually placed in polyethylene bags (or other suitable container) to preclude extensive drying and/or contamination. In some cases, it may be necessary to place the primary sample-containing bag in a second bag for transport.

Samples should be kept cool in a container and not exposed to direct sunlight. For

Table 1
COMMON SYMPTOMS OF BACTERIAL PLANT DISEASES

Symptom	Causal bacteria (genera)
Angular leafspots delimited by veins with or without distinct chlorotic or necrotic spots	*Clavibacter, Pseudomonas, Xanthomonas*
Fruit spots	*Clavibacter, Erwinia, Pseudomonas, Xanthomonas*
Blighted stems, petioles, fruits	*Arthobacter, Erwinia, Pseudomonas, Xanthomonas*
Cankers	*Clavibacter, Erwinia, Pseudomonas, Xanthomonas*
Vascular wilts, gumming, black rot	*Clavibacter, Curtobacterium, Erwinia, Xanthomonas*
Fasciation	*Rhodococcus*
Dry necroses	*Erwinia*
Gumming of inflorescences	*Corynebacterium*
Cutting rot, bulb rot, black rot	*Xanthomonas*
Soft-rots	*Erwinia, Pseudomonas, Xanthomonas, (Acetinobacter, Clostridium, Bacillus)*
Ring rot	*Clavibacter*
Galls	*Agrobacterium, Nocardia, Pseudomonas*
Bud proliferation	*Nocardia*
Hairy root	*Agrobacterium*
Declines, stunting, yellows-type diseases, proliferation disorders	Mollicutes (spiroplasmas, mycoplasma-like organisms), fastidious vascular bacteria (gracilicutes; rickettsia-like bacteria)
Scabs and/or rots	*Streptomyces*

phytopathogenic bacteria, it is preferable that the diseased material be partially dry rather than wet to minimize multiplication of secondary microorganisms that can interfere with subsequent isolation of bacterial plant pathogens.

When possible, diseased material such as leaf spots, blights, cankers, galls, and fasciations should be pressed and dried or otherwise preserved for herbarium specimens. Preserved herbarium specimens are useful reference sources for future diagnoses. In addition, pressed and dried herbarium specimens are useful for isolation of bacterial plant pathogens and for preservation of some phytopathogenic bacteria (e.g., most xanthomonads).

C. Symptoms

A complete description or record of symptoms and accurately identified host are critical for accurate disease diagnosis.[35] Symptoms of presumptive bacterial plant diseases include watersoaked spots (or tissues) that appear translucent with transmitted light, oozing of bacterial masses from affected tissues, soft rots, and blighted twigs. It is also helpful to check host indices, descriptions of plant diseases (e.g., text books and original publications), photographs, and herbarium specimens to determine what bacterial diseases have been described for the host plant. Symptoms of some selected common bacterial plant diseases are summarized in Table 1.

Oozing or streaming of bacterial masses from cut lesions is almost invariably associated with bacterial plant diseases, except those caused by agrobacteria.[35] Absence of detectable ooze is not an absolutely reliable criterion that bacteria are not involved.

Dried exudates of bacterial ooze may occur on the lower surfaces of leaf lesions or on the surfaces of stems and petioles with lesions. Droplets of bacterial ooze may develop on freshly infected material after incubation for several days in a moist chamber. Alternatively, small portions (3 to 5 mm) of material (e.g., leaf lesions) are excised from the boundary between diseased and healthy tissue, mounted in a droplet of water under a coverslip on a microscope slide, and examined at 100 to 400×. Oozing or streaming of bacterial masses is visible; however, the rate of egress and the amount of bacteria present depend upon the age and condition of the diseased material. Particulate material, such as latex, plastids, and starch granules, may be confused with bacterial cells.[35]

D. Pathogen Isolation

1. General

Appropriate samples in good condition are essential for successful isolation of bacterial plant pathogens. Isolations are made as soon as possible after samples are collected. Initial isolations on suitable selective or semiselective media may be made in the field. Surface disinfestation of tissue from which pathogen isolations are to be made is not always necessary; however, washing samples in running tap or distilled/demineralized water is desirable. If surface disinfestation is necessary, the material may be immersed in 0.5 to 1% sodium hypochlorite or 70 to 95% ethanol for 1 to 2 min, followed by extensive rinsing with sterile distilled/demineralized water.

Generally, portions of diseased material are excised aseptically from the boundary between diseased and healthy tissue, and suspended in a small volume (e.g., 1 to 5 mℓ) of appropriate suspending medium (sterile water, saline, phosphate-buffered saline, dilute nutrient medium). After incubation for 30 to 60 min at room temperature to allow separation of bacteria from plant tissue debris and diffusion from the diseased material into the suspending medium, aliquots of the suspension are streaked onto the surface of a suitable general or selective nutrient agar medium. Alternatively, the excised diseased portion may be aseptically triturated in a small volume of suspending medium and the macerate allowed to stand for 5 to 20 min before streaking aliquots onto the nutrient agar medium. Dilutions (e.g., 1:10 to 1:100) of turbid, opaque suspensions should also be streaked.

Small aseptically excised pieces of diseased plant tissue exposed by cutting through the boundary between diseased and healthy tissue may be allowed to grow out directly onto the medium from the tissue piece. Alternatively, a small droplet of sterile water may be placed on the agar surface adjacent to the tissue piece near the edge of the plate, allowed to stand for 15 to 30 min, and then streaked over the surface of the isolation medium.[35]

A simplified technique for isolating phytopathogenic bacteria from leafspots involves aseptically stabbing a flame-sterilized needle through the margin of a young lesion directly into a appropriate isolation medium.[30] In an adaptation of this method, the margins of lesions on pods, stems, and other fruit are aseptically pierced with a sterile needle. The needle is withdrawn and stabbed into the isolation medium. With a sterile inoculating loop, streaks may be made aseptically from the stabs in the agar over the surface of the isolation medium.

The surface of the isolation medium must be dry in order to ensure subsequent development of distinct, well-separated colonies. When the agar medium is cooled to about 45 to 50°C before pouring, the surfaces of the set medium will dry in about 24 to 48 hr. Open plates can be aseptically dried at 37 to 40°C if inverted, or at room temperature in a laminar air flow hood for 15 to 20 min.

Streaks are made on nutrient agar media with a platinum loop in four right-angle directions, with the loop being flame-sterilized after each directional streak.[58] After streaking, plates are generally incubated at 25 to 30°C for isolation of most phytopathogenic bacteria. For isolation of corynebacteria, incubation at 23 to 25°C may be more suitable.[35] Primary isolation plates are examined daily for appearance of bacterial colonies. Common saprophytes generally grow faster than phytopathogenic bacteria. Bacteria that produce macroscopic colonies within 24 to 36 hr are likely to be saprophytes. A dissecting microscope is used to determine colony characteristics such as morphology, elevation, surface texture, viscosity, striations, crater formations, and margin features. Other colony and cultural characteristics to note are pigmentation, production of diffusible pigments, and fluorescence.

In order to establish a pure culture of an isolated bacterial strain, a single well-isolated colony is usually selected and restreaked on fresh medium. Further purification can be achieved by repeating the single colony selection process from the subculture medium.

2. Nonselective Isolation Media for Common Plant Pathogenic Bacteria

For most routine isolations, a few nonselective isolation media are adequate. Recipes for these media are given elsewhere.[26,58] A few selected examples of these media include Difco nutrient agar supplemented with yeast extract and yeast extract-mannitol agar for *Agrobacterium;* glucose-yeast extract-calcium carbonate agar for *Arthobacter, Clavibacter, Corynebacterium, Erwinia, Rhodococcus* and *Xanthomonas;* sucrose-peptone agar for *Erwinia, Pseudomonas* and *Xanthomonas;* tetrazolium medium and King's medium B for *Pseudomonas;* and yeast extract nutrient agar and glucose-peptone-yeast extract agar for *Xanthomonas.*[35] Difco® nutrient agar supplemented with glucose or sucrose and Wilbrink's medium are also useful for routine isolations of some xanthomonads (e.g., *X. albilimeans*).[35]

3. Selective Isolation Media for Common Bacterial Plant Pathogens

Presence of saprophytic microorganisms in soil, decomposing plant material, and old lesions can make isolation of common plant pathogenic bacteria difficult. Saprophytes generally are present in larger numbers and grow faster than phytopathogenic bacteria to be isolated. Unfavorable conditions also caused by saprophytes in primary isolation plates for phytopathogenic bacteria include competition for nutrients, production of antibiotic substances, and adverse pH effects.[35] Some or all of these problems can be overcome by the use of selective or semiselective isolation media that exclude or preclude growth of saprophytes while permitting growth of the pathogen to be isolated. Exclusion of saprophytes on a specific selective isolation medium may be based on lack of appropriate nutrients, inhibition of growth by antibiotics or pH, or a combination of these. Selectivity of the medium is usually achieved by use of specific carbon and nitrogen sources in defined or complex media, alone or in combination with dyes, antibiotics, or other inhibitors which favor the pathogen.[35] Specific recipes of selective or semiselective isolation media for bacterial plant pathogens are given elsewhere.[14,26,58,60,61]

Generally, selective isolation media for plant pathogenic bacteria are complex, and may be expensive to prepare.[35] Nevertheless, these are useful for isolating specific bacterial plant pathogens for disease diagnostic purposes and for epidemiological studies involving pathogen survival, occurrence, and distribution.

4. Enrichment

Direct isolation of common phytopathogenic bacteria is often difficult because the organism to be isolated is present in a very small fraction of a mixed microbial population. Many nonpathogenic microorganisms in the mixture grow more rapidly on primary isolation plates than inhibit growth of the bacterial plant pathogen to be isolated. In some cases, selective increase of the bacterium to be isolated can be achieved in enrichment culture. A form of phytopathological enrichment consists of inoculating crude, infectious preparations from soil samples or diseased plant material into healthy host plant tissue or plant parts such as stems, attached or detached leaves or fruit, or slices of tissue of a suitable host (e.g., carrot and potato). Following a suitable incubation period under appropriate conditions for infection and selective growth, the bacterial plant pathogen can be isolated in axenic culture.

5. Fastidious Vascular Prokaryotes

a. Spiroplasmas

Plant pathogenic spiroplasmas occur in sieve tube elements of the phloem. Young plant tissues in the early stages of disease development usually contain high populations of spiroplasmas that are readily isolated. These tissues are not likely to contain other microorganisms, and sap expressed from these tissues are likely to contain few factors inhibitory to the growth of primary isolates.[10]

Diseased plant tissue is surface disinfested, rinsed with sterile water, and placed in an appropriate culture medium. Cuts can be made cross-wise through the vascular tissue to release organisms from the phloem. The primary culture can be subcultured directly, to avoid growth inhibition, by diluting into fresh medium. Alternatively, the primary culture can be centrifuged and the organisms in the pellet resuspended in fresh medium. Plant pathogenic spiroplasmas can also be isolated from the hemolymph or whole body extracts of infected insect vectors.[10]

SMC medium[27] and modified SMC medium, formulated as BSR, are useful for primary isolation of *Sprioplasma citri* from plants and leafhopper vectors.[10] A simple medium (C-3G) consisting of PPLO broth base, horse serum, and sucrose and a slightly modified medium (R_2) containing PPLO broth, fetal bovine or horse serum, and sucrose are easily prepared and have been used for primary isolations of several spiroplasmas.[10,42] Medium MIA,[37] a modification of medium Ml,[71] will support growth of several different spiroplasmas, including the more fastidious strains of the corn stunt spiroplasma.

b. Xylem-Inhabiting Bacteria

Isolation of fastidious xylem-inhabiting bacteria is usually made from host plants which contain high populations of organisms. Generally, tissues are stripped of bark or outer epidermis, disinfested (e.g., immersion in sodium hypochlorite with ethanol) and rinsed in sterile water. After exposing fresh xylem tissue aseptically, sap is expressed by crushing and then blotted onto the surface of an agar medium. Sap can also be extracted in a few milliliters of filter-sterilized phosphate-buffered saline (PBS) by centrifugation at 20,000 *g* for 10 to 15 min. An agar medium can then be inoculated with the supernatant. Tissue can also be macerated in sterile PBS and serial dilutions of the homogenate inoculated onto an agar medium. Intact wedges or sections of tissue can be placed directly on an agar medium or in a broth medium.

Primary cultures are usually incubated at 20 to 25°C in an unmodified normal atmosphere. Bacterial isolates must be subcultured by streaking onto fresh medium approximately every 5 to 30 days, depending upon the strain and medium.

Numerous semisynthetic and chemically defined media are available for isolation and cultivation of xylem-inhabiting bacteria.[10] The filamentous, rod-shaped, Gram-positive ratoon stunting disease and bermuda grass bacteria have been isolated on SMH, SC, and S8. The Gram-negative plum leaf scald, phony peach disease, ragweed, Pierce's disease of grapevine, almond leaf scorch, periwinkle wilt, clove wilt, and alfalfa dwarf bacteria have been isolated and cultivated on S8, JD-1, JD-2, JD-3, BCYE, BZYE, and PW media.

c. Citrus Greening Disease Organism

Citrus greening (and greening-like) disease(s) is(are) caused by a unique, Gram-negative, catalase- and oxidase-positive bacterium with a distinctive cell wall structure.[6,7,27-29] There may be at least two distinct physiological types of the pathogen in various parts of the world. Asian greening disease (including Likubin disease in Taiwan, tree decline or dieback in India, leaf mottle-yellow disease in the Phillipines, and yellow shoot disease in the People's Republic of China) symptoms develop at 35°C, and the pathogen is transmitted primarily in nature by the citrus psylla *Diaphorina citri*.[7,9,28,44,49,50] African greening disease symptoms develop at 24°C,[59] and the pathogen is transmitted in nature primarily by the citrus psyllid *Trioza erytreae*. The organisms associated with both African and Asian greening diseases occur in the sieve tubes of infected citrus tissue and replicate in the phloem, as well as in the hemolymph of the insect vectors.[29,50]

A putative greening diesase bacterial pathogen has been isolated using a medium consisting of the major components of citrus phloem and insect vector hemolymph.[28] Antiserum prepared against a strain of this organism may be useful for serodiagnosis of this disease by enzyme-linked immunosorbent assay (EIA) and radioimmunoassay (RIA) tests.[25]

E. Pathogenicity Tests

1. General

A pathogenicity test based on host inoculation is required to fulfill Koch's postulates for disease diagnosis. Generally, a susceptible host, preferably the same host and cultivar from which the pathogen was isolated, is used. Unusual or atypical symptoms or hypersensitive reactions (see Section E.5) may develop on a resistant cultivar. Natural infection should be simulated as closely as possible. Inoculations should be made on immature, rapidly developing plant tissue, as mature tissues become resistant to plant infection by pathogenic bacteria. Results of pathogenicity tests are affected by inoculum concentration, method of inoculation, and pre- and postinoculation conditions under which the plants are maintained. Inocula containing reference strains that induce known reactions should be included in each test if possible. Nonpathogens can also induce a variety of host reactions, especially when high inocula concentrations are used.

2. Inoculum Preparation

Bacterial growth from a fresh (e.g., 24- to 72-hr-old) nutrient agar medium culture is usually suspended in sterile water or other suitable medium. Alternatively, cells are collected from a log-phase liquid medium culture by low-speed centrifugation (e.g., 5000 to 10,000 *g* for 15 to 20 min) and resuspended in sterile water or other suitable medium. The cells can be further washed by two or three successive centrifugations and resuspension of the pelleted cells. The cell suspension is adjusted to a turbidity equivalent to 0.1 to 0.3 absorbance units at 590 to 620 nm (1 cm light path) photometrically or visually by comparison with turbidity standards. Depending upon the strain, this is usually equivalent to approximately 1 to 5 $\times$ 10^8 colony-forming units (CFU) per milliliter. The cell suspension is then diluted appropriately to contain approximately 10^4 to 10^7 CFU per milliliter of suspending medium.

3. Pre- and Postinoculation Conditions

Maintenance of plants to be inoculated at high relative humidity prior to inoculation is generally not necessary for a successful pathogenicity test for bacterial plant pathogens.[26] However, this may be helpful in some cases. Selection and inoculation of appropriate plant tissue at the proper stage of development, and maintenance of inoculated plants under high relative humidity conditions after inoculation are important for infection by phytopathogenic bacteria. Inoculated plants are usually held for about 24 to 48 hr in a cabinet or growth chamber maintained with high relative humidity, or covered with a plastic bag to increase and maintain high humidity. The incubation period for symptom development may be shortened and the severity of symptoms may be increased by increasing the length of time inoculated plants are held under high-humidity conditions after inoculation.[26]

4. Inoculation Methods

a. Foliar, Stem, and Fruit Pathogens

Inocula containing approximately 10^6 to 10^8 CFU per milliliter are sprayed at low pressure onto both surfaces of leaves to runoff without infiltration of the tissue. Small needle or pin puncture wounds made through droplets of inoculum may be made after spraying if the pathogen is known to or suspected of infecting only through wounds. In addition, to expanded leaves, young emerging shoots and leaf whorls can be similarly spray inoculated.

Infiltration of incoulum under the epidermis of a leaf is generally not recommended for inoculations for routine diagnosis.[26] With high inoculum concentration, discrete lesions may develop at the margin of the infiltrated area; however, extensive tissue damage to the infiltrated tissue may occur. Nevertheless, discrete lesions without extensive tissue damage will often develop following inoculation with some foliar plant pathogenic bacteria if inoculum containing less than 10^5 CFU per milliliter is infiltrated into susceptible leaf tissue.

Stems and petioles are usually inoculated by placing droplets of inoculum on the surface. The inoculum is then pricked into the tissue through the droplet with a needle or pin.

b. Vascular Pathogens

Test host plants for bacterial plant pathogens that affect the vascular system of their host may be inoculated by stem wounding as above. Alternatively, the root system may be inoculated. The roots of seedlings or plants removed from soil and rinsed free of soil particles are usually wounded by lightly trimming before or after dipping in an inoculum suspension. The seedlings or young plants are then replanted. Roots of seedlings or plants in soil can be inoculated *in situ* by cutting through the soil with a scalpel to wound the roots, and then pouring an inoculum suspension into the soil.

c. Excised Plant Tissues and Detached Organs

Use of excised plant tissues and detached organs are convenient for screening suspected isolated strains for pathogenicity. Detached whole immature fruit leaves and bean pods are useful for screening for fruit- and leafspot-causing *X. campestris* pathovars, *P. syringae* pathovars, and *E. amylovora*. Potato slices as well as whole intact vegetables such as eggplant, squash, and cucumber are useful for testing the macerating and pectolytic activity of soft-rotting erwinias, pseudomonas, and xanthomonads. Inverted carrot slices are useful for testing for *A. tumefaciens* and *A. rhizogenes*.

Generally, the detached tissue or organ is thoroughly washed and surface disinfested with sodium hypochlorite or ethanol. Droplets of inocula suspension are usually deposited into small needle puncture holes in intact whole fruit or vegetables or deposited on the cut surfaces of fruit, potato, or carrot slices. Inocula may also be infiltrated into thin leaves, such as those of *Prunus* species, deposited on the unwounded surfaces of intact leaves, or deposited on needle puncture wounds of intact leaves. Positive inocula and negative controls consisting of uninoculated wounds and nonpathogenic strain-containing inocula should be used whenever possible. Inoculated exised plant tissues and detached organs are usually incubated in sealed containers or Petri dishes with moist filter paper at 25 to 30°C for 1 to several days depending upon the bacterial strains.

5. Hypersensitivity Test

The hypersensitivity test using tobacco, tomato, or pepper is very useful as a general screening method for potential pathogenicity for most fruit- and leaf-spotting bacteria.[26,35] Nevertheless, a negative hypersensitive reaction (HR) is not conclusive, as the ability to induce HR is variable among strains of the same pathogen and some pathogens do not induce HR reactions. The HR reaction is affected by a variety of factors, including the host species and pre- and postinoculation growth temperature and light. In addition, HR induction by different groups of pathogens may be affected in different ways.[26] To be useful as a diagnostic aid, standardization of this test needs to be developed.

In general, bacterial suspensions containing 10^8 to 10^9 CFU per milliliter are infiltrated into the lower surface of tobacco (*Nicotiana tabacum* cv. White Burley or Hicks) or tomato leaves with a syringe equipped with a 25G needle. A dry necrosis, characteristic of HR, occurs rapidly within 8 to 24 hr. In contract, susceptible host plants will develop typical disease symptoms after several days rather than rapidly depending upon the environmental conditions.

F. Recent Technological Advances

1. General

Classical methods of microbial isolation and identification are not always conducive to rapid disease diagnoses. Thus, more sensitive, specific, and reliable techniques have been developed for pathogen detection and identification in clinical samples. Significant advances have been made in development of serodiagnostic techniques and molecular probes for bacterial plant diseases.

2. Serodiagnostic Techniques

Serodiagnostic techniques for phytopathogenic bacteria are generally based on the use of polyclonal antibodies in precipitin tests, agglutination (or microagglutination) tests, direct and indirect immunofluoresence (IF) assays,[57] and variations of enzyme-linked immunosorbent assays (EIA).[12,13,15] Increased sensitivity and specificity of these tests can potentially be increased by development of improved serological reagents such as monoclonal antibodies,[1,20,22,43A] and procedures such as biotin-avidin or other means of amplification of enzyme-labeled immunoassays. Other immunosorbent methods, such as immunosorbent immunfluoresence microscopy and immunosorbent dilution plating are potentially useful for improving the detection, identification, and selective isolation of phytopathogenic bacteria.[67] Cells of individual strains of phytopathogenic bacteria can also be detected and identified by IF staining of tissue extracts of clinical samples collected on membrane filters and by adaptation of immunogold and immunogold-silver staining techniques.[19,66]

IV. IDENTIFICATION OF PHYTOPATHOGENIC BACTERIA

A. General

A variety of general physiological and biochemical tests are available for identification of an isolated phytopathogenic bacterium at the generic and species level. Recent advances have also been made in development and use of more specialized techniques. Confirmation of the identification of phytopathogenic bacteria is usually dependent upon a pathogenicity test.

B. Morphology

Morphological features such as cell size and shape and arrangement are usually observed by phase contrast microscopy of fresh cultures. Motility of the cells is also usually determined. The number, position, and arrangement of flagella are best determined by electron microscopic examination of negatively stained actively motile cells; however, light microscopy techniques are available. The presence of refractile polymer inclusions, such as poly-β-hydroxybutyrate (PHB) inclusions, are produced by many aerobic Gram-negative bacteria. While these can be observed by phase contrast microscopy in stained cells, their identity can be confirmed by extraction and chemical analysis.

C. Gram Reaction

The Gram stain reaction is an important and essential feature for identifying an isolated bacterium. Numerous modifications of the Gram stain are available. Generally, a smear of young culture cells on a glass slide is air dried and gently heat fixed by light flaming of the underside of the slide. The heat-fixed smear is flooded with crystal violet for 1 min, washed for 3 to 4 sec in a gentle stream of tap water, decolorized with 95% ethanol, washed, and counterstained with a contrasting stain such as safranin. Bacteria whose cells retain the primary violet dye and appear blue are Gram positive. Bacteria whose cells lose the violet coloration and are counterstained, appearing red, are Gram negative.

At least two alternatives to the Gram stain are available. They are the KOH solubility test and aminopeptidase activity. The KOH solubility test is rapid and easy to perform. One or two drops of freshly prepared 3% KOH are placed on a clear glass slide. A colony, or a few colonies, are aseptically picked from 24 to 48 cultures on a solid nutrient medium and transferred to the drop of KOH. The material is stirred in the KOH for several (5 to

15) sec. If the KOH becomes viscous and a mucoid thread of slime appears when the mixture is lifted with an inoculating loop or toothpick, this is a KOH-positive reaction and the bacterium is concluded to be Gram negative. If the KOH does not become viscous, remains watery, and no mucoid threads develop, the reaction is KOH-negative characteristic of Gram-positive bacteria.

D. Physiological and Biochemical Tests

Commonly isolated phytopathogenic bacteria can usually be differentiated by cultural characteristics on a few artificial nutrient media and by selected determinative bacteriological tests based in physiological and biochemical properties.[26,35,41,58,60,61] Identification of bacteria of uncertain affiliation or minor importance as plant pathogens, as well as fastidious prokaryotic phytopathogens, may require specialized diagnostic tests. Methodology generally used for conventional determinative bacteriological tests for phytopathogenic bacteria are described elsewhere.[26,41,58]

E. Bacteriophages

In some cases, bacteriophages are useful for rapid, presumptive identification of phytopathogenic bacteria for disease diagnosis,[3,4,51,52] especially when used in conjunction with other diagnostic tests. Bacteriophages vary in their spectrum of activity. Some bacteriophages have broad host ranges while others are highly host specific. Highly host-specific phages generally occur in naturally infected plant material, and phages with very broad host ranges occur in soil.[3,4,31,62] Host-specific bacteriophages for identifying and distinguishing strains, pathovars, pathotypes, and biovars of the major genera of phytopathogenic bacteria have been described.[5,16,17,32,43,53,62,69] It is important that the host range of any phage be established before it can be used for bacterial identification. Phages for diagnostic and identification purposes should be tested against a wide range of strains of the specific pathogen from different hosts and geographical areas, other closely related bacteria, and epiphytic bacteria commonly isolated with the pathogen.

Phage techniques for phytobacterial pathogen detection and identification are generally easy to perform. Nevertheless, these techniques have not been widely used for these purposes. There is clearly more potential for bacteriophage typing systems to be developed in regard to phytopathogenic bacteria detection and identification for disease diagnosis.[5,51]

F. Recent Advances

The sensitivity and specificity of bacterial plant pathogen identification tests based on conventional determinative bacteriological tests and serological analyses might be improved in several ways. Treatments could be developed to maximize the release of intact bacterial cells or specific bacterial components.[68] Alternatively, techniques could be developed to identify and isolate pathogenic determinants and/or disease-specific components. Progress is currently being made in the development and application of diagnostic identification tests for phytopathogenic bacteria based on protein and nucleic acid (chromosome and plasmid DNA)[11,34,39A] analyses, cellular fatty acid composition analyses,[21,45,55,65,70] specific enzyme (aminopeptidase activity, isozyme patterns) analyses, and specific molecular (DNA) probes.[54]

REFERENCES

1. **Alvarez, A. M., Benedict, A. A., and Mizumoto, C. Y.,** Identification of *Xanthomonas* and grouping of *Xanthomonas campestris* pv. *campestris* strains with monoclonal antibodies, *Phytopathology*, 75, 722, 1985.

2. **Behncken, G. M.,** Mollicutes and rickettsia-like bacteria in plants, in *Plant Bacterial Diseases: A Diagnostic Guide,* Fahy, P. C., and Persley, G. J., Eds., Academic Press, North Ryde, Australia, 1983, chap. 10.
3. **Billing, E.,** The value of phage sensitivity tests for identification of phytopathogenic *Pseudomonas* spp., *J. Appl. Bacteriol.,* 26, 193, 1963.
4. **Billing, E.,** Isolation, growth, and preservation of bacteriophages, in *Methods in Microbiology,* (Series 3B), Norris, J. R. and Robbins, D. W., Eds., Academic Press. London, 1969, 315.
5. **Billing, E. and Garrett, C. M. E.,** Phages in the identification of plant pathogenic bacteria, in *Microbial Classification and Identification,* Goodfellow M. and Board, R. G., Eds., Academic Press, London, 1980, 319.
6. **Bove, J. M.,** Greening in the Arabian Peninsula: toward new techniques for its detection and control, *FAO Plant Prot. Bull.,* 34, 7, 1986.
7. **Bove, J. M., Bonnet, P., Garnier, M., and Aubert, B.,** Penicillin and tetracycline treatments of greening disease-affected plants in the glasshouse, and the bacterial nature of the prokaryote associated with greening, in *Proc. 8th IOCV Conf.,* Calavan, E. C., Garnsey, S. M., and Timmer, L. M., Eds., Riverside, Calif., 1980, 91.
8. **Carlson, R. R. and Vidaver, A. K.,** Taxonomy of *Corynebacterium* plant pathogens, including a new pathogen of wheat, based on polyacrylamide gel electrophoresis of cellular proteins, *Int. J. Syst. Bacteriol.,* 32, 315, 1982.
9. **Chen, M., Miyakawa, T., and Matsui, C.,** Citrus likubin pathogens in salivary glands of *Diaphorina citri, Phytopathology,* 63, 194, 1973.
10. **Chen, T. A., Wells, J. M., and Liao, C. H.,** Cultivation in vitro: spiroplasmas, plant mycoplasmic, and other fastidious, walled prokaryotes, in *Phytopathogenic Prokaryotes,* Mount, M. S. and Lacy, G. H., Eds., Academic Press, New York, 1982, chap. 20.
11. **Civerolo, E. L.,** Indigenous plasmids in *Xanthomonas campestris* pv. *citri, Phytopathology,* 75, 524, 1985.
12. **Civerolo, E. L. and Fan, F.,** *Xanthomonas campestris* pv. *citri* detection and identification by enzyme-linked immunosorbent assay, *Plant Dis.,* 66, 321, 1982.
13. **Civerolo, E. L. and Helkie, C.,** Indirect enzyme-linked immunosorbent assay of *Xanthomonas campestris* pv. *citri, Proc. 5th Int. Conf. Plant Path. Bact.,* CIAT, Cali, Colombia, 1981, 105.
14. **Civerolo, E. L., Sasser, M., Helkie, C., and Burbage, D.,** Selective medium for *Xanthomonas campestris* pv. *pruni, Plant Dis.,* 66, 39, 1982.
15. **Clark, M. F.,** Immunosorbent assays in plant pathology, *Annu. Rev. Phytopathol.,* 19, 83, 1981.
16. **Crosse, J. E. and Garrett, C. M. E.,** Relationship between phage type and host plant in *Pseudomonas mors-prunorum* Wormald, *Nature,* 192, 379, 1961.
17. **Cuppels, D. A.,** Isolation and characterization of phages useful for identifying *Pseudomonas syringae* pv. *tomato, Phytopathology,* 73, 1376, 1983.
18. **Davis, M. J., Gillaspie, A. G., Jr., Vidaver, A. K., and Harris, R. W.,** *Clavibacter:* A new genus containing some phytopathogenic coryneform bacteria, including *Clavibacter xyli* subsp. *xyli* sp. nov., subsp. nov. and *Clavibacter xyli* subsp. *cynodonotis* subsp. nov., pathogens that cause ratoon stunting disease of sugarcane and bermudagrass stunting disease, *Int. J. Syst. Bacteriol.,* 34, 107, 1984.
19. **De Boer, S. H.,** Enumeration of two competing *Erwinia carotovora* populations in potato tubers by a membrane filter-immunofluorescence procedure, *J. Appl. Bacteriol.,* 57, 517, 1984.
20. **De Boer, S. H. and McNaughton, M. E.,** Production of monoclonal antibodies to *Erwinia carotovora* subsp. *atroseptica* lipopolysaccharide, *Phytopathology,* 76, (Abstr.), 1138, 1986.
21. **De Boer, S. H. and Sasser, M.,** Differentiation of *Erwinia carotovora* ssp. *carotovora* and *E. carotovora* ssp. *atroseptica* on the basis of cellular fatty acid composition, *Can. J. Microbiol.,* 32, 796, 1986.
22. **De Boer, S. H. and Wieczorek, A.,** Production of monoclonal antibodies to *Corynebacterium sepedonicum, Phytopathology,* 74, 1431, 1984.
23. **Dianense, J. C.,** Proteinas celulares nacaracterizacaodas bacterias fitopatogenicas, *Fitopatol. Brasil.,* 11, 55, 1986.
24. **Dianese, J. C. and Schaad, N. W.,** Characterization of cell membranes of *Xanthomonas campestris* pv. *campestris, Phytopathology,* 72, 1284, 1984.
25. **Duncan, F. and Garnett, H. M.,** Serological studies of the greening organism, *Citrus Sub-Trop. Fruit J.,* 611, 9, 1985.
26. **Fahy, P. C. and Hayward, A. C.,** Media and methods for isolation and diagnostic tests, in *Plant Bacterial Diseases: A Diagnostic Guide,* Fahy, P. C. and Persley, G. J. Eds., Academic Press, North Ryde, Australia, 1983, chap. 12.
27. **Fudl-Allah, A. E. A., Calavan, E. C., and Igwegbe, E. C. K.,** Culture of a mycoplasmalike organism associated with stubborn disease of citrus, *Phytopathology,* 61(Abstr.), 1321, 1971.
28. **Garnett, H. M.,** Isolation of the greening organism, *Citrus Sub-Trop. Fruit J.,* 611, 4, 1985.
29. **Garnier, M., Latrille, J., and Bove, J. M.,** *Spiroplasma citri* and the organism associated with likubin: comparison of their envelope systems, in *Proc. 7th IOCV Conf.,* Calavan, E. C., Ed., Athens, Greece, 1976, 13.

30. **Goth, R.,** Puncture method for isolating bacterial blight of beans, *Phytopathology,* 55, 930, 1965.
31. **Goto, M. and Starr, M. P.,** Phage-host relationships of *Xanthomonas citri* compared with those of other xanthomonads, *Ann. Phytopathol. Soc. Jpn.,* 38, 26, 1972.
32. **Goto, M., Takahashi, T., and Messina, M. A.,** A comparative study of the strains of *Xanthomonas campestris* pv. *ctri* isolated from citrus canker in Japan and cancrosis B in Argentina, *Ann. Phytopathol. Soc. Jpn.,* 46, 329, 1980.
33. **Grogan, R. G.,** The science and art of plant-disease diagnosis, *Annu. Rev. Phytopathol.,* 19, 333, 1981.
34. **Hartung, J. S. and Civerolo, E. L.,** Genomic fingerprints of *Xanthomonas campestris* pv. *citri* strains from Asia, South America and Florida, *Phytopathology,* 77, 282, 1987.
35. **Hayward, A. C.,** Preliminary diagnosis of plant diseases caused by bacteria, in *Plant Bacterial Diseases: A Diagnostic Guide,* Fahy, P. C. and Persley, G. J., Eds., Academic Press, North Ryde, Australia, 1983, chap. 1.
36. **Hirano, S. S. and Upper, C. D.,** Ecology and epidemiology of foliar bacterial plant pathogens, *Annu. Rev. Phytopathol.,* 21, 243, 1983.
37. **Jones, A. L., Whitcomb, R. F., Williamson, D. L., and Coan, M. E.,** Comparative growth and primary formulations, *Phytopathology,* 67, 738, 1977.
38. **Joshi, B. and Vidaver, A. K.,** Evaluation of cellular protein profiles for identification of *Corynebacterium michiganense* ssp. *nebraskense* and *C. m.* ssp. *tessullarius* from infected plants, *Phytopathology,* 74(Abstr.), 879, 1984.
39. **Koike, H.,** The aluminum-cap method for testing sugarcane varieties against leaf scald disease, *Phytopathology,* 55, 317, 1965.
39A. **Lazo, G. R. and Gabriel, D. W.,** Conservation of plasmid DNA sequences and pathorar identification of *Xanthomonas campestris, Phytopathology,* 77, 448, 1987.
40. **Leben, C.,** Epiphytic microorganisms in relation to plant disease, *Annu. Rev. Phytopathol.,* 3, 209, 1965.
41. **Lelliott, R. A., Billing, E., and Hayward, A. C.,** A determinative scheme for the fluorescent plant pathogenic pseudomonads, *J. Appl. Bacteriol.,* 29, 470, 1966.
42. **Liao, C. H. and Chen, T. A.,** Culture of corn stunt spiroplasma in a simple medium, *Phytopathology,* 67, 802, 1977.
43. **Liew, K. W. and Alvarez, A. M.,** Phage typing and lysotype distribution of *Xanthomonas campestris, Phytopathology,* 71, 274, 1981.
43A. **Lin, C. P., Chen, T. A., Wells, J. M., and van der Zwet, T.,** Identification and detection of *Erwinia amylorora* with monoclonal antibodies, *Phytopathology,* 77, 376, 1987.
44. **Martinez, A. L. and Wallace, J. M.,** Citrus leaf-mottle-yellows disease in the Phillipines and transmission of the causal virus by a psyllid, *Diaphorina citri, Plant Dis. Rep.,* 51, 8, 1967.
45. **Miller, L. T. and Sasser, M.,** Fatty acid profiling of plant related bacteria, *Phytopathology,* 75(Abstr.), 757, 1984.
46. **Minsavage, G. V. and Schaad, H. W.,** Characterization of membrane proteins of *Xanthomonas campestris* pv. *campestris, Phytopathology,* 73, 747, 1983.
47. **Moline, H. E.,** Differentiation of postharvest soft rotting bacteria with two-dimensional polyacrylamide gel electrophoresis, *Phytopathology,* 75, 549, 1985.
48. **Moline, H. E., Johnson, K. S., and Anderson, J. D.,** Evaluation of two-dimensional polyacrylamide gel electrophoresis of acidic proteins of ribosome preparations for identifying plant pathogenic soft-rotting bacteria, *Phytopathology,* 73, 224, 1983.
49. **Moll, J. N. and Martin, M. M.,** Electron microscope evidence that citrus psylla *(Trioza erytreae)* is a vector of greening disease in South Africa, *Phytophylactica,* 5, 41, 1973.
50. **Moll, J. N. and Martin, M. M.,** Comparison of the organism causing greening disease with several plant pathogenic gram negative bacteria, rickettsia-like organisms and mycoplasma-like organisms, *Inserm,* 33, 89, 1974.
51. **Okabe, N. and Goto, M.,** Bacteriophages of plant pathogens, *Annu. Rev. Plant Pathol.,* 1, 397, 1963.
52. **Persley, G. J.,** Bacteriophages for the identification of plant pathogenic bacteria, in *Plant Bacterial Disease: A Diagnostic Guide,* Fahy, P. C. and Persley, G. J., Eds., Academic Press, North Ryde, Australia, 1983, chap. 10.
53. **Persley, G. J. and Crosse, J. E.,** A bacteriophage specific to race 2 of the cherry strain of *Pseudomonas morsprunorum, Ann. Appl. Biol.,* 89, 219, 1978.
54. **Roth, D. A. and Johnson, J.,** DNA hybridization probes for the identification of phytobacteria, Proc. 6th Int. Conf. Plant Path. Bact., College Park, Md., June 2 to 7, 1985.
55. **Santos, R. M. D. B. and Dianense, J. C.,** Comparative characterization of *Xanthomonas campestris* pv. *cassavae* and *X. campestris* pv. *manibotis, Phytopathology,* 75, 581, 1985.
56. **Sasser, M., Fieldhouse, D. J., and Carter, C. N.,** Computer assisted identification of bacteria based on fatty acid analysis, *Phytopathology,* 74(Abstr.), 882, 1984.
57. **Schaad, N. W.,** Serological identification of plant pathogenic bacteria, *Annu. Rev. Phytopathol.,* 17, 123, 1979.

58. **Schaad, N. W.,** *Laboratory Guide for Identification of Plant Pathogenic Bacteria,* American Phytopathological Society, St. Paul, Minn., 1980.
59. **Schwarz, R. E. and Green, G. C.,** Das 'Citrus Greening' und der citrusblatt-floh *Trioza erytreae* -ein temperatur-abhangiger erregerubertragerkamplex, *Z. Pflanzenkr. Pflanzenschutz,* 9, 490, 1970.
60. **Starr, M. P., Stolp, H., Truper, H. G., Balows, A., and Schlegel, H. G., Eds.,** *The Prokaryotes: A Handbook of Habitats, Isolation, and Identification of Bacteria* Vol. 1, Springer-Verlag, New York, 1981, 1.
61. **Starr, M. P., Stolp, H., Truper, H. G., Balows, A., and Schlegel, H. G., Eds.,** *The Prokaryotes: A Handbook of Habitats, Isolation, and Identification of Bacteria,* Vol. 2, Springer-Verlag, New York, 1981, 1103.
62. **Stolp, H. and Starr, M. P.,** Bacteriophage reactions and speciation of phytopathogenic xanthomonads, *Phytopathol. Z.,* 51, 442, 1964.
63. **Teakle, D. S.,** The sugarcane ratoon stunting disease bacterium, in *Plant Bacterial Diseases: A Diagnostic Guide,* Fahy, P. C. and Persley, G. J., Eds., Academic Press, North Ryde, Australia, 1983, chap. 11.
64. **Tully, J. G., Rose, D. L., Garcia-Jurado, O., Vingault, J. C., Saillard, C., Bove, J. M., McCoy, R. E., and Williamson, D. L.,** Serological analysis of a new group of spiroplasmas, *Curr. Microbiol.,* 3, 369, 1980.
65. **van der Zwet, T. and Sasser, M.,** Characterization of *Erwinia amylovora* through fatty acid profiling, *Phytopathology,* 75(Abstr.), 1281, 1985.
66. **Van Laere, O., De Wael, L., and De Mey, J.,** Immunogold staining (IGS) and immunogold silver staining (IGSS) for the identification of the plant pathogenic bacterium *Erwinia amylovora* (Burrill) Winslow et al., *Histochemistry,* 83, 397, 1985.
67. **Van Vuurde, J. W. L.,** Immunosorbent immunofluorescence microscopy (ISIF) and immunosorbent dilution plating (ISDP): new methods for detection of plant pathogenic bacteria, *Seed Sci. Technol.,* 11, 523, 1983.
68. **Vidaver, A. K.,** Bacteria, in *Challenging Problems in Plant Health,* Kommedahl, T. and Williams, P. H., Eds., American Phytopathological Society, St. Paul, Minn., 1983, 249.
69. **Wakimoto, S.,** Classification of strains of *Xanthomonas oryzae* on the basis of their susceptibility against bacteriophages, *Ann. Phytopathol. Soc. Jpn.,* 25, 193, 1960.
70. **Wells, J. M. and Raju, B. C.,** Cellular fatty acid composition of six fastidious, Gram-negative, xylem-limited bacteria from plants, *Curr. Microbiol.,* 10, 231, 1984.
71. **Williamson, D. L. and Whitcomb, R. F.,** Plant mycoplasmas: a cultivable spiroplasma causes corn stunt disease, *Science,* 188, 1018, 1975.

Chapter 15

PLANT DIAGNOSTIC PROBLEMS: INSECTS AND MITES

Lloyd Knutson

TABLE OF CONTENTS

I. INTRODUCTION AND DEFINITION OF OBJECTIVES

The purpose of this chapter is to describe insect and mite identification (diagnostic) problems in relation to plant quarantine. Current identification capabilities will be related to taxonomic research and to collections, literature, and other resources necessary to conduct the research. It will be shown that the quality of identifications depends on the resources. Emphasis will be placed on the identification of major problem areas, specific problems, methodologies, and current and potential solutions. The chapter title refers to insects and mites, but for brevity, the word "insects" will also include mites.

The basic objective of plant quarantine diagnosis is to provide rapid and authoritative identifications. These identifications are needed at several levels — at the inspector level, the regional diagnostic level, and the taxonomic specialist level (the ultimate authority). Not all countries have three levels of personnel, but some type of "screening" process exists in most countries with national plant quarantine services. A primary objective in plant quarantine diagnosis should be to place as much capability as possible at the first level, i.e., the point of earliest detection, the inspectors at the ports of entry.

II. REVIEW OF LITERATURE

Only a few publications address problems in insect and mite diagnosis and plant quarantine. These are useful papers, but since they are not readily available, certain sections are quoted in order to give the reader a better indication of their contents.

The functional relationships among (1) Plant Protection and Quarantine (PPQ), Animal and Plant Health Inspection Service (APHIS), USDA; (2) The U.S. National Collections of Insects, a major resource of diagnostic information; and (3) the Systematic Entomology Laboratory, USDA (SEL), the group of taxonomic specialists that use these resources to provide diagnostic information were analyzed by Foote.[12] He focused on the flow of information from the collection of the taxonomist to the user group. Foote characterized these interactions as follows:

> Accurate and precise identifications are indispensable to the conduct of quarantine programs, but the research that makes these determinations possible must take place in museums that offer adequate reference collections, catalogs, and libraries as working tools. These activities, identification and research, must go hand in hand for two reasons: (1) the taxonomist delivers information much more effectively to the users if he is thoroughly familiar with the action programs involved, and (2) the user often requires identification-associated information (overall distribution, biology, economic importance, etc.) that the taxonomist is not likely to deliver on a routine basis. The taxonomist, in complete command of his resources, is in a unique position to conduct inventories of the groups on which he specializes. Such inventories will eventually form the backbone of comprehensive information systems that will be used extensively by those engaged in action programs and by museum personnel alike. The resources of museums, devoted perforce mostly to museum objectives, must be augmented in many ways before the full potentials of museums can be brought to bear on problems of the non-museum world, and such augmentation should be provided by those organizations involved most closely with action programs.

Foote's table of the relationships between programs of PPQ and SEL and the kinds of available biosystematic services includes both domestic and foreign quarantine programs (Table 1).

Research taxonomists supply not just names of organisms, but diverse and extensive information to their user groups. These functions are referred to as "biosystematic" services rather than the less inclusive term "diagnostic" or "identification" services.

Fowler[13] presented a fine overview of the taxonomic support required by quarantine programs. His coverage was as follows:

1. Operations at ports of entry, airports, maritime/land border crossings, mail
2. Operations at foreign ports, preclearance

Table 1
INFORMATION NEEDS OF PROGRAMS OF THE ANIMAL AND PLANT HEALTH INSPECTION SERVICE (APHIS), U.S. DEPARTMENT OF AGRICULTURE, IN INSECT REGULATION AND DETECTION

APHIS program	What is the extent of problem?	What information is needed?	Who needs the information?	What is best source of the information?	To what extent are taxonomic identification services needed?
Domestic surveys and pest control:					
Program pests	5—6 species of primary importance	Extent of spread into new areas (new county records, etc.), extent of damage, effectiveness of control	Control personnel	Survey personnel, field and published records	Verification of specimens from new localities (program pest species only)
Surveys	All insects	Type and extent of damage, seasonal population fluctuations	Survey and control personnel	Survey personnel, records	Precise (species level) to general (e.g., "grasshoppers")
Cooperative Plant Pest Report (CPPR)	Up to 10,000 species affecting man and his activities	Type and extent of damage, seasonal population fluctuations	Survey and control personnel	Survey personnel, records	Precise (species level) to general (e.g., "grasshoppers")
Quarantines:					
Interceptions at port of entry	Exotic insects worldwide	Point of origin, economic importance at point of origin, biology at point of origin	Customs and USDA quarantine officials	Literature, previous records of importation, specimen data	As precise as possible, especially if insect is known to be important at source
New pest detection	Exotic insects worldwide	Home range, economic importance in home range, biology in the home range	Control and survey personnel	Literature, previous records of importation, specimen data	As precise as possible, especially if insect is known to be important at source, and indication of closely related species
Insects not known to occur in the U.S. (INKTO)	300—400 exotic insects of economic importance worldwide	Home range, economic importance in home range, biology (all aspects) in home range, means for identification	Control and survey personnel	Literature, specimen data	Precise identification (to species level) and indication of closely related species

From Foote, R. H., in *Beltsville Symposia in Agricultural Research 2. Biosystematics in Agriculture*, Romberger, J. A., Ed., Allanheld, Osmun, Montclair, N.J., 1978, 261. With permission.

Table 2
REGULATORY/QUARANTINE NEEDS AND EFFECTS OF THE LACK OF TAXONOMIC INFORMATION

Example (program specific)	Effect of the lack of taxonomic information
Ports of entry/training	PPQ officer cannot enforce the U.S. plant quarantine laws
Foreign ports/export certification	PPQ officer cannot assist in the development of biologically sound plant quarantine regulations
Pest risk analysis	Permits for the entry of agricultural products may be denied
Survey and monitoring	Entry of pest insects without detection
Emergency programs	Nonproductive control programs
Grapholita molesta vs. *Grapholita* sp.	Infested plant material not allowed to enter the U.S. unless treated
Microgryllus pallipes vs. *Hoplosphyrum*	Stimulated control program and behavior studies
Africanized vs. European honeybee	Hampered development of adequate control program
Gryllus spp.	Promotion of biologically unsound regulations and the use of unregistered pesticides
Anastrepha spp.	Stopped or altered the international movement of agricultural products

From Fowler, J. F., in Biosystematic Services in Entomology, Knutson, L., Harris, K. M., and Smith, I. M., Eds., Agricultural Research Service, U.S. Department of Agriculture, Washington, D.C., in press.

3. Pest risk analysis, permit issuance, new pest status
4. Surveying and monitoring domestic, new pest status, foreign
5. Emergency programs
6. Training programs

Fowler included a listing of specific examples of regulatory/quarantine needs and the impacts when taxonomic information is not available to meet these needs (Table 2). He also presented a table showing what types of taxonomic information are required to deal with specific quarantine needs (Table 3).

Shannon[31] described the current system in the U.S. for meeting the timely, authoritative identification needs of APHIS-PPQ. He emphasized the value of close, effective cooperation and communication between quarantine and taxonomic units.

In a previous paper,[20] this author has described the broader area of biosystematic services of which diagnostic services are a part:

What are biosystematic services? I refer to biosystematic services, not simply identification services. Many persons have a simplistic impression of taxonomic or biosystematic services. They think primarily of identifications, and consider these as routine and easily provided. Identifications are one product of a biosystematic service; however, many identifications are not routine but are actually mini-research projects. That is, the specimens may represent an undescribed species, *or* the species may have been incorrectly or insufficiently described or diagnosed when orginally published, *or* additional collections may have revealed the existence of related, similar species. Solution of these problems often requires that the previous literature and material be reexamined in detail. Furthermore, a great deal of biosystematic information is provided by systematists — such as on distribution, hosts, prey, natural enemies, biology, and habitat. Most taxonomists have knowledge about the biology of the organisms they study, and they know the published literature on a broad range of information on their insect or mite groups. A comprehensive biosystematic program such as that of the SEL provides another essential service to users, that of verifying the validity of scientific names in manuscripts and other documents. A program such as this can be useful by providing authoritative reference specimens, for example, reference specimens of foreign pests can be supplied to plant quarantine inspectors which assists them in making rapid identifications to facilitate commodity processing.

What is an identification? The nature and complexity of taxonomic identification is occasionally overlooked by many users of biosystematic services because they lack knowledge of certain basic facts: (1) The class Insecta is tremendously diverse — about one million species have been described and an estimated 3—30 times more remain to be described; (2) most species are known from only one of the various life stages, usually the adult, and often

Table 3
TAXONOMIC INFORMATION NEEDED TO MEET REGULATORY/ QUARANTINE NEEDS

Effect	Taxonomic need (increase in that available)
Inability to enforce quarantine laws	Identification training, reference collection, rapid indentification of survey material
Development of nonbiologically sound plant quarantine regulations	Identification training, reference collections, transfer of information on related species, synonomy lists, host preference information, keys for all life stages, new taxonomic methodologies
Denial of permit to enter	Identification keys for insects from foreign countries, transfer of taxonomic information, worldwide reference collections, range, and economic importance
Entry of pest insects	Specialized identification services, information on related species, identification keys for larval forms and for insects from foreign countries, host preferences, behavior patterns, innovative identification methodologies, local and worldwide reference collections
Nonproductive control projects	Identification training, reference collections, information on related species, and host preferences

From Fowler, J. F., in Biosystematic Services in Entomology, Knutson, L., Harris, K. M., and Smith, I. M., Eds., Agricultural Research Service, U.S. Department of Agriculture, Washington, D.C., in press.

it is the damaging or beneficial immature stage that needs to be identified; (3) many of the one million described species are, in fact, poorly described and many of the currently used classifications and keys are grossly inadequate; and (4) good classifications have tremendous predictive powers.

Whitehead,[38] weevil specialist in the SEL, has described an identification as follows:

Many biologists think of taxonomy primarily as a service for providing names of organisms, usually free of charge. They assume that the taxonomist is an expert, but most do not understand the limitations of taxonomic expertise. They assume that correct identifications are easy and inexpensive and that names of organisms carry invariable and interpretable meaning.

A taxonomic identification provided by an expert is, by definition, authoritative. However, having comprehensive knowledge does not mean that the expert has detailed knowledge throughout the subject area. The identification may seem unequivocal yet serve only to perpetuate errors from previous analyses. The identification may be correct in terms of current knowledge, but it may confuse species that are similar or even apparently identical in physical structure yet biologically very different. Even a taxonomically correct identification may be an oversimplification because different populations of a species may be biologically and thus, critically, different.

Users tend to assume that names can be quickly and unequivocally applied—and that they have particular biological meaning. However, the particular meaning often is taxonomic or nomenclatural rather than biological and therefore of limited use to the client. For instance, a subspecies name may disguise complex problems of biological relationships and normally should not be used unless there is a precise understanding of what is meant. On the other hand, the name applied to a specimen may be insufficient because additional information may be hidden under other vernacular names, synonyms, published incorrect identifications, or even other species.

Johnson,[17] in describing the U.S. Cooperative National Plant Pest Survey and Detection Program, noted:

It is important to remember that national pest surveys and their identification support services must be maintained since few quarantine programs are 100 percent effective. While some of these quarantine and survey programs monitor established pests, many determine the presence of exotic or newly introduced pests. In programs aimed at exotics, the identification needs may be concerned with one specific insect as in the case of *Ceratitis capitata* (Mediterranean fruit fly); or the focus may be directed toward a wide variety of pest species, as in the case of citrus. The survey and monitoring efforts of PPQ are involved with programs of both types. In these programs, plant pest information is gathered using many trapping and monitoring techniques. Experienced taxonomists from cooperating State and Federal agencies collect, identify, and record the organisms obtained from these field operations. The information is then processed and stored by automated data systems for use by other national and international scientists and agricultural communities. Precise taxonomic determination is required to define any actions that will be taken as a result of these programs. . .

. . . it is always best to know the biota which exists at a given point in time as a reference point from which to conduct surveys. As we know from our program, much data exists, but it is scattered, hard to find, and difficult to retrieve. In addition to helping to find new pests that are not now known to occur here, as seems likely, a national biological survey would establish that base point from which to conduct survey programs.

Johnson's[17] concerns about diagnostic services for survey work apply equally well to quarantine needs:

The taxonomic aspects of the [survey] program are a problem in several ways. First of all the identification capabilities are very limited in many States, at least for certain families and genera of organisms. The program will certainly impact adversely on the taxonomists who must identify the additional specimens that are submitted as a result of the coordinated survey. Additionally, as we request or conduct special surveys for certain exotic species, such as a new pheromone trapping project we have initiated, we find that the number of taxonomists able to identify those species is very limited. Identification keys and aids are needed along with reference specimens. Frequently these are difficult to obtain or are unavailable.

Muesebeck[26] observed nearly half a century ago the need for quarantine diagnosticians to recognize to which functional group an unidentifiable specimen belongs:

Of particular importance in quarantine and nursery inspection is an understanding of the correlation of structure and habits in insects. The inspector, whose facilities and taxonomic training are usually limited, can expect to make definite identifications for only a comparatively small number of species and only when these occur in restricted associations; but recognition of the more or less well-defined group to which an insect belongs may suffice to indicate the immediate action to be taken. Superficial differences, sometimes bizarre and very striking, between closely allied species, or between polymorphic forms of the same species, may tend to confuse the picture until the really significant characters have been determined; but as taxonomic work grows more critical, it becomes increasingly evident that normally a rather close correlation exists between taxonomic affinities and habits in insects.

III. THE OVERALL INSECT SYSTEMATICS ENTERPRISE IN RELATION TO PLANT QUARANTINE DIAGNOSIS

Taxonomic research and identification capabilities exist to meet a broad range of purposes, the primary one of which is basic research in systematics and evolutionary biology. The user groups of biosystematics research outside the disciplines of systematics and evolutionary biology include (1) agricultural research and action programs, especially for pest management (and most especially for biological control by natural enemies); (2) public health; (3) national and local plant health interests in quarantine and survey programs; and (4) basic organismal research in various disciplines, e.g., ecology and behavior. Diagnostic needs of plant quarantine services are met by (1) in-house specialists, (2) taxonomic units of other organizations, and (3) to some extent by unrelated organizations and by individual professional and ''amateur'' taxonomists in various situations. The functional relationships between plant quarantine diagnostic needs and identification capabilities are considered against this background of different origins, funding, missions, and interests of the units involved.

The flow of specimens from the points of interception to the points of identification ranges from a well-defined system to a state of chaos. The overall picture of the identification ''system'' is that of a diffuse and changing network, due to the variety of needs, insufficient taxonomic expertise to meet these needs, changing staffs and thus changing capabilities/interests of personnel, and realignments in priorities of both supplier and user organizations. The picture varies depending upon the organizational structure of the plant protection agency and the extent and diversity of taxonomic expertise in the country.

As long as the intercepted specimens are handled within the various levels of the plant quarantine agency or a closely related agency, the day-to-day physical location and current state of identification should be easy to determine. But because most plant quarantine agencies do not have all of the identification capability needed in-house, much material has to be forwarded to other agencies and individuals (usually in the same country) for authoritative identification. A problem may result if the taxonomist on whose desk the specimen appears does not understand or appreciate the need for timely identification. Fortunately, most experienced taxonomists understand plant quarantine diagnostic needs and priorities and are highly responsive to requests for assistance, even if they are not paid to provide such service. This can be a dangerous situation for a quarantine agency. The quarantine agency has the responsibility to arrange for, or provide for, the taxonomic expertise it needs, should know where such expertise is located, and should take a proactive role in nurturing relationships with the general taxonomic community in the country (and outside the country to some extent). This author's overall evaluation of this problem is that with ever increasing demands being placed on current or decreasing levels of capability in diagnostic work, difficulties will persist in the rapid flow and prompt identification of material. Thus, quarantine agencies and identification services need to be in effective and frequent communication to discuss the details of the needs/capabilities and plan for gaps in coverage *before they occur*. It may not be possible to fill all gaps, but at least by addressing the problems together, the agencies can plan for alternatives or temporary solutions.

IV. THE KNOWLEDGE BASES FOR INSECT IDENTIFICATION

To understand the problems facing identification services for plant quarantine, one must have an appreciation for the knowledge bases on which resolution of the problems ultimately depend. We simply do not know enough about the organisms. Because of the deficiencies in the quality of the knowledge base, it is essential that the quarantine agency be in good communication with the taxonomic research community. The quality and extent of taxonomic information now on the shelf is seriously inadequate for plant quarantine identification needs. It is incumbent on the plant quarantine community to work with the taxonomic community to improve the knowledge base. Plant quarantine should think of its needs partly in terms of research needs, not simply in terms of specimen identification needs.

For insects and mites, the first fact of life is simply the vast numbers of species. With 3 to 30 million species, the insects are the most numerous of the larger organisms. The numbers of insect and mite species of importance as plant pests or potential plant pests is similarly very high. An overview of numbers of species (Table 4) may be useful in putting these numbers in perspective. Note especially the columns "Potential pests not known to occur in the U.S." and "Estimated total U.S. spp.". Although the numbers refer to the U.S., they also are probably representative of other regions. The point is, there are many *potential* pests. This is a significant problem indeed when one is confronted with such numbers in relation to the large number of endemic species. Similar insights on the extent of the problem can be obtained by comparing other columns of numbers in Table 4.

The problem of numbers is further complicated by the fact that insects have four or more developmental stages and, in most species, some, if not all, are as good as or better commercial travelers than the adult stages. Also, most immature stages are smaller than the adult stages, hidden in the commodity, or are otherwise cryptic and thus more difficult to detect and identify than are the adults. The critical point is that our knowledge of the identity of immature stages is exceedingly rudimentary. For most groups, the immature stages of less than 10% of the known species have been described.

Another problem concerns the unpredictability of diagnostic needs. Although Shannon[31] pointed out that the degree of repetition of organisms transported in international commerce is high, the changing agricultural practices, the increased variety of commodities shipped,

Table 4
NUMBERS OF SPECIES IN MAJOR TAXA THAT INCLUDE PESTS,[a] AND IN THE BEST KNOWN MAJOR TAXA, FOR COMPARISON

						Pest and nonpest species[b]			
	Native pests in U.S.	Immigrant pests established in U.S.	Pests not known to occur in U.S.	Potential pests not known to occur in the U.S.	Total world pests	Estimated total world spp.	Estimated described world spp.	Estimated total U.S. spp.	Total no. U.S. described spp.
				Numbers of Species in Major Taxa that Include Pests					
All insects	5,000	865	600	6,000	10,000	10—30,000,000	751,000		87,100
Mites	350	150	300	400	1,200	1,000,000+	30—40,000	20—30,000	7,000
Molluscs	55	50	100	900	950	46,800	42,000	6,000	5,500
Animal disease microorganisms[c]	114	300	113	500	527	100,000—160,000[d]	24,000[d]	20—30,000[d]	6,000[d]
Plant pathogenic fungi	8,000	2,000	4,000	5,000	14,000	250,000	70,000	100,000	29,000
Plant parasitic nematodes	950	50	2,000	13,000	30,000	108,000[e]	18,200	35,000	6,000
Plant pathogenic microbes[f]	400	100	100	500	2,000	2,000	700	1,000	500
Vascular plants	500	1,900	4,700	2,000	8,100	238,000	215,775	21,750	21,600
Vertebrates	120	86—111				86,300	85,900	6,069	5,219
				Numbers of Species in Best-Known Major Taxa, for Comparison					
Butterflies	7+	1				17,500	16,500	763[f]	763[f]
Birds	7	3	30—100	30—100	30—100	9,000	8,600	1,106	1,106
Mammals	25	8				4,300	4,170	500	490
Reptiles	0	0				7,000	6,000	400	369
Amphibians	0	2				6,000	4,300	300	223
Fish	88	75—100				60,000	40,000	3,000	2,268

[a] Pest is defined as a species deleterious (causing economic loss or other negative impact, e.g., competitive exclusion of native species) to man, useful plants or domestic animals, or likely to be deleterious. To qualify as a ''pest,'' a species does not need to be serious enough to warrant control actions on a regular basis. Sources of information are available from the compilers, L. Knutson, D. W. S. Sutherland, and R. L. Johnson.
[b] For columns 6—9, the total number of pests and nonpests are included.
[c] Domestic animals (cattle, sheep, goats, horses, swine, poultry). Includes only Protozoa and helminths.
[d] Includes only helminths and protozoan parasites of mammals and birds.
[e] Includes plant parasites, free-living and predaceous forms in soil, parasites and associates of insects, and marine nematodes.
[f] United States and Canada.

From Klassen, W., Agricultural research: the importance of a national biological survey to food production, in *Foundations for a National Biological Survey*, Kim, K. C. and Knutson, L., Eds., Association of Systematics Collections, Lawrence, Kan., 1986, 65. With permission.

and new trading partners mean that the unpredictability of identification needs becomes more significant.

One of the most important problems is our inadequate knowledge of identification characters for even the known pests, and the generally inadequate development of classification systems that provide the level of predictive capability needed. We have very little capability to predict whether a species other than a well-known pest is of quarantine significance without organized knowledge of the natural relationships of major groups, that is, without classification systems based on relatedness.

V. IDENTIFICATION RESOURCES — PUBLICATIONS

A. Research Publications

Problems with access to and availability of publications, especially publications appropriate to specific needs, are key problems in plant quarantine. The literature that enables the most authoritative identifications is essentially the taxonomic *research* literature. This literature is generally not prepared primarily to meet the needs of first-line quarantine personnel. Since this situation is not likely to change, it behooves the quarantine establishment to ensure that the training of its diagnostic staff includes the ability to use the extant literature. Again, good communication between identification staffs in plant quarantine and taxonomists producing the literature should help to further sensitize taxonomists to consider the needs of plant quarantine specialists in preparation of keys or taxonomic revisionary studies — especially the kinds and extent of illustrations, and the unspecialized level of expertise that often must be assumed for the user.

The problem of providing sufficient literature at the initial diagnostic points is probably unsolvable, due to the extent of diversity involved, the changing nature of kinds of and origins of commodities (and the associated plant pests), and the immense size of the world diagnostic literature. Access to the world literature is needed in order to place the intercepted plant pests in relation to the native species. Yet the world literature is too extensive for it to be widely available. This fact alone results in the continuing need for the assistance of major taxonomic research units.

B. Specialized Publications

In addition to the basic research literature, especially needed are specialized publications geared to meet the requirements of nonspecialist identifiers. Robinson[27] commented on the taxonomists' responsibilities to produce useful publications for the nontaxonomist. There are several problems in producing such literature. A taxonomic specialist is required to produce identification aids for the nonspecialist user. However, the researcher has more incentive to publish taxonomic research that meets peer and supervisor approval than routine taxonomic publications for nonspecialist use. An original, basic taxonomic study, i.e., monographic research, requires certain kinds of extensive documentation not required for the production of an identification manual. Another important problem in producing diagnostic publications for the nonspecialist is the need for extensive illustrations. Most taxonomic units have difficulty in maintaining a sufficient staff of scientific illustrators.

C. Translations

A major deficiency in plant quarantine diagnostics, again related more to funding difficulties than to unavailability of expertise, is the grossly inadequate extent of translation of diagnostic literature from one language to another. This represents a serious deficiency in the overall plant quarantine enterprise because many publications, with translation and occasionally with moderate modification, would meet the needs of researchers in various countries.

In the U.S. Department of Agriculture, for example, only a few avenues are available for securing translations of foreign-language material. The National Agricultural Library in Beltsville, Md. maintains a translations file and collaborates with the John Crerar Library's National Translation Center in Chicago. Through these sources, approximately 800,000 English translations of foreign-language material in all subject areas are available, but most of them are more than 5 years old, and neither facility provides active translation services. The Translations Liaison of the National Agricultural Library provides free translations to USDA staff through their Special Foreign Currency program, but the turn-around time averages 1 year between submittal and receipt of the translation. Finally, commercial translations firms have about a 1-month turn-around time for most publications, but charge $30 to $100/1000 words, depending upon the source language. A problem common to all technical translations, moreover, is the uneven quality of translations, which range from highly usable to useless.

VI. IDENTIFICATION RESOURCES — COLLECTIONS

In addition to problems involving a lack of specialized taxonomic expertise and appropriate publications, a closely related, major problem is the lack of availability of collections for researchers and diagnosticians. Collection needs may be characterized as general reference collections, research collections, and special reference collections.

A. General Reference Collections

Reference collections at ports of entry should be relatively easy to develop by the retention of authoritatively identified intercepted material. Reference collections at the intermediate levels in the screening process need to be more extensive and should also be built, in part, by the return of properly prepared and labeled specimens from the taxonomic specialists.

B. Research Collections

The collections needed by taxonomic specialists include both research and reference collections; usually one collection serves both purposes. Such collections should have priority for deposition of interesting and important material intercepted by the quarantine agency. Partly because of the unpredictability of the interception of pest insects and mites, such collections need to be large and continuously developed. Even very large collections, e.g., the U.S. National Collection with over 25 million specimens, frequently do not include specimens of many foreign species necessary for comparison and identification.

C. Special Reference Collections

A few types of special collections can be developed to help solve quarantine diagnostic problems. Reference collections of potential pest species that *may* become introduced and established are useful at ports of entry and at intermediate levels in the screening process. Such collections are useful not only as sources of information, but also should serve to increase the awareness of diagnostic staff to be on the alert for new pests. For several years, the Biosystematics and Beneficial Insects Institute has made a concerted attempt to develop and distribute such reference collections.[21] For the vast number of foreign plant-feeding insects and mites, the current ability to predict the potential feeding behavior, host range, and geographic range is limited. Predictability of such aspects for stored-product pests, animal pests, and some other categories of pests is somewhat higher. Of course, certain well-known foreign plant pests can be singled out and ranked as to pest potential; however, the effects of many others cannot be determined until they become established in new areas. Yet some entire higher taxonomic categories (tribes, families, and even orders) of nonpest groups (e.g., Trichoptera) can be excluded, and other higher taxa can be included as potential pests.

It is not possible to have reference specimens of all species of potential pests in all locations where such reference material would be useful because of the large number of known insect pests, the diversity of crops, the huge geographic area involved in plant pest quarantine, and the difficulties of prediction. However, since some information is better than none, our Institute distributes reference collections of "Pest Insects Not Known to Occur" in the U.S. or of limited distribution (PINKTO) to the 24 PPQ area entomology identifiers supporting the quarantine officers at the international ports of entry, and to at least one identification center in each of the 50 states and Puerto Rico.

All PINKTO specimens are marked with a dye visible only under ultraviolet light, to prevent any subsequent false report of the establishment of an immigrant. Each specimen bears a red label, "PINKTO COLLECTIONS — Insect Identification and Beneficial Insect Introduction Institute, ARS" (subsequently, "Biosystematics and Beneficial Insects Institute, ARS)". Species of known economic importance, particularly species treated in the PPQ series "INKTO" or "PKNTO" published in the *Cooperative Economic Insect Report,*[8] the *Cooperative Plant Pest Report,*[9] or the *APHIS 81* series,[5] have been emphasized, although availability of material has influenced selection of species.

Certain other countries are also developing such special reference collections and, as a way of cooperating with them and of obtaining material of interest, a stockpile or collection base of Common North American Insect Pests (CNAIP) has been developed in the Biosystematics and Beneficial Insects Institute for distribution and exchange. We have found it difficult to obtain the long series (100 specimens is the general preference) of well-prepared specimens of adult and immature stages of foreign species needed for PINKTO collections. The Institute would very much appreciate receiving or exchanging material for both the CNAIP and PINKTO collections, particularly arthropod species included in the PPQ PNKTO publications. We are also interested in providing sets of the collections to appropriate users. In communicating about these collections, please write to Research Leader, Systematic Entomology Laboratory, Building 003, Beltsville Agricultural Research Center West, Beltsville, Md. 20705.

Finally, voucher specimens of newly established pests should be deposited in permanent reference collections. Specimens from the original finds should be maintained in the country's primary research collection and/or in specialized central collections such as that being developed with the Western Hemisphere Immigrant Arthropod Database in the Biosystematics and Beneficial Insects Institute.

VII. IDENTIFICATION RESOURCES — PERSONNEL

With reference to the major problem of lack of expertise, there are relatively few statistics available, particularly information on skilled personnel in comparison with information on diagnostic needs. An evaluation of insect systematics personnel needs in the U.S. has recently been completed,[24] and the procedures for carrying out that analysis should be of interest to other countries. The personnel needs are excruciatingly obvious. One of the major solutions to this long-term problem is effective dialogue between suppliers and users, especially at the higher levels of management where funds are controlled.

Training needs, especially those required to upgrade the knowledge of incumbent staff, are continuing problems. Again, the research taxonomists will likely be a good source of assistance, and short courses provided by universities and professional societies can be important means of improvement.

Maintaining the integrity of the diagnostic service depends to a large extent upon the plant quarantine organization's ability to maintain knowledge of the current capability of its staff. This can be done in part by the research taxonomists carrying out occasional verification of identifications made by port inspectors and other nonspecialist identifiers. Appropriate review of the inspector's or identifier's authority to provide final identification of intercepted pests is an effective procedure. Fowler[13] has noted, "The confirmed identifications of SEL [Systematic Entomology Laboratory] serve as a mechanism to substantiate the ability of the PPQ identifiers. On the average, identifiers are able to give definitive determinations on more than 90% of intercepted insects. . . . After having correctly identified a plant pest a specified number of times (three or more, depending on the difficulty of the identification), the identifier is given 'discard authority'. Discard authority is defined as the responsibility for final identification without verification by others." This is another example of the importance of effective communication between quarantine taxonomic units, because the taxonomist, knowing the identification authority and capability of the inspector or area identifier, is in a unique position to alert the quarantine agency of new research information requiring changes in identification authority. Review of identifications made by taxonomic specialists, usually including all identifications upon which regulatory actions are taken, is somewhat more of a problem as frequently no peer reviewers are available.

There frequently is a need to locate diagnostic expertise for special programs (such as "high-hazard surveys"), special quarantine alerts, and alternative sources of information when a key taxonomic specialist has retired or is otherwise no longer available, or when there is a crisis. Lack of specific expertise is a problem because taxonomic expertise is scattered in many institutions throughout a country. The taxonomic expertise is dispersed and generally already overcommitted. Surprisingly, the relatively simple expedient of maintaining directories of identification capability is not used more widely to meet these problems.

A potential solution to special expertise problems is for the quarantine agency to contract for identifications. This approach has several inherent difficulties, most importantly the questionable credibility of the information provided. What usually happens is that a budget officer selects the contractor and the quarantine staff may have little technical input in selecting or evaluating contract identifiers. Such contractual services generally should be used only in the preliminary screening processes, not for providing identifications on which quarantine actions depend.

VIII. KEY OPERATIONS IN DIAGNOSTIC UNITS

Effective management of the diagnostic unit will prevent many problems. Only three large, multipurpose, internationally oriented diagnostic units for insects and mites exist: the Biosystematics Research Centre, Agriculture Canada; the Commonwealth Institute of Entomology, U.K.; and the Biosystematics and Beneficial Insects Institute, USDA. These organizations provide biosystematics research and services for many areas of activity in addition to plant quarantine.

Although the role of the Biosystematics and Beneficial Insects Institute diagnostic service is much broader than plant quarantine, I will refer to few of the management practices of the Institute that relate, at least in part, to the management of plant quarantine diagnostic services.

A. Proper Documentation

The PPQ port interception form (PPQ 309, Figure 1), backed up by a specific protocol, has been used effectively for many years in documenting interceptions and the identification of the material. With the proper directions for use of the form communicated to the quarantine staff and to the taxonomists, this specialized form works well.

In some situations, a more general-purpose documentation form may be needed. In these cases, the identifier needs certain information about the request: the purpose of the request, how soon the identifications are needed, if specimens may be retained by the identifier, and to what taxonomic level the identification should be carried out. We began to experiment

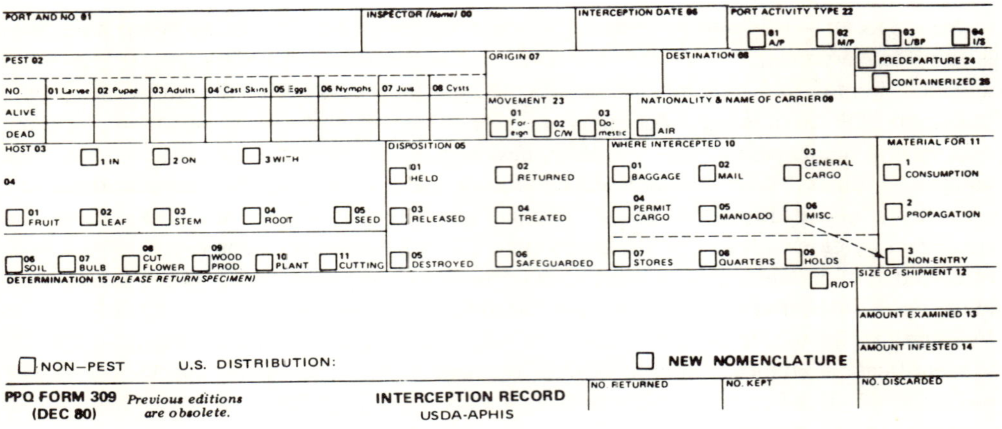
PORT AND NO 01 | INSPECTOR (Name) 00 | INTERCEPTION DATE 06 | PORT ACTIVITY TYPE 22
01 A/P | 02 M/P | 03 L/BP | 04 I/S
PEST 02 | ORIGIN 07 | DESTINATION 08 | PREDEPARTURE 24 | CONTAINERIZED 25
NO. | 01 Larvae | 02 Pupae | 03 Adults | 04 Cast Skins | 05 Eggs | 06 Nymphs | 07 Juvs | 08 Cysts
ALIVE
DEAD
MOVEMENT 23: 01 Foreign | 02 C/W | 03 Domestic
NATIONALITY & NAME OF CARRIER 09 | AIR
HOST 03: 1 IN | 2 ON | 3 WITH
04
01 FRUIT | 02 LEAF | 03 STEM | 04 ROOT | 05 SEED
06 SOIL | 07 BULB | 08 CUT FLOWER | 09 WOOD PROD | 10 PLANT | 11 CUTTING
DISPOSITION 05: 01 HELD | 02 RETURNED | 03 RELEASED | 04 TREATED | 05 DESTROYED | 06 SAFEGUARDED
WHERE INTERCEPTED 10: 01 BAGGAGE | 02 MAIL | 03 GENERAL CARGO | 04 PERMIT CARGO | 05 MANDADO | 06 MISC. | 07 STORES | 08 QUARTERS | 09 HOLDS | R/OT
MATERIAL FOR 11: 1 CONSUMPTION | 2 PROPAGATION | 3 NON-ENTRY
DETERMINATION 15 (PLEASE RETURN SPECIMEN)
SIZE OF SHIPMENT 12
AMOUNT EXAMINED 13
AMOUNT INFESTED 14
NON-PEST | U.S. DISTRIBUTION: | NEW NOMENCLATURE
NO. RETURNED | NO. KEPT | NO. DISCARDED
PPQ FORM 309 (DEC 80) *Previous editions are obsolete.*
INTERCEPTION RECORD
USDA-APHIS

FIGURE 1. Animal and Plant Health Inspection Service, Plant Protection Quarantine, port interception form (PPQ 309).

with a general-purpose request form in 1976 and finalized the ARS form 748 a few years ago (Figure 2). The form serves as a checklist of needed information for the user and identifier. Such a standardized format makes it easier for both the clerical staff and taxonomists to scan the request, and is essential for the ''smoothest'' operation of a computerized inventory and reporting system such as ours. A one-sheet set of instructions for submittals and a blank form ARS-748 are included with most of our identification reports. Our general request form can also be modified for special purposes, e.g., for in-country surveys maintained by some quarantine organizations (PPQ 391, Figure 3).

In plant quarantine diagnosis, there is a special need to know the geographical origin of the material and any transshipment that has occurred en route, and this kind of information should be included in the documentation.

B. Identification Reports

Identification requests submitted with our general-purpose request form are entered into a database in a Wang Virtual Storage-100 computer for effective tracking (inventory) and subsequent preparation of reports. This work, for about 13,000 identification requests per year, is performed by a clerical/support staff of four, not by the taxonomists. Whenever possible, standardized sentences are used in reports. An important aspect of our reports is the use of an ''Incomplete Identification'' (II) code (NER 626, Figure 4). If the taxonomist cannot identify the specimen to species, he or she simply enters the appropriate II code. The clerk strikes the number of the code on the terminal keyboard in preparing the report and, very efficiently, critical information of value to the user in interpreting the identification is provided. We also encourage our taxonomists to use standardized terminology for citing incomplete identifications, avoiding question and quotation marks.

C. Difficulties in Maintaining Diagnostic Services

It is convenient, indeed important, that users of biosystematic services have a source of services that is comprehensive, especially for the many projects that involve diverse taxa. Even with 27 research taxonomists, the Systematic Entomology Laboratory has major gaps in its current capability (Tortricoidea, Ichneumonidae, and Coleoptera of importance to forest entomology). However, it is difficult to maintain funds for such a large scientific staff, and in the past there has been an inclination to retain scientific positions at the expense of support staff. Currently, there are only eight technicians for the 27 SEL taxonomists.

Our Institute is a part of the Agricultural Research Service (ARS) of the USDA, and our

U.S. DEPARTMENT OF AGRICULTURE
AGRICULTURAL RESEARCH SERVICE
BIOSYSTEMATICS AND BENEFICIAL INSECTS INSTITUTE

IDENTIFICATION REQUEST

NOTE:
- Please *type or print* all information.
- Do not write in shaded areas.
- Give explanations where requested in "Remarks" section at bottom of form.
- Attach additional pages if (and only if) more space is needed.

BBII LOT NO.	BBII PRIORITY
DATE	Sender's Reference No.
DATE IDENTIFICATION REQUIRED *(month, day, year)* If less than two months, explain below.	
TOTAL NUMBER SENT Pinned: Vials: Slides: Other:	

NAME & COMPLETE MAILING ADDRESS OF SENDER *(Include Zip code)*

RETURN TO *(If other than sender) (Include Zip code)*

SOURCE

AR ☐ ARS
AP
AQ ☐ APHIS-PPQ
FS ☐ FS
DD ☐ U.S. Military
OF ☐ Other Federal
SA ☐ State Agricultural Agency
SU ☐ State University
OS ☐ Other State
PU ☐ Private University
IN ☐ Individual
CO ☐ Commercial Organization
FN ☐ Non-U.S.
CI ☐ CICP

LEVEL/TYPE OF IDENTIFICATION NEEDED

☐ Family ☐ Genus ☐ Species

☐ Positive or negative verification of ecological group:

☐ phytophagous ☐ parasitic ☐ predaceous

☐ saprophagous ☐ aphidophagous ☐ other:

OTHER INFORMATION REQUESTED — Will be supplied as conditions allow, as determined by taxonomist.

a ☐ Pest status
b ☐ Geographical distribution
c ☐ Usual hosts (prey)
d ☐ Best reference
d ☐ Identifying characteristics
f ☐ Other (Specify)

SOURCE OF PROJECT SUPPORT

☐ ARS ☐ APHIS ☐ FS ☐ CSRS Regional project no.:
☐ Hatch ☐ EPA ☐ DOI ☐ NIH ☐ NSF ☐ FAO
☐ USAID ☐ Other *(Specify)*

REASON FOR IDENTIFICATION *(Check and complete as appropriate)*

a Biological control
1 ☐ Scientific name of target pest:
2 ☐ General quarantine or biocontrol research
3 ☐ Identity of host of natural enemy
4 ☐ Recovery of released natural enemy
5 ☐ Suspected contaminant in culture
6 ☐ Quarantine reference collection
7 ☐ Voucher specimen of field release
8 ☐ Holding living material pending identification
b ☐ Damaging crop, plants — *identify host plants:*
c ☐ Suspected pest of regulatory concern — *give details below*
d ☐ Stored product pest — *commodity affected:*
e ☐ Livestock, wildlife, or domestic animal pest — *host:*
f ☐ Danger to human health
g ☐ Household pest — *damage:*
h ☐ Possible immigrant — *new to:*
i ☐ Reference collection — *for:*
j ☐ Survey — *explain in detail below*
k ☐ Thesis problem — *describe project below* ☐ M.S. ☐ Ph.D.
l ☐ Other — *explain below*

SPECIMEN DISPOSITION

☐ **Return** ☐ **Place in National Collection**
☐ **Keep part of series** ☐ **Other** *(Specify)*

VOUCHER SPECIMEN(S) INCLUDED

How marked?
Will be deposited in:

TELEPHONE REPORT REQUESTED

If yes, give number — include area code and extension
Requests are handled at the discretion of BBII

DESCRIPTION OF PROJECT — Include Project Title and name of Project Leader. **(Reference previous communications pertaining to this submittal)**

REMARKS *(Explanations, tentative identification, etc.)*

FOR BBII USE
DATE RECEIVED
NO.
LABEL
SORTED
PREPARED
DATE ACCEPTED
CC's

ARS-748 (5/86) **PART I—BBII COPY**

FIGURE 2. Biosystematics and Beneficial Insects Institute, identification request form (ARS 748).

primary responsibility is to produce research results. Our taxonomists are classified as research entomologists. Identifications and other services (requiring about 25% of our time) and curatorial work must be secondary responsibilities. It is difficult to maintain a balance among these competing demands. Because the classifications of insects and mites are incomplete and inadequate, a great deal of research is required for taxonomists to be able to provide identifications. Indeed, many identifications themselves are research efforts that

This report is authorized by law (7 U.S.C. 147a). While you are not required to respond your cooperation is needed to make an accurate record of plant pest conditions.

FORM APPROVED
OMB NO. 0579-0010

U.S. DEPARTMENT OF AGRICULTURE
ANIMAL AND PLANT HEALTH INSPECTION SERVICE

SPECIMENS FOR DETERMINATION

Instructions - Type or print information requested. Press hard and print legibly when handwritten. Item 1 assign number for each collection beginning with year, followed by collector's initials and collector's number. Example (collector, John J. Dingle); 83-JJD-001. Pest Data Section - Complete items 14, 15 and 16 or 19 or 20 and 21 as applicable. Complete items 17 and 18 if a trap was used.

FOR IIBIII USE
LOT NO.
PRIORITY

1. COLLECTION NUMBER
2. DATE — MO | DA | YR
3. SUBMITTING AGENCY — ☐ STATE ☐ PPQ ☐ OTHER ....

SENDER AND ORIGIN
4. NAME OF SENDER
6. ADDRESS OF SENDER
ZIP

INTERCEPTION SITE
5. TYPE OF PROPERTY *(Farm, Feedmill, Nursery, etc.)*
7. NAME AND ADDRESS OF PROPERTY OR OWNER
COUNTY

PURPOSE
8. REASON FOR IDENTIFICATION *("X" Applicable Items)*

A. ☐ Biological Control (Target Pest Name)	E. ☐ Livestock, Domestic Animal Pest
B. ☐ Damaging Crops/Plants	H. ☐ Possible Immigrant
C. ☐ Suspected Pest of Regulatory Concern	J. ☐ Survey *(Explain in remarks)*
D. ☐ Stored Product Pest	L. ☐ Other *(Explain in remarks)*

9. ☐ TELEPHONE RESPONSE REQUESTED ☐ FTS (..........................)

HOST DATA
10. HOST INFORMATION — NAME OF HOST *(Scientific name when possible)*
11. QUANTITY OF HOST — NUMBER OF ACRES/PLANTS — PLANTS AFFECTED *(Insert figure & indicate number or percent)* ☐ Number ☐ Percent

12. PLANT DISTRIBUTION
☐ LIMITED
☐ SCATTERED
☐ WIDESPREAD

13. PLANT PARTS AFFECTED

☐ Leaves, Upper Surface	☐ Trunk/Bark	☐ Bulbs, Tubers, Corms	☐ Seeds
☐ Leaves, Lower Surface	☐ Branches	☐ Buds	
☐ Petiole	☐ Growing Tips	☐ Flowers	
☐ Stem	☐ Roots	☐ Fruits or Nuts	

PEST DATA
14. PEST DISTRIBUTION
☐ FEW
☐ COMMON
☐ ABUNDANT
☐ EXTREME

15. ☐ INSECTS ☐ NEMATODES ☐ MOLLUSKS

NUMBER SUBMITTED	LARVAE	PUPAE	ADULTS	CAST SKINS	EGGS	NYMPHS	JUVS.	CYSTS
ALIVE								
DEAD								

16. SAMPLING METHOD
17. TYPE OF TRAP AND LURE
18. TRAP NUMBER

19. PLANT PATHOLOGY - PLANT SYMPTOMS *("X" one and describe symptoms)*
☐ ISOLATED ☐ GENERAL

20. WEED DENSITY — ☐ FEW ☐ SPOTTY ☐ GENERAL
21. WEED GROWTH STAGE — ☐ SEEDLING ☐ VEGETATIVE ☐ FLOWERING/FRUITING ☐ MATURE

22. REMARKS

23. TENTATIVE DETERMINATION

24. DETERMINATION AND NOTES *(Not for Field Use)*

FOR IIBIII USE
DATE RECEIVED
NO.
LABEL
SORTED
PREPARED
DATE ACCEPTED
RR

PPQ FORM 391 (NOV 83) *Replaces PPQ 390 (Aug 73) and PPQ 391 (Oct 81) which are obsolete.*

PART 1 - PPQ, HYATTSVILLE AFTER IDENTIFICATION

FIGURE 3. Animal and Plant Health Inspection Service, Plant Protection Quarantine, domestic collection form (PPQ 391).

cannot be made by use of the existing literature. A large, well-curated research collection is essential, particularly for these identifications, as well as for confirmation of published information. Even with the 25+ million specimens in the U.S. National Collection, there are occasions every day when we find our collection inadequate, particularly for Afrotropical, Australasian, and Oriental material. The expertise of the staff on insects and mites of various geographic regions also provides some limitations.

Page of

IIBIII IDENTIFICATION REPORT

TAXONOMIST

SENDER

LOT NO.

MATERIAL FORWARDED

Pinned: Vials: Slides: Other:

DATE FORWARDED

PRIORITY

TEL. REPORT
Tax.:
TSU:

Incomplete Identification Code (II)

1 — Species is apparently undescribed
2 — Group now under revision *(Optional — enter name and address of revisor below identification)*
3 — Group needs revision; species are not identifiable
4 — More precise determination depends on characters found in other sex or caste
5 — More precise determination depends on characters found in other life stage
6 — More precise determination depends on genital characters that must be dissected and/or mounted on a slide to be seen
7 — Specimen *(Code)*: C — Critical parts missing
P — In poor condition I — Improperly prepared
8 — Known habits of this group suggest specimen is irrelevant to your project; if relevant, resubmit with details
9 — More precise determination depends on knowledge of host
10 — Other *(Specify)*:
11 — No comparable identified specimens in U.S. National Collection

Specimen/ Vial/Slide/ No.	Family — Genus — Species — Author	Total Number Specimens	No. Specimens Kept	II Code	DB

IDENTIFIER

DATE

Form NER-626 (Jul 83)

USDA-ARS

FIGURE 4. Biosystematics and Beneficial Insects Institute, identification report form (NER 626).

IX. TAXONOMIC PROBLEM AREAS

It is difficult to identify taxonomic problem areas on a world basis because of differences in pest exclusion philosophies and approaches from country to country, the varying nature of international trade in agricultural commodities, and the wide range of availability of taxonomic expertise and identification resources. It is instructive to analyze the historical record of immigrant species, by higher taxonomic category, for specific regions. Figure 5

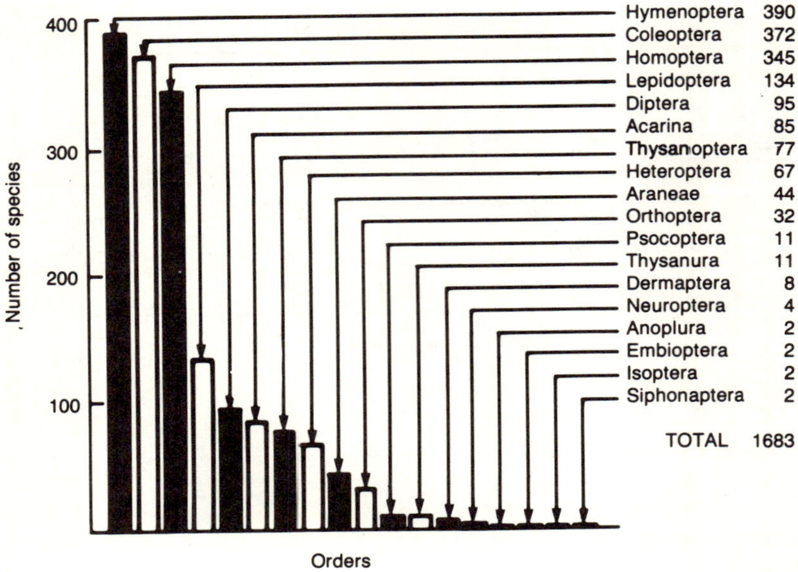

FIGURE 5. Number of species in each of the orders represented in the immigrant arthropod fauna of the 48 contiguous U.S. states. (From Sailer, R. I., in *Exotic Plant Pests and North American Agriculture,* Wilson, C. L., and Graham, C. L., Eds., Academic Press, New York, 1983. With permission. Illustration by Limhout Nong.)

shows the numbers of species in each order represented in the immigrant arthropod fauna of the 48 contiguous states in the U.S. Of course, one needs to discount the large numbers of introduced biological control agents in relating these data to plant quarantine. Sailer[30] summarized this information as follows:

Sixteen insect orders, together with Acarina (mites) and Araneae (spiders), are represented among the 1683 species presently included in the file [of immigrant arthropods in the 48 contiguous U.S. states] developed over the past 12 years (Fig. 2). Although this represents an increase of only 298 above the number reported in 1978 (Sailer, 1978), a much larger number has actually been added. The difference is made up of species removed for various reasons. Also, publications of two major reference works, *Introduced Parasites and Predators of Arthropod Pests and Weeds: A World Review* (Clausen 1978) and the *Catalog of the Hymenoptera of America North of Mexico* (Krombein et al., 1979), have made available much previously inaccessible information regarding both purposely introduced and adventitious species. The result has been the addition of a disproportionate number of Hymenoptera to the list and a shift of Homoptera from first position to third. Also, special attention to the spider literature has resulted in moving the Araneae from 15th to 9th place. Otherwise, there has been little change in the proportion of species, as shown by Sailer (1978). Three orders account for 66% of the total number of species. If intentionally introduced species are removed, Homoptera becomes the most numerous order (345), with Coleoptera a close second (342) and Hymenoptera falling to third place (233). Of the 134 Lepidoptera, nine are intentional introductions. As might be expected in view of the parasitic and predacious habits of many Diptera, almost 30% of the 95 immigrant species have been intentionally introduced. Of the remaining orders, only one species of Thysanoptera and one species of Heteroptera are intentional introductions.

Changes in Systematic Composition through Time

There have been pronounced changes in the kinds of foreign insects that have gained entry and become established during the past 360 years. . . . These changes reflect changes in the technology of transport, with the time in transit reduced from several weeks to a few hours. Where once the most serious obstacle to colonization was the ability to survive a long sea voyage, today's insect traveler is more likely to find success limited by the availability of food or the ability to find a mate. From the early dominance of Coleoptera transported in ballast, and the shift toward predominance of Homoptera during the 50-year period of intensive unregulated plant introduction activity prior to passage of the 1912 Plant Quarantine Act, the order affinities of invading species have become much less predictable. Of those species not intentionally introduced, data for the period 1970—80 indicate that Coleoptera (25) and Homoptera (24) are still dominant orders; however, Heteroptera (13) has moved up to third place, closely followed by Lepidoptera (11). Most of the remaining immigrant species during this period are accounted for by Acarina (9) and Diptera (8).

Also, interception records, such as the *List of Intercepted Plant Pests*[25] could be analyzed to identify geographic, host, and taxonomic problem areas. Analysis of the interception of *pest* groups *not* identified to the species level could be especially useful in identifying groups that need further taxonomic study.

Another approach to identifying the problem areas is represented by analyses of taxa (Tables 5 and 6). Hodges[16] asked some of the systematists at SEL the following question, "On the basis of the [U.S. National] collection and the literature available to you, what percent of the names in the literature can you associate with [adult] specimens for a series of faunas with a fair degree of certainty?" From the data (Table 5), Hodges drew the obvious conclusion, "Currently, we cannot identify with certainty a relatively large percentage of the described world fauna."

It is difficult to arrange the problems in plant quarantine diagnostic capabilities on a world priority basis, because the needs of most countries are more or less specific to problems in identification of insects and mites on specific commodities imported from certain areas. Also, these needs change over time. However, the major international identification problems can be focused to some extent. Following the approach of Hodges,[16] this author asked certain SEL systematists, "What percent of the described economically important plant pest (or likely potential plant pest) species intercepted at ports of entry in the U.S. can you identify to the species level, based on the U.S. National Collection, the available literature, and your experience?" The results are shown in Table 6.

I also asked some of the same taxonomists the following questions, aimed at identifying international needs:

1. Of the more important plant pests intercepted or likely to be intercepted at ports of entry *thoughout the world,* where are some of the major identification problems (needs) for higher taxa (genera, tribes, subfamilies, or families)? That is, on an international level, which major groups need the most attention?
2. What kinds of systematics *research* are needed to solve these problems?
3. What special resources (e.g., reared and associated adults and immatures from Central America) are needed to conduct the research and produce the necessary publications?
4. What kinds of publications (e.g., color illustrations, pictorial keys, etc.) or other identification aids (e.g., reference specimens) would be most useful to port inspectors, area identifiers, and diagnostic centers?

The responses, by taxonomic group, were as follows:

1. *Curculionidae* (weevils) (D. R. Whitehead)
   a. Illustrated manuals of tropical weevils, by country/region
   b. Basic research — generic (and higher level) revisions of critical groups
   c. Quality material (long series, host and habitat information)
   d. Illustrated manuals to genera plus included pest species, and pertinent reference specimens, throughout the world
2. *Chrysomelidae* (leaf beetles) (R. E. White)
   a. Galerucinae and Alticinae (Chrysomelidae) of Central and South America
   b. Generic and species revisions; most importantly, keys to genera and keys to species

Table 5
PERCENT OF SPECIMENS ASSOCIABLE WITH DESCRIBED SPECIES BY MUSEUM WORKERS FOR A SERIES OF FAUNAS

	N. America		Palaearctic		Neotropical		Exotic	
	%	Described species	%	Described species	%	Described species	%	Described species
Diptera								
Culicidae	95	250					50	3,500
Mycetophilidae	5	612					8	
Cecidomyiidae	5	1,200					1	
Ceratopogonidae	75	400					25	4,000
Tipulidae	95	1,500					80	14,000
Syrphidae	40	1,000					25	6,000
Muscidae	90	622					40	
Sarcophagidae	95	327					60	
Calliphoridae	95	78					65	
Tachinidae	82	1,281			28	2,864		
Coleoptera								
Curculionidae	75	2,600	50—60		0—5		0—5	
Scarabaeidae	90	1,400	90	500	50	2,000	5—10	
Coccinelidae	99	400	95	200	50	600	25	
Bruchidae	100	100			90		40—50	900
Buprestidae	80	660					25	11,400
Colydiidae	95	110					25	1,375
Dermestidae	95	130					40	877
Histeridae	50	360					5	3,500
Heteroptera	75		60		50		25—50	
Homoptera								
Aphidae	60	1,500					45—50	2,000
Aleyrodidae	90	500						
Coccoidea	80	1,500	80+	1,500	20	1,000	40—50	2,000+
Cicadellidae (♂♂)	80	3,500					12	20,000
Hymenoptera								
Symphyta	75	1,000	60	2,500	25	1,000	40	2,500
Formicidae								
workers	80	650	30	2,000	15	2,700	10	3,000
♀♀	15		5		1		1	
♂♂	10		5		1		1	
Ichneumonidae	50	2,850					15	15,000
Chalcidoidea	50—60		50—60				5—15	
Braconidae	65	2,000					10	10,000
Isoptera								
soldiers	90	45			15	1,000	15	1,500
alates	90				15		15	
workers	5				5		1	
Lepidoptera								
Lycaenidae	95	300	98+		70		80—85	
Noctuoidea	50	3,100					30	50,000
Geometridae	90	1,200+			50		10—15	10,000
Gelechiidae	80	750					10-15	4,000

From Hodges, R. W., *J. Lepid. Soc.*, 30, 245, 1976. With permission.

Table 6
PERCENT OF ECONOMICALLY IMPORTANT PLANT PEST SPECIES[a] (ADULT STAGE, EXCEPT AS INDICATED) INTERCEPTED AT U.S. BORDERS THAT ARE IDENTIFIABLE AS A DESCRIBED SPECIES NOT KNOWN TO OCCUR IN THE U.S.

	NA[b]	NT	P	AT	O	AO
Acarina	90	1	1	1	1	1
Coleoptera						
Curculionidae	95	1	25	1	1	1
Chrysomelidae	80—90	20—30	70—80	20—30	30—40	20—30
Scarabaeidae	99	20	80	15	20	10
Coccinellidae	99	15	80	10	30	10
Bruchidae	95	15	10	2	10	80
Diptera						
Cecidomyiidae	75	2	85[c]	1	1	1
Anthomyiidae	60	30	75	10	10	—
Tephritidae	95	70	70	50	50	30
Agromyzidae	75	30	50	30	50	30
Homoptera						
Aphididae	10	10	10	5	5	5
Aleyrodidae	70	10	75	10	20	10
Coccoidea	70	30	40	25	35	30
Cicadellidae	95	60	75	30	30	30
Hymenoptera						
Symphyta	80	70	70	50	50	80
Lepidoptera						
Noctuidae[d]	70	60	90	60	50—60	70
Gelechiidae	70	40	50	15[e]	15—20	10
Pyralidae	85	50	50	10	20	10
Thysanoptera	80	60	75	50	50	60

Note: Abbreviations of regions: NA, Nearctic; NT, Neotropical; P, Palaearctic; AT, Afrotropical; O, Oriental; AO, Australasian-Oceanian.

[a] Species usually requiring at least some level of control.
[b] The column for Nearctic is included as a basis for comparison and refers to knowledge of the Nearctic fauna, not to interceptions.
[c] The economically important species are relatively well known.
[d] Interceptions seen are primarily larvae.
[e] Primarily South African; virtually nothing is known about the rest of the region.

c. More extensive collections than are now available
d. Pictorial keys and reference collections

3. *Scarabaeidae* (grubs) (R. D. Gordon)
   a. An overall revision of the higher classification of the family down to the species level (modern taxonomic practice requires that all such publications are adequately illustrated)
   b. Revisions of genera, particularly those containing known pests
   c. Of less importance than the preceding, but certainly helpful, would be illustrated manuals to known pests
4. *Coccinellidae* (lady bird beetles) (R. D. Gordon)
   a. A thorough revision to the species level of the Old World Epilachninae (plant feeders)

5. *Bruchidae* (seed weevils) (J. M. Kingsolver)
   a. Keys to genera and species of the 50 or so economic bruchids in the world, and to genera and species of the most important Dermestidae in the world
   b. Many of the species in each of these families never adequately illustrated or described; delineations of genera needed for bruchids
   c. More extensive collecting to try to pinpoint more accurately the geographic origins of economic bruchids to aid in searches for biocontrol agents and to determine the native hosts, if the bruchid has transferred to another host, so that a prediction can be made as to where the bruchid may become a pest
   d. Well-illustrated keys, handbooks, host lists for bruchids
6. *Cecidomyiidae* (gall flies) (Cecidomyiinae) (R. J. Gagné)
   a. Capability to incorporate new information to existing base of information
   b. Revisions of groups in the areas of origin of imported commodities; descriptions of new species
   c. Scientific illustrators and technical support; reliable host and damage information
   d. A new procedure/format for inclusion of information/descriptions/illustrations of new or additional species as the changing scene in importation results in needs for such species to be added to the base of information; up-to-date, well-illustrated guides to damage and host lists; keys to larvae
7. *Tephritidae* (fruit flies) (A. L. Norrbom)
   a. Subfamily Trypetinae (*s. lat.*), especially of tropical regions; subfamily Dacinae (all stages); immature stages — entire family; larvae are the most common stage intercepted and submitted for identification
   b. Comprehensive revisions and keys to entire taxa (most available keys are for single countries or restricted regions only) comprehensive biosystematic database, including host plant information, because host plant data are often very useful for identification
   c. Careful fieldwork to produce reared adults, with host-plant data and preserved immatures
   d. Pictorial keys and reference collections should provide the tools needed for identification
8. *Anthomyiidae* (root and leaf mining maggots) (F. C. Thompson)
   a. Basic research — synthesis of the various regional revisions of *Pegomya* (leaf miners) and *Delia* (root maggots)
   b. Illustrated keys to adults and maggots of the known *Delia* pest species
   c. More research material — reared series, host and habitat data
9. *Agromyzidae* (leaf mining flies) (A. L. Norrbom)
   a. Identification to species level possible only with adult male; female and immature stages badly need work
   b. See comments under Tephritidae
   c. See comments under Tephritidae
   d. Most identifications based on male genitalia; detailed illustrations needed, particularly surstyli and the entire aedeagus, from different views
10. *Aphidoidea* (aphids) (M. B. Stoetzel)
   a. Keys to genera, worldwide, and keys to species of selected genera, especially *Aphis*
   b. Basic research resulting in revisions of groups containing pest species
   c. Host transfer experiments required to establish specific relationships of some pests; reared series of the various aphid morphs would be ideal
   d. Well-illustrated manuals, handbooks, keys, etc. must be accompanied by more detailed host and morphological information so that the user can have a degree

of certainty that his/her identification is correct; with aphids, identifications must be limited to developing colonies; alates land everywhere, and one cannot assume a plant is a true host in the absence of a colony

11. *Coccoidea* (scale insects) (D. R. Miller)
    a. Armored scales, mealybugs, and soft scales from Asia, Neotropics, Europe, and Africa
    b. Basic revisionary work
    c. Resources for collection, preservation, analysis, and presentation of material in these groups
    d. Manuals and handbook on the economic species of the world and close relatives
12. *Cicadellidae and Fulgoroidea* (leafhoppers) (J. P. Kramer)
    a. Outside of continental U.S. and Western Europe, taxa at all levels need attention.
    b. Basic research needed to delimit genera and species; information should be presented in the form of well-illustrated keys
    c. Pest status, particularly of exotic species, needs to be better documented by direct observations of feeding and damage; in most cases, the information needed to designate pest status to an exotic leaf- or planthopper has not been gathered
13. *Symphyta* (saw flies) (D. R. Smith)
    a. Larvae on their host plants most commonly found and most difficult to identify
    b. Comprehensive study on a worldwide basis to determine family, subfamily, generic, and species characters that may be used to separate and more accurately identify larvae
    c. Series of adult-larval associations with associated host plant and life history data
    d. Manual to families — genera of larvae, worldwide, including illustrations and separation of the most commonly encountered and most economically important species
14. *Gelechioidea — Tortricoidea* (leaf-mining moths) (R. W. Hodges)
    a. All higher taxa need revision; tortricids better known of the two
    b. Basic revisions — generic for world; specific perhaps by zoogeographic region
    c. Strong emphasis on quality material of adults associated with immature stages
    d. Keys should be illustrated; care must be taken to prevent an identifier from incorrectly concluding that species at hand is in reference being used
15. *Pyralidae* (pyralid moths) (D. C. Ferguson)
    a. Worldwide, the frequently intercepted Pyralidae most in need of attention
    b. Basic research — generic revisions of critical groups for most regions of the world
    c. Quality material, accurately identified, from the most troublesome regions. This must include larvae, positively associated with reared adults, so that some progress can be made in identification of larvae.
    d. Color-illustrated manuals, with well-designed keys for both adults and larvae, covering each of the major source areas of intercepted species; for North America, a complete, color-illustrated, faunal work on moths, as an essential starting point for recognition of what is, or is not, already here
    e. A well-curated reference collection; critical for identification; keys with photographic illustrations of wings also would be useful
16. *Thysanoptera* (thrips) (S. Nakahara)
    a. Thripidae
    b. Generic revisions
    c. Association of adults and immatures — world; better collections from S.E. Asia
    d. Basic generic revisions and reference collections

X. SPECIAL PROBLEMS AND SOLUTIONS IN IDENTIFICATION

Over the next several decades the overwhelming need in insect and mite diagnosis will continue to be for identification by morphological characters. However, with the increasing recognition of the importance of intraspecific and population differences in genetic makeup (biotypes, races, strains, etc.),[11] there will also be increasing need to refine morphological techniques, apply existing biochemical and molecular methodologies, and develop new methodologies.

Refinement of morphological approaches is being accomplished by use of scanning electron microscopy, morphometric analysis, and image analysis. Currently, these are essentially tools and methodologies for improving the quality of taxonomic research. Use of the stereoscan microscope over the past 20 years is providing new, microstructural characters for building classifications, and is of value in illustrating identification aids. Morphometric analyses at varying levels of complexity are being increasingly used in taxonomic research at subspecific levels. Simpler morphometric procedures, based on quantitative research, are being developed for broader use. Image analysis equipment is beginning to be used to solve classification problems at the intraspecific level (e.g., Africanized bees) and likely will find broader application as the methodology is improved and the costs of equipment decrease.

Several biochemical and molecular methodologies have been used over recent years in characterizing species and populations of species.[6,7] Some of these are likely to become routine methods at various levels of diagnostics. The methodologies include isozyme-allozyme analysis (electrophoresis), protein analysis (high-performance liquid chromatography and gas chromatography), cuticular hydrocarbon analysis, immunology-microcomplement fixation, analyses of mitochondrial DNA variability, and, less likely to be of practical value, nucleic acid sequencing. The methodologies most likely to be used in the near future in more or less routine diagnosis are enzyme-linked immunosorbent assay (ELISA or EIA) and cuticular hydrocarbon analysis. Dead, preserved specimens can be used in both of these techniques.

These advances in methodologies provide us with potentially useful techniques for resolving certain problems in identification, especially in the development of "fingerprints" to determine the geographical origin of biotypes.[28,29] Currently, studies of these kinds are being carried out on a few major groups of pests — *Heliothis,* fruit flies, boll weevils, the honey bee tracheal mite, and Africanized bees. In insects, only one molecular technique, gel electrophoresis of enzymes, has been widely used. Although gel electrophoresis remains to be further utilized as part of routine systematic techniques among systematists, other molecular approaches require equal attention in insect systematics. So far, only limited research involving insects has been carried out in the areas of DNA hybridization, nucleic acid sequencing, and microcomplement fixation, and this research has been limited to only a few species of flies.

For some groups, gel electrophoresis has proven useful in biotype discrimination, species identification, and hierarchical classification. However, the comparative study of the characteristics of mitochondrial DNA allows greater resolution than does gel electrophoresis in determining phylogenetic and evolutionary relationships to animals, both within and among species. The ability of monoclonal antibodies to distinguish minor antigenic differences has made them useful probes with which to differentiate the "strains" of organisms. The inclusion of mtDNA mapping, ELISA for detecting low concentrations of one or a few species-specific proteins, and radioimmunoassay (RIA), along with electrophoresis in insect systematics research will complement current identification capability. Since both fresh and dry specimens can be used in ELISA and RIA, the application of these immunological techniques will greatly increase our capability to use material collected under any circumstances for systematic investigations. The development of field (port) identification kits for insect pests using molecular technologies such as ELISA will be especially important in providing rapid identification of certain quarantine pests.

XI. SPECIAL NEEDS AND CAPABILITIES

Part of the future of meeting plant diagnostic problems lies in the application of computer methodologies for the development, transmission, and use of information. Decision support systems, information management systems, and expert systems all have a role to play. Fundamental to the proper application of such technology is the need to determine when traditional hard copies are most useful and most economical, and when electronic formats are best. The economics and capabilities of electronic data processing (EDP) will be a deciding factor, and thus managers of diagnostic services will need to keep informed of current and near-future developments in EDP technology. Utilization of the proper kind of hardware is an important consideration. The increasing capability and decreasing costs of micro or "personal" computers (PCs) is especially significant.[4]

A. Taxonomic and Biosystematic Databases

Applying the correct scientific name to the described species is a major problem because only one third to one thirtieth of all insect species have been described and named. The problem exists and will continue to exist for several reasons, basically because of the poor state of development of the classification of insects. There are perhaps 2,000,000 species-group names, and 100,000 generic names in the literature. These have been published, beginning in 1758 (tenth edition of *Systema Naturae* by C. Linneaus), in all major and many minor languages and in a plethora of journals, books, etc. throughout the world. There are many synonyms and homonyms in the taxonomic literature, because many species (and genera) have been inadequately defined and reported in the literature, because of some (especially earlier) taxonomists' lack of knowledge of the work of others, and because of lack of access to the world literature and to type or reference collections. Corrections are generally published first in the specialized literature, and therefore it may take a considerable amount of time for them to find their way into working documents and general treatments available to nonspecialists.

Taxonomic (nomenclatural) catalogs or checklists of higher taxa, e.g., orders, for major regions are the most comprehensive sources of information on the correct names and the overall classification. World catalogs should be the most authoritative sources, as such treatments should uncover most nomenclatural problems. However, there are inherent problems even with comprehensive catalogs: they are out of date as soon as they are published because of the lengthy production time; they include errors made or perpetuated because of the incompetence of some authors; they quickly become out of print because they are printed in rather limited editions of one or two thousand copies; and they are usually expensive for the same reason.

Key solutions to some of these problems include computerization of catalog production, more frequent and wider distribution of hard copies as well as soft copies to be used on personal computers, and remote access to computerized catalog-databases. The first major taxonomic catalog initiated (and continuing) in a computerized mode was the *Coleoptera of America North of Mexico;*[18] 22 fascicles have been distributed to date. The second major taxonomic catalog for insects produced in a computerized mode was the three-volume "*Catalog of Hymenoptera in America North of Mexico.*"[23] However, the computer program (SELGEM) used in the production of that catalog was awkward to work with, the database could not be searched remotely, and, although the database was rich in information, the procedures for obtaining searches were difficult to use and not well known. The *Biosystematic Database of Files of the World* was recently initiated by the USDA, with the first step being an update of the nomenclature in the *Catalog of Diptera of America North of Mexico*.[32] The *Biosystematic Database of Flies of the World* promises to be a major advance in taxonomic "catalogs" in that it will actually be an in-depth database including distributional, host, and other biological information. The *Biosystematic Database* will be produced with state-of-the-art software, in fascicles to ensure relatively prompt publication, and will make use of the world's expertise.

In addition to problems with access to basic nomenclatural information, access to the broad range of related information in the taxonomic literature, in collections, and in the various unpublished files that are typically situated in taxonomic units is a major problem. Such information, as noted above under the comments on the *Biosystematic Database of Flies of the World,* goes beyond basic diagnostic needs, but is closely related and supportive of diagnostic activities. Developments in and availability of hardware and software are expected to enable data entry and search by a broad spectrum of contributors and users from remote locations. Other improvements in producing and making use of taxonomic "catalogs" and databases are to be expected as technology advances and availability of computerization increases. Optical disc storage, laser printing, flat-screen terminals, and satellite-based communications networks allowing workers everywhere to access local databases are actual or potential improvements. As the production of "catalogs" proceeds from the traditional mode to that of modern information systems, these needs can be expected to be met.

B. Immigrant Species Databases

Further steps can be taken in "packaging" special databases, certain kinds of taxonomically related information of special interest to plant quarantine diagnosis activities. The Western Hemisphere Immigrant Arthropod Database (WHIAD), being produced by the Biosystematics and Beneficial Insects Institute, USDA, is one example of this kind of special database.

The objectives of WHIAD are to develop a computerized database of essential information on insects, mites, and related arthropods immigrant to or introduced into the Western Hemisphere, with emphasis on taxa in North America, especially in the U.S.; to develop a system for obtaining, maintaining, analyzing, and documenting this information; and to make this information available to the broad spectrum of users, including regulatory agencies, research and state extension units, agribusinesses, international organizations, and private individuals.

A card file of essential data on the species known to be immigrant to North America, initiated by the late R. I. Sailer in 1968, is being expanded to include 51 fields of information and updated by reviewing publications and investigating unpublished sources of information. Taxonomic position and nomenclature of the taxa are verified periodically through consultation with taxonomists. A searchable, computerized database is being developed and maintained on a Wang Virtual Storage-100 system. The hard-copy file now includes about 1660 species, but we estimate a rapid increase to about 3000 species during the first 2 to 3 years of computerization. Thereafter, 10 to 25 species per year may be added and 2 to 3 deleted. Information will also be incorporated from the "Release of Beneficial Organisms Database" being developed in the Beneficial Insects Laboratory, BBII, on introduced beneficial arthropods established in the U.S. after 1981. After review and editing of the database, a first compilation of the entire list will be reproduced and sent to members of the entomological community, requesting that they contribute information to fill in the many gaps that will undoubtedly appear. Revised compilations subsequently will be published for broader distribution. It is expected that a continuing series on exotic arthropods new to North America, presented in the WHIAD format, will be established by an entomological journal.

C. Common Names

Common names of pests are useful in general communication. Effective, international communication is essential at many levels, and common names can be important in this regard, particularly when specialists communicate with the public. Also, common names can be useful in the development of more universal methods for retrieving information. There are a few major lists (e.g., Sutherland and Brassard,[34] Greiff,[15] and Werner[37]) of common names of insects and mites and many specialized lists of various kinds. An essential requirement for common-names lists is to have the proper equivalent scientific name. This is difficult to accomplish as scientific names generally change more frequently than common names. In order to achieve some consensus on the formation of common names and stability in their use, a continually updated list, institutionalized in an entomological society or some other organization, appears to be the best approach. Computerization of such lists, with the addition of certain key information (higher taxonomic category, code numbers, etc.), such as the list of about 4500 names maintained by the U.S. Environomental Protection Agency[34] is very useful. Furthermore, computer abbreviations of pest names could be useful; the European Plant Protection Organization (EPPO)[2] has recommended that the Bayer system of computer abbreviations of pest names be made known to the EPPO member countries as the system to be used for preference. From an international point of view, multilingual, synonymical lists would be particularly useful.[35] An excellent English-Spanish list has recently been published.[15] An "International Register of Lists of Common Names" is an identified need for the more effective and broader use of common names of pests of all groups.

D. National and International Pest Registers

One of the major problems in plant quarantine is the lack of knowledge of native pests. In some countries, there is a lack of basic information about the occurrence of pests and their hosts and distributions. In most countries, the available information is not well organized nor readily available. Obviously, computerization of information would be a major advantage with this problem. The problem has been recognized. For example, Section XII of the U.S. Federal Insecticide, Fungicide, and Rodenticide Act of 1968 calls upon the Department of Agriculture and Environmental Protection Agency to "prepare a list of pests that need to be controlled." At the 1983 International Meeting for the Strengthening of Regional Plant Protection Organizations with Special Emphasis on Latin America and the Caribbean[3] it was recommended. "1) that IICA [Inter-American Institute for Cooperation in Agriculture] coordinate the development of a list of pests of the major crops of the Western Hemisphere for transmittal to FAO, showing the distribution in each country, 2) that regional organizations in the Western Hemisphere develop a list of pests of plant quarantine concern. The definition of a plant quarantine pest shall be the one developed at the FAO IPPC [International Plant Protection Congress] meeting in Rome in 1976. The list shall be divided into two groups: one for organisms not present within a region and the other for those that occur within the region but for which organized control of quarantine activities are being conducted, 3) that the lists of pests be forwarded to FAO through IICA and that FAO be requested to develop a worldwide list of pests of major crops, 4) that each regional organization form committees so that the lists are reviewed and modified as necessary on an annual basis, 5) that a similar approach could be followed by regional plant protection organizations in other parts of the world, and that FAO develops a computer system to relay information on pest lists worldwide."

Although some steps have been taken in this direction, a comprehensive source of information is not available. A computerized National Agricultural Pest Register for the U.S. has been proposed.[22] This would provide certain core data (nomenclature, host-vector relationships, pest status, distribution, and key references on biology) on pest species to all users and would identify the diverse supporting or related databases that could be used to rapidly compile data needed to evaluate pest management problems. As an index or directory to the species in various databases, such a register could be of significant value to plant quarantine activities. Major difficulties relative to compatibility and telecommunication between diverse databases will need to be solved. Technological problems are gradually being overcome, but information about the existence of databases and protocols for their use are expected to be continuing problems.

Databases in national biological survey activities could also be primary elements[17] for plant quarantine and national plant pest survey and detection databases.

E. Expert Systems

A recent paper on expert systems in entomology[33] defines ''expert systems'' as ''. . . computer programs that offer solutions to complex problems by mimicking known reasoning processes (heuristics) and employing a knowledge base of information extracted from human experts.'' The author goes on to point out that, ''In thinking of an appropriate expert system application, one should look for a problem that is well defined in scope, but which requires scarce human expertise to solve. Problems that involve incomplete data and require judgment, approximation, or opinion in their solution are well suited to an expert system approach. Agricultural pest management decisions, insect and general pest identification, and crop production are all good examples of such problems.''

The hierarchical nature of animal and plant classification lends itself well to the use of computerized expert systems for decision-making in diagnosis. Specialized software such as DELTA (*DE*scriptive *L*anguage for *TA*xonomy) is available for the entry, manipulation, and output in various forms of taxonomic information.[10,36] DELTA was designed to aid taxonomists in the efficient and systematic gathering, editing, storing, presenting, and updating of taxonomic data. The system is flexible, easy to use, and allows scientists to collect and present data in a systematic and comparative manner. Taxonomic data coded in DELTA format can be automatically converted for natural language descriptions, identification keys, numerical analysis, on-line identification, and information retrieval. Natural language descriptions as produced from the programs can be fully comparative and consistent in punctuation, character sequence, and terminology. In addition, DELTA allows for flexibility in the descriptive format, taxa described, and characters used. DELTA-produced keys are mechanically well-designed and agree with the descriptive data. The taxonomist interacts with the programs in the key-making process to produce a key that reflects his concepts of character importance and key design. Additional advantages of the method include the production of special-purpose keys and descriptions, production of treatments in other languages, and the direct linkage of output to automatic typesetting facilities. DELTA includes three programs: CONFOR, KEY, and ONLINE. CONFOR checks data and converts data in DELTA format to various other formats, KEY analyzes data and converts data in identification keys, and ONLINE analyzes data for on-line identification and information retrieval.

Further developments of special systems such as DELTA will predictably lead to networks of research taxonomists contributing to the development and improvement of databases that at the same time are being used by quarantine personnel at remote locations throughout the world for decision making.

Integration of the identification process with the quarantine decision-making process has been proposed by Dr. F. R. Gaeta, Interamerican Institute for Cooperation in Agriculture of the Organization of American States,[14] for use on PCs in southern South America. Separation between such bodies of information can complicate and protract decision making. Ready access to and understanding of these areas of information should enhance quarantine programs, eventually leading to a truly integrated plant quarantine and should result in improved international trade.

ACKNOWLEDGMENTS

I thank J. F. Fowler and D. R. Whitehead for permitting me to quote from their unpublished manuscripts; M. D. Huettel, A. C. F. Hung, and R. L. Roehrdanz for providing information on biochemical and molecular biological methodologies; W. L. Murphy for his usual excellent editorial assistance; and V. L. Blackburn, R. A. Bram, L. M. Chilson, E. Civerolo, J. R. Coulson, J. F. Fowler, W. Friedman, F. R. Gaeta, R. Gonzalez, E. R. Hoebeke, J. Kirkbride, M. A. Lacey, J. Lightfield, G. Mathys, C. E. Miller, D. R. Miller, S. Nakahara, A. Y. Rossman, R. Stewart, D. W. S. Sutherland, T. E. Wallenmaier, and C. L. Wilson for their technical review of the manuscript.

I thank the following taxonomists of the Systematic Entomology Laboratory for contributing information for Section IX, and for Table 6: E. W. Baker, D. C. Ferguson, R. J. Gagné, R. D. Gordon, R. W. Hodges, J. M. Kingsolver, J. P. Kramer, D. R. Miller, S. Nakahara, A. L. Norrbom, R. W. Poole, D. R. Smith, M. B. Stoetzel, F. C. Thompson, R. E. White, and D. R. Whitehead.

I thank the following for giving me permission to reproduce certain material: R. H. Foote (Table 1), J. F. Fowler (Tables 2 and 3), W. Klassen (Table 4), R. W. Hodges (Table 5), Animal and Plant Health Inspection Service, Plant Protection Quarantine (Figures 1 and 3), and Academic Press (pp. 24-25, and Figure 5).

REFERENCES

1. **Anon.,** Pests Not Known to Occur in the United States or of Limited Distribution, Plant Protection and Quarantine, U.S. Department of Agriculture, Washington, D.C., 1981 — 1986.
2. **Anon.,** Recommendations made by the EPPO Council, *EPPO Bull.*, 13, 609, 1983.
3. **Anon.,** Recommendations of the International Meeting for the Strengthening of Regional Plant Protection Organizations with Special Emphasis on Latin America and the Caribbean, *EPPO Bull.*, 13, 613, 1983.
4. **Anon.,** *Microcomputers and Their Applications for Developing Countries*, Westview Press, Boulder, Colo., 1986.
5. APHIS 81 Series, 1981.
6. **Berlocher, S. H.,** Insect molecular systematics, *Annu. Rev. Entomol.*, 29, 403, 1984.
7. **Bush, G. L. and Kitto, G. B.,** Application of genetics to insect systematics and analysis of species differences, in *Beltsville Symposia in Agricultural Research, 2. Biosystematics in Agriculture*, Romberger, J. A., Ed., Allanheld, Osmun, Montclair, N.J., 1978, 89.
8. Cooperative Economic Insect Report.
9. Cooperative Plant Pest Report.
10. **Dallwitz, M. J.,** A general system for coding taxonomic descriptions, *Taxon*, 29, 41, 1980.
11. **Diehl, S. R. and Bush, G. L.,** An evolutionary and applied perspective of insect biotypes, *Annu. Rev. Entomol.*, 29, 471, 1984.
12. **Foote, R. H.,** Museums, museum data, and the real world—the taxonomic connection, in *Beltsville Symposia in Agricultural Research. 2. Biosystematics in Agriculture*, Romberger, J. A., Ed., Allanheld, Osmun, Montclair, N.J., 1978, 261.
13. **Fowler, J. F.,** Regulatory/quarantine, in Biosystematic Services in Entomology, Knutson, L., Harris, K. M., and Smith, I. M., Eds., Agricultural Research Service, U.S. Department of Agriculture, Washington, D.C., in press.
14. **Gaeta, R. F.,** personal communication, 1986.
15. **Greiff, M.,** Spanish-English-Spanish Lexicon of Entomological and Related Terms, Commonwealth Agricultural Bureaux, Slough, U.K., 1985.
16. **Hodges, R. W.,** Presidential Address 1976 — What insects can we identify?, *J. Lepid. Soc.*, 30, 245, 1976.
17. **Johnson, R. L.,** Plant protection and a national biological survey, in *Foundations for a National Biological Survey*, Kim, K. C., and Knutson, L., Eds., Association of Systematics Collections, Lawrence, Kan., 1986, 77.
18. **Kingsolver, J. M. and Gordon, R. D., Eds.,** Coleoptera of America North of Mexico, Agricultural Research Service, U.S. Department of Agriculture, Washington, D.C., 1978.
19. **Klassen, W.,** Agricultural research: the importance of a natural biological survey to food production, in *Foundations for a National Biological Survey*, Kim, K. C. and Knutson, L., Eds., Association of Systematics Collections, Lawrence, Kan., 1986, 65.

20. **Knutson, L.,** Improving support for biosystematic service organizations, in Biosystematic Services in Entomology, Knutson, L., Harris, K. M., and Smith, I. M., Eds., Agricultural Research Service, U.S. Department of Agriculture, Washington, D.C., in press.
21. **Knutson, L. and Lacey, M. A.,** Reference collections for identification of immigrant pest insects, *FAO Plant Prot. Bull.,* 32, 95, 1984.
22. **Knutson, L., Sutherland, D. W. S., Dogger, J. R., and Johnson, R. L.,** The Concept of a National Agricultural Pest Register, paper presented at Entomological Society of America National Meeting, San Antonio, Tex., 1984.
23. **Krombein, K. V., Burks, B. D., Hurd, P. D., Jr., and Smith, D. R., Eds.,** *Catalog of Hymenoptera in America North of Mexico,* Vols. 1 to 3, Smithsonian Institution Press, Washington, D.C., 1979.
24. **Lattin, J. B.,** Availability of taxonomic specialists in the United States, with special reference to agriculture, in *Biosystematic Services in Entomology,* Knutson, L., Harris, K. M., and Smith, I. M., Eds., Agricultural Research Serv., U.S. Department of Agriculture, Washington, D.C., in press.
25. List of Intercepted Plant Pests, Plant Protection and Quarantine, U.S. Department of Agriculture, Washington, D. C., 1917-1986.
26. **Muesebeck, C. F. W.,** Fundamental taxonomic problems in quarantine and nursery inspection, *J. Econ. Entomol.,* 35, 753, 1942.
27. **Robinson, W. H.,** Improving taxonomic literature to meet regulatory needs, *Melsheimer Entomol. Ser.,* 32, 33, 1982.
28. **Roehrdanz, R. L. and Hung, A. C. F.,** in correspondence.
29. **Roehrdanz, R. L., Hung, A. C. F., Steiner, W., and Temayer, K.,** Molecular techniques in systematics and evolution, in Molecular Entomology Working Conf., Agricultural Research Service, United States Department of Agriculture, Washington, D.C., 1986.
30. **Sailer, R. I.,** History of insect introductions, in *Exotic Plant Pests and North American Agriculture,* Wilson, C. L. and Graham, C. L., Eds., Academic Press, New York, 1983.
31. **Shannon, M. J.,** Systematics: a basis for effective regulatory activities, *Bull. Entomol. Soc. Am.,* 29, 47, 1983.
32. **Stone, A., Sabrosky, C. W., Wirth, W. W., Foote, R. H., and Coulson, J. R., Eds.,** A Catalog of the Diptera of America North of Mexico, Agricultural Research Service, U.S. Department of Agriculture, Washington, D.C., 1965.
33. **Stone, N. D., Coulson, R. N., Frisbie, R. E., and Loh, D. K.,** Expert systems in entomology: three approaches to problem solving, *Bull. Entomol. Soc. Am.,* 32, 161, 1986.
34. **Sutherland, D. W. S. and Brassard, D. W.,** EPA List of Insects and Other Organisms, 3rd microfiche ed., U.S. Environmental Protection Agency, Washington, D.C., 1982.
35. **Sutherland, D. W. S. and Knutson, L.,** Proposal for a synonymical, multi-language listing of insect common names for the Western Hemisphere, Abstracts of Papers, IX Int. Congr. Plant Protection and 71st Annual Meet. American Phytopathological Society, No. 994, 1979.
36. **Taylor, R. W.,** Descriptive taxonomy: past, present, and future, in Australian Entomology: A Bicentennary Perspective, Highley, E. and Taylor, R. W., Eds., Commonwealth Scientific and Industrial Research Organization, Melbourne, 1983, 93.
37. **Werner, F. G.,** *Common Names of Insects and Related Organisms — 1982,* Entomological Society of America, College Park, Md., 1982.
38. **Whitehead, D. R.,** The interactions between producers and users of diagnostic taxonomic services, and comments on the nature of taxonomic questions and answers, in Biosystematic Services in Entomology, Knutson, L., Harris, K. M., and Smith, I. M., Eds., Agricultural Research Service, U.S. Department of Agriculture, Washington, D.C., in press.

Chapter 16

PLANT QUARANTINE DIAGNOSTIC PROBLEMS: FUNGI

J. M. Waller

TABLE OF CONTENTS

I. INTRODUCTION

Plant disease diagnosis requires an understanding of both the plant and the pathogen causing the disease. In order to recognize when a plant is diseased, it is necessary to know the healthy plant; only then is one able to recognize the unhealthy state and to discern the symptoms characteristic of a particular disease. Disease symptoms, however, can be caused by many factors and, from the plant quarantine point of view, it is the identity of these factors that is particularly important, because the objective of the plant quarantine exercise is to detect and exclude particular pathogenic organisms. Consequently, the distinction between disease and pathogen must be appreciated.

The diagnosis of plant diseases requires not only the recognition of a particular disease by its symptoms, but also the identification of the cause. This may be complicated by interactions with environmental factors and other organisms which can confuse the straightforward host/pathogen concept of disease. Therefore, meaningful diagnosis must also involve an appreciation of why the disease has occurred. This is particularly important when considering plant diseases caused by fungi. Many fungal pathogens are facultative necrotrophs, only causing serious disease problems on damaged or debilitated plants. Usually such organisms are widely distributed and of little concern in plant quarantine. Nevertheless, their effects must be known so that other more serious pathogens can be distinguished.

II. NOMENCLATURE

A. Diseases and Pathogens

Unfortunately, the usage of the names of pathogens and those of diseases has become very confusing. Names of diseases usually refer to particular symptoms, yet frequently particular diseases can be caused by several different pathogens and sometimes by pests or abiotic agencies. Generalized terms such as rust, mildew, leaf blight, etc. are quite inappropriate for describing pathogens. Thus, in the context of plant quarantine (and in many others where precision is required), the term ''pests and diseases'' is often misleading: it is not the disease which is of direct concern but the pathogen(s) able to cause it. ''Pests and pathogens'' is a more appropriate term, as they are the damaging organisms at which plant quarantine activities are targeted. A further problem related to this (and one which will be discussed later) is that many pathogens can only be recognized in the first instance by the symptoms of disease that they cause. If no symptoms are evident, it is usually assumed that the pathogen is absent — a fundamental problem for plant quarantine and many other aspects of plant disease control.

B. Pathogen Nomenclature

As it is the individual pathogens which are the targets of plant quarantine measures, we should be very clear about their names. The International Code of Botanical Nomenclature governs the use of generic and specific names given to fungi. This does not solve all problems, however, and nomenclatural difficulties frequently arise in plant quarantine work. There are three main reasons for this.

First, different countries may have different usages — either the same name may refer to different fungi (rare except where differences are slight) or, more usually, a particular pathogen may be known under different names. Such differences become highlighted in the sphere of international exchange.

Second, the taxonomy and hence the nomenclature of fungi is developing and expanding like any other science; new facets are continually being uncovered, particularly about microorganisms, and some of these may demand changes of classification and nomenclature. Unfortunately many plant pathogens were originally described and named by nontaxonomists, and subsequent examination has revealed the need to reclassify and rename them. Nomenclatural changes are often slow to filter out to the nontaxonomic world, and even slower to receive general acceptance.

Third, the life cycle of most fungi involves imperfect (or asexual) and perfect (or sexual) stages during which the species assumes different forms producing different types of spores. Because the type of spores and the way they are produced are of fundamental importance in fungal classification, this results in the teleomorph (perfect stage) being given a different generic name to that of the anamorph (imperfect stage). For most fungal pathogens, it is the anamorph which is the most common; the teleomorph may be produced only rarely or during "off-season" saprophytic survival or not at all in many plant pathogens. This has resulted in plant pathogenic fungi usually being called by their anamorphic name rather than their telemorph. However, according to the International Code, the correct name for the fungal species as a whole (i.e., applicable to all its forms), technically termed the holomorph, is usually that applied to the teleomorph.

Further information on these nomenclatural aspects can be obtained from Hawksworth[10] and Voss and Greater.[12] The correct naming of fungi, and indeed recognition of the names used, is therefore an exacting task, and those involved with plant quarantine need to be aware of the pitfalls and seek expert advice where appropriate.

III. PRINCIPLES OF DIAGNOSIS

A. Symptoms

The first step in diagnosing plant diseases is recognition of the symptoms. In some cases these may be sufficiently distinctive by themselves to allow a positive identification by those familiar with the disease, particularly where the characteristic sporulating structures of the pathogen are an integral part of the disease syndrome, as in rusts and smuts. However, it is always advisable to check the specific identification of the pathogen by direct microscopic examination. More usually, symptoms of disease are not a reliable guide to the identity of the pathogen as not only can the same symptom be produced by several different fungi or other agents, but the same pathogen may produce different symptoms on different host species or cultivars, at different stages of plant growth or under different environmental conditions. Nevertheless, recognition of a symptom type is a good indication of which pathogen species may be involved and where it is likely to be located.

Diseases characterized by local lesions on the aerial parts of plants such as leaf or fruit spots and stem cankers are usually relatively easy to diagnose, as the pathogen can either be found sporulating on the lesion or can be induced to do so by incubation in a moist chamber. Where systemic symptoms such as wilting, stunting, etiolation, or general chlorosis occur, these may either be the result of systemic infection such as occurs with the vascular wilt pathogens *Fusarium oxysporum (formae speciales)* and *Verticillium dahliae* or may be secondary symptoms resulting from diseases restricted to the root system or stem base of plants. Sometimes local lesions such as marginal leaf scorching and stem tip dieback may also be secondary symptoms induced by primary infection elsewhere, such as the roots, which hinder the proper physiological function of the plant. Therefore in attempting to diagnose the cause of such effects, some knowledge of whole-plant physiology is required in order to be aware of the type of effects pathogens can cause. It follows from this that pathogens do not always occur at the site of symptom expression in the plant, and it is essential to know where to look before the pathogen can be identified.

In plant quarantine work, particular problems exist with plant material which although infected may not show symptoms of disease, i.e., the disease is in a latent stage. The only guide as to whether plant material may be carrying latent infections is to know whether it has been previously exposed to infection. Incubation of latently infected material will often induce symptoms to appear, but with biotrophic pathogens, the symptoms may only appear if the plant is grown on to a particular growth stage. Containment facilities or external quarantine will be needed if plants are to be grown on to allow latent diseases to appear. Propagating material, including dormant storage organs and seeds, are well-known sources of fungal pathogens which do not show overt symptoms of disease or signs of pathogen presence. The detection of pathogens in propagating material is discussed elsewhere in this volume.

B. Causal Agent

Having established a provisional diagnosis, however broad, of a disease by symptom examination, the next step is to confirm it by determining the causal agent. For many well-known diseases it may be fairly obvious which pathogen is the accepted causal agent, and all that is required is confirmation of its presence. But there are a significant number of both local lesion and systemic diseases where a variety of agents may be responsible and it may not be at all obvious which type of organism, if any, is involved. This particularly applies where symptoms are not characteristic of known or familiar diseases. Fungal pathogens can often be detected sporulating on disease lesions either before or after incubation in a moist chamber by careful examination with a hand lens or, better, a low-power stereoscopic microscope. Where this is not so, lesions may have to be examined more carefully. Both fungi and bacteria can be seen microscopically in tissue mounts either in water (bacterial exudation) or by using clearing and staining techniques for fungi. If there are strong reasons to suspect a fungal pathogen either as the cause of a local lesion disease or a systemic disease and it cannot be readily identified on the material, then it may be necessary to isolate it into culture. However, before a fungal cause is assumed, other possible causes must be given due consideration.

Most disease diagnosis undertaken in connection with plant quarantine is on plant material removed from its growing situation. This severely limits the types of clues which pathologists seek when diagnosing diseases. These include patterns of disease spread in fields, prevailing or immediately preceeding climatic conditions, soil conditions, and many other factors which could either directly or indirectly be responsible for the observed symptoms or which may indicate the type of pathogen that is involved. Some common factors which can cause lesions resembling those due to fungal pathogens include (1) herbicide damage, which is a common cause of growth malformation and stem cankers; (2) necrosis caused by the feeding of hemipteran bugs such as *Helopeltis*, which can be readily confused with anthracnose symptoms; and (3) lightning damage, waterlogging, salinity and nematodes, all of which cause root damage that can be mistakenly attributed to pathogenic soil fungi.

IV. PATHOGENS AND SAPROPHYTES

The standard procedure for confirming the presence of a fungal pathogen which cannot be readily identified directly from diseased material is to isolate the fungus in culture. Problems here center around the fact that a wide range of fungi occurs naturally on plant surfaces, usually growing as harmless saprophytes on natural exudates and accumulated organic debris. Some also grow as nonpathogenic but parasitic endophytes or as symbiotic mycorrhizae. Others occur as hyperparasites of pathogens, e.g., *Darluca* and *Verticillium* spp. on rust pustules. The presence of these fungi interferes with attempts to isolate the pathogen, and appropriate isolation techniques must be used to select the pathogen. This may involve the use of baiting techniques or selective media as well as straightforward isolation onto plain agar after surface sterilization. The use of sugar-rich media such as PDA usually favors the rapid growth of saprophytes, whereas growth on plain agar requires the use of nutrients in the plant tissue and thus tends to favor the pathogen. Isolation from material excised aseptically from internal tissues is usually more successful in producing pure cultures on the pathogen. Burchill[3] gives a useful review of techniques used for isolating plant pathogens (see also Dhingra and Sinclair[4]).

Disease lesions and unhealthy plants usually become infected by ''secondary invaders'', weak parasites that can only invade moribund tissue, and eventually by saprophytes. These may totally obscure and replace the primary pathogen originally responsible for the disease lesion, and are frequently isolated in culture and mistakenly assumed to be primary pathogens. Some fungi are such common inhabitants of damaged tissue that the term ''imposter pathogen'' was coined by Garrett[8] for *Macrophomina phaseolina,* a common isolate from damaged roots in warm climates; *Botryodiplodia theobromae, Fusarium pallidoroseum,* and *Alternaria alternata* usually fall into the same category, although they can be important postharvest fruit pathogens in their own right. Such fungi are well distributed and of little relevance to plant quarantine, but they can nevertheless obscure the presence of other more virulent or nonfungal pathogens.

In the field situation it is often difficult to find a plant infected with only one pathogen or pest. Multiple infections by different pathogens are often the rule, particularly on plants growing in suboptimal conditions, where they may be predisposed to infection by nonspecific or weak pathogens such as the imposter pathogens referred to above.

Certain diseases, particularly those affecting roots, may be recognized as being of complex etiology where several pathogens and/or pests interact to cause the problem, but some such diseases have subsequently been shown to be due to a distinct primary pathogen, the presence of which had been obscured by a complex of secondary invaders. Those diseases whose true etiology remains obscure do constitute plant quarantine risks. Even when the interaction is well understood, problems can arise, as in the case of *Fusarium* wilt of cotton and root knot nematodes. Invasion of roots by *Meloidogyne* reduces the effectiveness of resistance of cotton cultivars to *Fusarium oxysporum* f. sp. *vasinfectum.* The disease develops late in the season and can become seed borne to a small but significant extent on ostensibly resistant cultivars. These and similar aspects of pathogen interaction have been reviewed by Waller and Bridge.[13]

The isolation and culturing of pathogens from plant material also takes time, and this is often of crucial importance in plant quarantine work. This is often a matter of weeks, and the problem is compounded when cultures have to be sent away for identification. Until more rapid methods such as those similar to ELISA tests for viruses or biochemical tests based on specific proteins or DNA characteristic of particular pathogen species have been developed, there is no short cut to fungal identification.

Isolation of particular fungi from plants may require the use of very fresh material. Pathogens such as *Phytophthora* species are difficult to isolate from older specimens in which the pathogen is not actively growing. The older the lesion is, or the more moribund the plant is, the greater is the chance that secondary invaders or saprophytes will have taken over. A major constraint to the adequate diagnosis of many plant diseases in developing countries is the time required for specimens to get to a laboratory or for pathologists to reach it in the field or port of entry.

V. IDENTIFICATION—TAXONOMY VS. PATHOGENICITY

A. Species Identification

Accurate identification of the pathogen provides the key to information about its biology and enables quarantine risks and problems to be assessed. Although most important fungal pathogens are now well known and can be identified relatively easily with access to microscopy and the relevant literature detailing their taxonomic characteristics, there are many instances where related species and even genera are confused. Identification to genus level should present few problems to those with reasonable mycological expertise — an increasingly uncommon attribute among pathologists. However, it is necessary to be able to distinguish parasites from saprophytes, and to do this one needs to know both — particularly where exotic and unfamiliar fungi are likely to be encountered as they are in quarantine work.

The accurate identification of fungi requires mycological expertise, and reference should be made to appropriate authorities who will undertake fungal identification such as the CAB International Mycological Institute, Kew, Surrey, U.K.; Centraalbureau voor Schimmelcultures, Oosterstraat, Baarn, Netherlands; and American Type Culture Collection, Parklawn Drive, Rockville, Md.

Although certain genera of fungi that contain well known pathogenic species are quite distinctive, there are often closely related species in the same genus which may be of little pathological significance. Confusion between such species is common, and can result either in failure to recognize the pathogen or attribution of pathogenic status to the saprophytic form. The genus *Alternaria* typifies this confusion. *A. alternata* and *A. tenuissima* are common secondary invaders or saprophytes often assuming the role of imposter pathogens. Yet pathogenic species such as *A. citri, A. longipes,* and others are close morphologically to these and difficult to separate.[6] *Fusarium* also creates many problems. The three species forming microspores commonly associated with diseased plants are *F. solani, F. oxysporum,* and *F. moniliforme* and are often confused. *F. solani,* although an important pathogen on some crops, is a very common secondary invader and saprophytic in warm climates, and is frequently confused with *F. oxysporum,* which again exists either as a saprophyte or in a number of *formae speciales* which are of limited distribution and cause serious vascular wilts of a number of crop species. *F. moniliforme* var. *subglutinans,* pathotypes of which cause sugarcane wilt and pineapple fruit rot, can only be distinguished from *F. oxysporum* by the lack of chlamydospores in mature cultures. Major features of the common pathogenic species have been summarized and keyed out by Booth.[2]

Biotrophic pathogens cannot be cultured in vitro and must be identified directly from their hosts. There is thus less flexibility for mycological manipulation, and similar pathogens which can occur on the same host can often be confused. A good example of this is the confusion which existed over sugarcane rusts some years ago. Differences between *Puccinia kuehnii* and *P. melanocephala* are slight, and past records have often confused the two. The situation was not helped by the fact that *P. erianthii* was the name used for *P. melanocephala* until the late 1970s. Only when sugarcane rust spread rapidly across the world was distinction made between the two species and synonymy between *P. erianthii* and *P. melanocephala* widely recognized.[5]

Graminaceous downy mildews (*Peronosclerospora* spp.) present a similar case. A range of very similar species can occur on maize, sorghum, and some related grasses. Yet each has different pathogenic characteristics and most are restricted to Asia. Distinction between the species relies on small differences in the morphology of the ephemeral conidia, which are difficult to assess in specimens that are not fresh. Williams[15] gives a succinct account of the problems associated with the taxonomy, pathogenicity, and distribution of these species.

An example of where two quite distinct pathogens are similar morphologically and in the symptoms they produce is the strong resemblance between *Synchytrium dolicholi* and *Phakopsora pachyrhizi*. The latter causes soybean rust, an important pathogen absent from most of the New World. The pathotype of this species which causes soybean rust is also absent from Africa, but the overtly similar *S. dolicholi* also affects soybean and other legumes and is more widely distributed.

B. Intraspecific Variation

Probably the most important problem with current morphological methods of identifying fungi is that virtually no weight is given to behavioral characteristics such as pathogenicity, yet they are of fundamental importance of plant pathologists and other concerned with plant health. Isolates with very different ecological and pathogenic characteristics are often included in the same species because of their morphological similarities. The terminology used to describe pathogenic variation within species is also very unsatisfactory. The terms race, strain, pathotype, and *formae speciales* have all been used to mean different degrees of pathogenic specialization in different circumstances, and there are no generally accepted rules governing their usage although guidelines have been suggested.[1]

The pathogenic variants within a species often have different geographic distribution. Within the species *Fusarium oxysporum,* there is not only the saprophytic unspecialized "wild type" but many *formae speciales.* Some are restricted to single host genera while others, such as *f. sp. vasinfectum,* can infect several. At present it is only possible to distinguish most of these by pathogenicity testing, although it has been reported that races within *f. sp. cubense* causing Panama wilt of banana can be morphologically distinguished in culture.[11] The link between cultural morphology differences and pathogenicity is not uncommon. Virulent and normal strains of *Ceratocystis ulmi*, the Dutch elm disease pathogen, can be distinguished in this way.[9]

The coffee berry disease (CBD) pathogen, which is restricted to Africa, demonstrates the problems of both nomenclature and intraspecific variability of pathogenicity. Species of *Colletotrichum* occur worldwide on coffee and many other tropical plants, causing anthracnose diseases. Isolates from coffee can be grouped according to several morphological criteria, and the pathogen falls within the broad concept of *C. gloeosporioides,* which is regarded as an anamorph of *Glomerella cingulata.* However, the greenish-gray cottony mycelium of fresh CBD isolates coupled with its pathogenicity to young coffee berries are distinctive characters, and the teleomorph *G. cingulata* has never been produced from the CBD isolate despite many attempts. On subculturing, CBD isolates rapidly lose their morphological characteristics and come to resemble the saprophytic strains, a fact which has caused confusion about its true taxonomic status. The original name *Colletotrichum coffeanum* was applied by Noack to isolates from coffee in Brazil, where CBD does not occur, and it is fairly clear that this name is a synonym of *C. gloeosporioides,* so it is strictly incorrect to use the name *C. coffeanum* for any isolates from coffee.[7] Recent work at CMI has shown that CBD isolates possess distinct biochemical characteristics which distinguish them from other *Colletotrichum* species. It therefore appears that the CBD pathogen is a distinct taxon from the *Colletotrichum* conidial state of *G. cingulata.*

VI. THE DYNAMICS OF PATHOLOGY

Apart from the technical difficulties associated with diagnosis, there are those which reflect the changing status of fungi in the agricultural context. Pathogens and their hosts are in a continuing state of change which is aided and abetted by agricultural development. Not only do pathogens continue to spread to new areas despite plant quarantine restrictions, but new diseases also arise. There are usually about 100 to 150 reports of both new country records of disease and of new host/pathogen combinations (i.e., new diseases) published each year in *Review of Plant Pathology.* The various aspects that influence this continuing development of diseases are too extensive to be discussed here, but have been reviewed by Waller[14] in relation to agricultural development. As far as plant quarantine work is concerned, this means that there is an ever-expanding volume of information which needs to be taken into account. Due to faulty diagnosis, nomenclatural changes, and other factors, older information about the occurrence and symptoms of disease can be misleading. The literature abounds with records of diseases attributed to imposter pathogens and secondary invaders. However, such fungi can produce pathogenic forms under the influence of selection pressure;

furthermore, new crop varieties have often been shown to have new susceptibilities to old, often discounted pathogens.

Diagnosis in relation to plant quarantine is the more difficult because it often has to be made away from the field situation where epidemiological and environmental factors can give important clues. Furthermore, it is usually concerned with detecting exotic and unfamiliar pathogens, often on plant material which may be very limited in amount and itself exotic. High-value ornamentals are an increasing commodity and pose a real danger because they may be alternative hosts to agriculturally dangerous pathogens. Add to this the problems caused by information both changing and increasing, and the task of detection and accurate diagnosis becomes difficult indeed. It is therefore particularly important that centers of expertise such as the CAB International Mycological Institute should be able to cooperate with plant quarantine authorities in providing both a reliable service for the accurate identification of fungal pathogens on a world-wide basis and a source of up-to-date information.

REFERENCES

1. **Anon.,** Guide to the Use of Terms in Plant Pathology, Phytopathological Paper No. 17, Commonwealth Agricultural Bureaux, Slough, U.K., 1973.
2. **Booth, C.,** Fusarium: A Laboratory Guide to the Identification of the Major Species, Commonwealth Agricultural Bureaux, Slough, U.K., 1977.
3. **Burchill, R. T.,** Methods in Plant Pathology, Phytopathological Paper No. 26, Commonwealth Agricultural Bureaux, Slough, U.K., 1981.
4. **Dhingra, O. S. and Sinclair, J. B.,** *Basic Plant Pathology Methods,* CRC Press, Boca Raton, Fla., 1986.
5. **Egan, B. T.,** Susceptible indicator varieties for the rust disease *(Puccinia melanocephala), Sugarcane Pathol. Newsl.,* 22, 10, 1979.
6. **Ellis, M. B.,** Dematiaceous Hyphomycetes, Commonwealth Agricultural Bureaux, Slough, U.K., 1977.
7. **Firman, I. D. and Waller, J. M.,** Coffee Berry Disease and Other Colletotrichum Diseases of Coffee, Phytopathological Paper No. 22, Commonwealth Agricultural Bureaux, Slough, U.K., 1977.
8. **Garrett, S. D.,** *Biology of Root-Infecting Fungi,* Cambridge University Press, Cambridge, 1956.
9. **Gibbs, J. N. and Brasier, C. M.,** Correlation between cultural characters and pathologenicity of *Ceratocystis ulmi* isolates from Britain, Europe, and North America, *Nature* (London), 241, 381, 1973.
10. **Hawksworth, D. L.,** Mycologist's Handbook, Commonwealth Agricultural Bureaux, Slough, U.K., 1974.
11. **Sun, E. J., Su, H. J., and Ko, W. H.,** Identification of *Fusarium oxysporum* f. sp. *cubense,* race 4 from soil or host tissue by cultural characteristics, *Phytopathology,* 68, 1672, 1978.
12. **Voss, E. G. and Greater, W.,** Synopsis of proposals on botanical nomenclature, Sidney, 1981, *Taxon,* 30, 95, 1981.
13. **Waller, J. M. and Bridge, J.,** Effects of pathogen interactions on tropical crop production, in *Plant Diseases: Infection, Damage, and Loss,* Wood, R. K. S. and Jellis, G. J., Eds., Blackwell Scientific, London, 1984, 311.
14. **Waller, J. M.,** The influence of agricultural development on crop diseases, *Trop. Pest Manage.,* 30, 86, 1984.
15. **Williams, R. J.,** Downy mildew of tropical cereals, in *Advances in Plant Pathology,* Vol. 2, Ingram, D. S., and Williams, P. H., Eds., Academic Press, New York, 1984, 1.

INDEX

A

B

C

D

E

I

J

K

L

W

X

Y